Lecture Notes in Physics

The Lecture Notes in Physics

The series Lecture Notes in Physics (LNP), founded in 1969, reports new developments in physics research and teaching – quickly and informally, but with a high quality and the explicit aim to summarize and communicate current knowledge in an accessible way. Books published in this series are conceived as bridging material between advanced graduate textbooks and the forefront of research and to serve three purposes:

- to be a compact and modern up-to-date source of reference on a well-defined topic

- to serve as an accessible introduction to the field to postgraduate students and nonspecialist researchers from related areas

- to be a source of advanced teaching material for specialized seminars, courses and schools

Both monographs and multi-author volumes will be considered for publication. Edited volumes should, however, consist of a very limited number of contributions only. Proceedings will not be considered for LNP.

Volumes published in LNP are disseminated both in print and in electronic formats, the electronic archive being available at springerlink.com. The series content is indexed, abstracted and referenced by many abstracting and information services, bibliographic networks, subscription agencies, library networks, and consortia.

Proposals should be sent to a member of the Editorial Board, or directly to the managing editor at Springer:

Christian Caron
Springer Heidelberg
Physics Editorial Department I
Tiergartenstrasse 17
69121 Heidelberg / Germany
christian.caron@springer.com

A. Kapustin
M. Kreuzer
K.-G. Schlesinger (Eds.)

Homological Mirror Symmetry

New Developments and Perspectives

 Springer

Anton Kapustin
California Institute of Technology
Mail Code 452-48
Pasadena CA 91125
USA
kapustin@theory.caltech.edu

Maximilian Kreuzer
TU Wien
Fak.Tech.Naturwiss./Inform
Inst. Theoretische Physik
Wiedner Hauptstr. 8-10
1040 Wien
Austria

Karl-Georg Schlesinger
Universität Wien
Formal- u. Naturwiss. Fakultät
Inst. Theoretische Physik
Boltzmanngasse 5
1090 Wien
Austria
kgschles@esi.ac.at

Kapustin, A. et al. (Eds.), *Homological Mirror Symmetry: New Developments and Perspectives*, Lect. Notes Phys. 757 (Springer, Berlin Heidelberg 2009), DOI 10.1007/978-3-540-68030-7

ISBN: 978-3-642-08767-7   e-ISBN: 978-3-540-68030-7

DOI 10.1007/978-3-540-68030-7

Lecture Notes in Physics ISSN: 0075-8450

Cover design: eStudio Calamar S.L., F. Steinen-Broo, Pau/Girona, Spain

Printed on acid-free paper

9 8 7 6 5 4 3 2 1

springer.com

Preface

In his 1994 ICM talk Kontsevich has put forward the Homological Mirror Symmetry (HMS) conjectures as a proposal to study the newly discovered duality of certain quantum field theories, coined mirror symmetry, in a mathematically rigorous form. HMS has since developed into a flourishing subject of its own. Eventually time was considered ripe for a workshop devoted solely to this topic, which was held at the Erwin-Schrödinger Institute, Vienna, Austria in June 2006.

It is at this occasion that we also felt the need for a set of representative, well-edited tutorial reviews that would both introduce and document the state of the art, and we set ourselves to the task, as a kind of ambitious follow-up project.

For introductions to the subject of mirror symmetry, we would like to refer to the two essay collections – S.T. Yau, *Essays on mirror manifolds*, International Press, Hong Kong 1991, and K. Hori et al. *Mirror symmetry*, AMS, Providence 2003. Unfortunately, there does not exist a general textbook introduction on homological mirror symmetry. For technically getting into the subject by starting from the simplest, yet highly involved, example, the case of complex tori, we refer to the excellent lecture course A. Polishchuk, *Abelian varieties, theta functions, and the Fourier transform*, Cambridge University Press, Cambridge 2002.

Two of the contributions might very well point the way to two possible future main roads of the subject: A. Kapustin gave one of the very first reviews of his ground-breaking joint work with E. Witten on S-duality and the Geometric Langlands program (the original paper with Witten appeared less than two months before the workshop). This work has already now started a considerable amount of follow-up activity on both the mathematics and the physics side, but the general impression is that we are still only perceiving the tip of an iceberg. The potential for supersymmetric gauge theory, as well as for the Langlands program in all its different versions in pure mathematics, is clearly enormous.

In a joint contribution M. Kontsevich and Y. Soibelman start to develop in detail a grand perspective on A_∞-algebras and A_∞-categories, considering them as a version of noncommutative algebraic geometry. This provides the basics for an extension of HMS to noncommutative spaces (a necessary extension from the perspective of physics, e.g., concerning resolution of singularities). The consideration of modules over A_∞-algebras ties HMS to one of the central themes in the important emerging field of (higher and) derived algebraic geometry. The presently rapidly developing study of Calabi-Yau categories is one of the central themes behind this contribution. Finally, the (noncommutative) geometric perspective on A_∞-algebras and A_∞-categories brings HMS into close contact with more general investigations involving deformation theory of quantum field theories, e.g., the recent mathematical approaches to the structure of the BV-quantization scheme. This contribution surely has the potential to become another future classic in the field of HMS.

The rigorous mathematical construction of the Fukaya category and its detailed study in examples is one of the central themes in HMS. K. Fukaya, P. Seidel, and I. Smith in their contribution give an overview of several approaches to the study of the Fukaya category of cotangent bundles and present a new approach in terms of family Floer cohomology.

Another important topic in HMS is the study of equivalences of derived categories of D-branes. For linear sigma-models corresponding to toric Calabi-Yau varieties, M. Herbst, K. Hori, and D. Page show how to construct a whole family of such equivalences on the Kähler moduli space.

That HMS has developed into an important field of its own is last but not least underlined by the fact that one can meanwhile truly speak of applications of HMS. An example of an important application is provided by the contribution of L. Katzarkov, showing the great use of the ideas of HMS for the study of rationality questions and the Hodge conjecture in algebraic geometry.

On the other hand, there are the deep roots of HMS in physics: HMS is inevitably tied to the question of the calculation of topological string partition functions and Gromov–Witten invariants. This is the topic of the contribution of M. Huang, A. Klemm, and S. Quackenbush, culminating in the specification of the partition function up to genus 51 for the quintic.

Another important point where HMS relates mathematics and physics is the fact that the identification of the derived categories appearing in HMS with the D-brane categories from physics relies upon the use of the renormalization group flow. This is reviewed in the contribution of E. Sharpe where it is also discussed how a similar application of the renormalization group flow appears if stacks are used as targets of sigma-models.

Finally, from the very beginning HMS has been closely related to the topic of deformation quantization. The contribution of Y. Soibelman is devoted to the study of noncommutative analytic spaces over non-archimedian fields. As is known from previous – already classical – work of Kontsevich and

Soibelman, noncommutative spaces over non-archimedian fields appear in examples of deformations which are not formal in the deformation parameter.

In conclusion, we would like to sincerely thank a number of people: Many thanks go to the Erwin-Schrödinger-Institute Vienna and its directors and president K. Schmidt, J. Schwermer, and J. Yngvason for hospitality and financial support. We thank I. Alozie, I. Miedl, and M. Windhager from the ESI administrative office for their very efficient support and handling of all the practical needs. Special thanks go to W. Beiglböck for his constant support of our book project at Springer and for providing his perfect organizational skills. We also acknowledge the efficient support by C. Caron, G. Hakuba and the staff of Springer in the final publishing process. Last but not least we thank all the authors for spending many hours on preparing detailed overviews of the field which should hopefully be of long-term value.

Pasadena and Vienna   *Anton Kapustin, Maximilian Kreuzer*
October 2007   *and Karl-Georg Schlesinger*

Contents

Derived Categories and Stacks in Physics

The Symplectic Geometry of Cotangent Bundles from a Categorical Viewpoint

K. Fukaya[1], P. Seidel[2], and I. Smith[3]

[1] Department of Mathematics, Kyoto University, Kyoto 6606-01, Japan
fukaya@math.kyoto-u.ac.jp
[2] M.I.T. Department of Mathematics, 77 Massachusetts Avenue, Cambridge MA
02139, USA
seidel@math.mit.edu
[3] Centre for Mathematical Sciences, Wilberforce Road, Cambridge CB3 0WB, UK
i.smith@dpmms.cam.ac.uk

Abstract. We describe various approaches to understanding Fukaya categories of cotangent bundles. All of the approaches rely on introducing a suitable class of non-compact Lagrangian submanifolds. We review the work of Nadler-Zaslow [30, 29] and the authors [14], before discussing a new approach using family Floer cohomology [10] and the "wrapped Fukaya category". The latter, inspired by Viterbo's symplectic homology, emphasizes the connection to loop spaces, hence seems particularly suitable when trying to extend the existing theory beyond the simply connected case.

1 Overview

A classical problem in symplectic topology is to describe exact Lagrangian submanifolds inside a cotangent bundle. The main conjecture, usually attributed to Arnol'd [4], asserts that any (compact) submanifold of this kind should be Hamiltonian isotopic to the zero-section. In this sharp form, the result is known only for S^2 and $\mathbb{RP}^2$, and the proof uses methods which are specifically four-dimensional (both cases are due to Hind [16]; concerning the state of the art for surfaces of genus > 0, see [17]). In higher dimensions, work has concentrated on trying to establish topological restrictions on exact Lagrangian submanifolds. There are many results dealing with assorted partial aspects of this question (see [7, 26, 35, 37, 38] and others; [26] serves as a good introduction to the subject), using a variety of techniques. Quite recently, more categorical methods have been added to the toolkit, and these have led to a result covering a fairly general situation. The basic statement, which we will later generalize somewhat, is as follows:

Theorem 1.1 (Nadler, Fukaya–Seidel–Smith) *Let Z be a closed, simply connected manifold which is spin, and $M = T^*Z$ its cotangent bundle.*

Fukaya, K., et al.: *The Symplectic Geometry of Cotangent Bundles from a Categorical Viewpoint.* Lect. Notes Phys. **757**, 1–26 (2009)
DOI 10.1007/978-3-540-68030-7_1   © Springer-Verlag Berlin Heidelberg 2009

Suppose that $L \subset M$ is an exact closed Lagrangian submanifold, which is also spin, and additionally has vanishing Maslov class. Then (i) the projection $L \hookrightarrow M \to Z$ has degree ± 1; (ii) pullback by the projection is an isomorphism $H^*(Z; \mathbb{K}) \cong H^*(L; \mathbb{K})$ for any coefficient field $\mathbb{K}$ and (iii) given any two Lagrangian submanifolds $L_0, L_1 \subset M$ with these properties, which meet transversally, one has $|L_0 \cap L_1| \geq \dim H^*(Z; \mathbb{K})$.

Remarkably, three ways of arriving at this goal have emerged, which are essentially independent of each other, but share a basic philosophical outlook. One proof is due to Nadler [29], building on earlier work of Nadler and Zaslow [30] (the result in [29] is formulated for $\mathbb{K} = \mathbb{C}$, but it seems that the proof goes through for any $\mathbb{K}$). Another one is given in [14], and involves, among other things, tools from [36] (for technical reasons, this actually works only for char($\mathbb{K}$) $\neq 2$). The third one, which is a collaborative work of the three authors of this chapter, is not complete at the time of writing, mostly because it relies on ongoing developments in general Floer homology theory. In spite of this, we included a description of it, to round off the overall picture.

The best starting point may actually be the end of the proof, which can be taken to be roughly the same in all three cases. Let L be as in Theorem 1.1. The Floer cohomology groups $HF^*(T_x^*, L)$, where $T_x^* \subset M$ is the cotangent fibre at some point $x \in Z$, form a flat bundle of $\mathbb{Z}$-graded vector spaces over Z, which we denote by E_L. There is a spectral sequence converging to $HF^*(L, L) \cong H^*(L; \mathbb{K})$, whose E_2 page is

$$E_2^{rs} = H^r(Z; End^s(E_L)), \tag{1}$$

$End^*(E_L) = Hom^*(E_L, E_L)$ being the graded endomorphism vector bundle. Because of the assumption of simple connectivity of Z, E_L is actually trivial, so the E_2 page is a "box" $H^*(Z) \otimes End(HF^*(T_x^*, L))$. The E_2 level differential goes from (r, s) to $(r + 2, s - 1)$

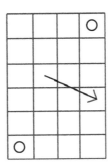

and similarly for the higher pages. Hence, the bottom left and top right corners of the box necessarily survive to E_∞. Just by looking at their degrees, it follows that $HF^*(T_x^*, L) \cong \mathbb{K}$ must be one-dimensional (and, we may assume after changing the grading of L, concentrated in degree 0). Given that, the spectral sequence degenerates, yielding $H^*(L; \mathbb{K}) \cong H^*(Z; \mathbb{K})$. On the other hand, we have also shown that the projection $L \to Z$ has degree $\pm 1 = \pm \chi(HF^*(T_x^*, L))$. This means that the induced map on cohomology is injective, hence necessarily an isomorphism. Finally, there is a similar spectral sequence for a pair of Lagrangian submanifolds (L_0, L_1), which can be used to derive the last part of Theorem 1.1.

At this point, we already need to insert a cautionary note. Namely, the approach in [14] leads to a spectral sequence which only approximates the one in (1) (the E_1 term is an analogue of the expression above, replacing the cohomology of Z by its Morse cochain complex, and the differential is only partially known). In spite of this handicap, a slightly modified version of the previous argument can be carried out successfully. The other two strategies (namely [29] and the unpublished approach) do not suffer from this deficiency, since they directly produce (1) in the form stated above.

From the description we have just given, one can already infer one basic philosophical point, namely the interpretation of Lagrangian submanifolds in M as (some kind of) sheaves on the base Z. This can be viewed as a limit of standard ideas about Lagrangian torus fibrations in mirror symmetry [11, 25], where the volume of the tori becomes infinite (there is no algebro-geometric mirror of M in the usual sense, so we borrow only half of the mirror symmetry argument). The main problem is to prove that the sheaf-theoretic objects accurately reflect the Floer cohomology groups of Lagrangian submanifolds, hence in particular reproduce $HF^*(L, L) \cong H^*(L; \mathbb{K})$. Informally speaking, this is ensured by providing a suitable "resolution of the diagonal" in the Fukaya category of M, which reduces the question to one about cotangent fibres $L = T_x^*$. In saying that, we have implicitly already introduced an enlargement of the ordinary Fukaya category, namely one which allows noncompact Lagrangian submanifolds. There are several possible ways of treating such submanifolds, leading to categories with substantially different properties. This is where the three approaches diverge.

1. **Characteristic cycles.** Reference [30] considers a class of Lagrangian submanifolds which, at infinity, are invariant under rescaling of the cotangent fibres (or more generally, asymptotically invariant). Intersections at infinity are dealt with by small perturbations (in a distinguished direction given by the normalized geodesic flow; this requires the choice of a real analytic structure on Z). An important source of inspiration is Kashiwara's construction [22] of characteristic cycles for constructible sheaves on Z; and indeed, Nadler proves that, once derived, the resulting version of the Fukaya category is equivalent to the constructible-derived category. (A similar point of view was taken in the earlier studies of Kasturirangan and Oh [23, 24, 31].) Generally speaking, to get a finite resolution of the diagonal, this category has to be modified further, by restricting the behaviour at infinity; however, if one is only interested in applications to closed Lagrangian submanifolds, this step can be greatly simplified.

2. **Lefschetz thimbles.** The idea in [14] is to embed the Fukaya category of M into the Fukaya category of a Lefschetz fibration $\pi : X \to \mathbb{C}$. The latter class of categories is known to admit full exceptional collections, given by any basis of Lefschetz thimbles. Results from homological algebra (more precisely, the theory of mutations, see for instance [15]) then ensure the existence of a resolution of the diagonal, in terms of Koszul dual bases. To apply this machinery one has to construct a Lefschetz fibration, with an antiholomorphic involution, whose real part $\pi_\mathbb{R}$ is a Morse function on $X_\mathbb{R} = Z$. This can be done easily, although not in a canonical way, by using techniques from real algebraic geometry. Roughly speaking, the resulting Fukaya category looks similar to the category of sheaves constructible with respect to the stratification given by the unstable manifolds of $\pi_\mathbb{R}$, compare [21]. However, because the construction of X is not precisely controlled, one does not expect these two categories to agree. Whilst this is not a problem for the proof of Theorem 1.1, it may be aesthetically unsatisfactory. One possibility for improving the situation would be to find a way of directly producing a Lefschetz fibration on the cotangent bundle; steps in that direction are taken in [20].

3. **Wrapping at infinity.** The third approach remains within M, and again uses Lagrangian submanifolds which are scaling-invariant at infinity. However, intersections at infinity are dealt with by flowing along the (not normalized) geodesic flow, which is a large perturbation. For instance, after this perturbation, the intersections of any two fibres will be given by all geodesics in Z connecting the relevant two points. In contrast to the previous constructions, this one is intrinsic to the differentiable manifold Z, and does not require a real-analytic or real-algebraic structure (there are of course technical choices to be made, such as the Riemannian metric and other perturbations belonging to standard pseudo-holomorphic curve theory; but the outcome is independent of those up to quasi-isomorphism). Conjecturally, the resulting "wrapped Fukaya category" is equivalent to a full subcategory of the category of modules over $C_{-*}(\Omega Z)$, the dg (differential graded) algebra of chains on the based loop space (actually, the Moore loop space, with the Pontryagin product). Note that the classical bar construction establishes a relation between this and the dg algebra of cochains on Z; for $\mathbb{K} = \mathbb{R}$, one can take this to be the algebra of differential forms; for $\mathbb{K} = \mathbb{Q}$, it could be Sullivan's model; and for general $\mathbb{K}$ one can use singular or Cech cohomology. If Z is simply connected, this relation leads to an equivalence of suitably defined module categories, and one can recover (1) in this way. In fact, we propose a more geometric version of this argument, which involves an explicit functor from the wrapped Fukaya category to the category of modules over a dg algebra of Cech cochains.

Remark 1.2 It is interesting to compare (1) with the result of a naive geometric argument. Suppose L is closed and exact; under fibrewise scaling $L \mapsto cL$, as $c \to \infty$, cL converges (in compact subsets) to a disjoint union of cotangent fibres $\bigcup_{x \in L \cap i(Z)} T_x^*$, where $i : Z \hookrightarrow M$ denotes the zero-section. In particular, for large c there is a canonical bijection between points of $L \cap cL$ and points of $\{L \cap T_x^* \,|\, x \in L \cap i(Z)\}$. Starting from this identification, and filtering by energy, one expects to obtain a spectral sequence

$$\bigoplus_{x \in L \cap i(Z)} HF(L, T_x^*) \Rightarrow H^*(L) \tag{2}$$

using exactness to identify $HF(L, cL) \cong H^*(L)$. This would (re)prove that $L \cap i(Z) \neq \emptyset$ and that $HF(L, T_x^*) \neq 0$; and, in an informal fashion, this provides motivation for believing that L is "generated by the fibres". It seems hard, however, to control the homology class of L starting from this, because it seems hard to gain sufficient control over $L \cap i(Z)$.

Each of the following three sections of the chapter are devoted to explaining one of these approaches. Then, in the concluding section, we take a look at the non-simply-connected case. First of all, there is a useful trick involving the spectral sequence (1) and finite covers of the base Z. In principle, this trick can be applied to any of the three approaches outlined above, but at the present state of the literature, the necessary prerequisites have been fully established only for the theory from [30]. Applying that, one arrives at the following consequence, which appears to be new:

Corollary 1.3 The assumption of simple-connectivity of Z can be removed from Theorem 1.1. This means that for all closed spin manifolds Z, and all exact Lagrangian submanifolds $L \subset T^*Z$ which are spin and have zero Maslov class, the conclusions (i–iii) hold.

From a more fundamental perspective, the approach via wrapped Fukaya categories seems particularly suitable for investigating cotangent bundles of non-simply-connected manifolds, since it retains information that is lost when passing from chains on ΩZ to cochains on Z. We end by describing what this would mean (modulo one of our conjectures, 4.6) in the special case when Z is a $K(\Gamma, 1)$. In the special case of the torus $Z = T^n$, Conjecture 4.6 can be sidestepped by direct geometric arguments, at least when char($\mathbb{K}$) $\neq 2$. Imposing that condition, one finds that an arbitrary exact, oriented and spin Lagrangian submanifold $L \subset T^*T^n = (\mathbb{C}^*)^n$ satisfies the conclusions of Theorem 1.1, with no assumption on the Maslov class.

Finally, it is worth pointing out that for any oriented and spin Lagrangian submanifold $L \subset T^*Z$, and any closed spin manifold Z, the theory produces a $\mathbb{Z}/2\mathbb{Z}$-graded spectral sequence

$$H(Z; \mathrm{End}(E_L)) \Rightarrow H(L) \tag{3}$$

which has applications in its own right, for instance to the classification of "local Lagrangian knots" in Euclidean space. Eliashberg and Polterovich [8] proved that any exact Lagrangian $L \subset \mathbb{C}^2$ which co-incides with the standard $\mathbb{R}^2$ outside a compact subset is in fact Lagrangian isotopic to $\mathbb{R}^2$. Again, their proof relies on an exclusively four-dimensional machinery.

Corollary 1.4 Let $L \subset \mathbb{C}^n$ be an exact Lagrangian submanifold which co-incides with $\mathbb{R}^n$ outside a compact set. Suppose that L is oriented and spin. Then (i) L is acyclic and (ii) $\pi_1(L)$ has no non-trivial finite-dimensional complex representations.

Using the result of Viterbo [38] and standard facts from 3-manifold topology, when $n = 3$ this implies that an oriented exact $L \subset \mathbb{C}^3$ which co-incides with $\mathbb{R}^3$ outside a compact set is actually diffeomorphic to $\mathbb{R}^3$. To prove Corollary 1.4, one embeds the given L (viewed in a Darboux ball) into the zero-section of $T^* S^n$, obtaining an exact Lagrangian submanifold L' of the latter which co-incides with the zero-section on an open set. This last fact immediately implies that $E_{L'}$ has rank one, so we see $\mathrm{End}(E_{L'})$ is the trivial $\mathbb{K}$-line bundle, and the sequence (3) implies $\mathrm{rk}_\mathbb{K} H^*(S^n) \geq \mathrm{rk}_\mathbb{K} H^*(L')$. On the other hand, projection $L' \to S^n$ is (obviously) degree ± 1, which gives the reverse inequality. Going back to $L \subset L'$, we deduce that L is acyclic with $\mathbb{K}$-coefficients. Since $\mathbb{K}$ is arbitrary, one deduces the first part of the Corollary. At this point, one knows *a posteriori* that the Maslov class of L vanishes; hence so does that of L', and the final statement of the Corollary then follows from an older result of Seidel [35]. When $n = 3$, it follows that L' is simply connected or a $K(\pi, 1)$, at which point one can appeal to Viterbo's work. (It is straightforward to deduce the acyclicity over fields of characteristic other than two in the other frameworks, for instance that of [14], but then the conclusion on the Maslov class does not follow.)

2 Constructible Sheaves

This section should be considered as an introduction to the two papers [29, 30]. Our aim is to present ideas from those papers in a way which is familiar to symplectic geometers. With that in mind, we have taken some liberties in the presentation, in particular omitting the (non-trivial) technical work involved in smoothing out characteristic cycles.

2.1 Fukaya Categories of Weinstein Manifolds

Let M be a Weinstein manifold which is of finite type and complete. Recall that a symplectic manifold (M, ω) is Weinstein if it comes with a distinguished Liouville (symplectically expanding) vector field Y, and a proper bounded below function $h : M \to \mathbb{R}$, such that $dh(Y)$ is positive on a sequence of level sets $h^{-1}(c_k)$, with $\lim_k c_k = \infty$. The stronger finite type assumption is that $dh(Y) > 0$ outside a compact subset of M. Finally, completeness means that the flow of Y is defined for all times (for negative times, this is automatically true, but for positive times it is an additional constraint). Note that the Liouville vector field defines a one-form $\theta = i_Y \omega$ with $d\theta = \omega$. At infinity, (M, θ) has the form $([0; \infty) \times N, e^r \alpha)$, where N is a contact manifold with contact one-form α, and r is the radial coordinate. In other words, the end of M is modelled on the positive half of the symplectization of (N, α). The obvious examples are cotangent bundles of closed manifolds, $M = T^*Z$, where Y is the radial rescaling vector field, and N the unit cotangent bundle.

We will consider exact Lagrangian submanifolds $L \subset M$ which are Legendrian at infinity. By definition, this means that $\theta|L$ is the derivative of some compactly supported function on L. Outside a compact subset, any such L will be of the form $[0; \infty) \times K$, where $K \subset N$ is a Legendrian submanifold. Now let (L_0, L_1) be two such submanifolds, whose structure at infinity is modelled on (K_0, K_1). To define their Floer cohomology, one needs a way of resolving the intersections at infinity by a suitable small perturbation. The details may vary, depending on what kind of Legendrian submanifolds one wants to consider. Here, we make the assumption that (N, α) is real-analytic, and allow only those K which are real-analytic submanifolds. Then,

Lemma 2.1 *Let (ϕ_R^t) be the Reeb flow on N. For any pair (K_0, K_1), there is an $\epsilon > 0$ such that $\phi_R^t(K_0) \cap K_1 = \emptyset$ for all $t \in (0, \epsilon)$.*

This is a consequence of the Curve Selection Lemma [28, Lemma 3.1], compare [30, Lemma 5.2.5]. Recall that, when defining the Floer cohomology of two Lagrangian submanifolds, one often adds a Hamiltonian perturbation $H \in C^\infty(M, \mathbb{R})$ (for technical reasons, this Hamiltonian is usually also taken to be time-dependent, but we suppress that here). The associated Floer cochain complex is generated by the flow lines $x: [0; 1] \to M$ of H going from L_0 to L_1; equivalently, these are the intersection points of $\phi_X^1(L_0) \cap L_1$, where X is the Hamiltonian vector field of H. We denote this cochain complex by

$$CF^*(L_0, L_1; H) = CF^*(\phi_X^1(L_0), L_1). \tag{4}$$

In our case, we take an H which at infinity is of the form $H(r, y) = h(e^r)$, where h is a function with $h' \in (0; \epsilon)$. Then, X is $h'(e^r)$ times the Reeb vector field R, hence $\phi_X^1(L_0) \cap L_1$ is compact by Lemma 2.1. Standard arguments show that the resulting Floer cohomology group $HF^*(L_0, L_1) = HF^*(L_0, L_1; H)$ is independent of H. It is also invariant under compactly supported (exact Lagrangian) isotopies of either L_0 or L_1. Note that in the case where $K_0 \cap K_1 = \emptyset$, one can actually set $h = 0$, which yields Floer cohomology in the ordinary (unperturbed) sense. Finally, for $L_0 = L_1 = L$ one has the usual reduction to Morse theory, so that $HF^*(L, L) \cong H^*(L)$, even for non-compact L.

At this point, we need to make a few more technical remarks. For simplicity, all our Floer cohomology groups are with coefficients in some field $\mathbb{K}$. If $char(\mathbb{K}) \neq 2$, one needs (relative) *Spin* structures on all Lagrangian submanifolds involved, in order to address the usual orientation problems for moduli spaces [13]. Next, Floer cohomology groups are generally only $\mathbb{Z}/2$-graded. One can upgrade this to a $\mathbb{Z}$-grading by requiring that $c_1(M) = 0$, and choosing gradings of each Lagrangian submanifold. For the moment, we do not need this, but it becomes important whenever one wants to make the connection with objects of classical homological algebra, as in Theorem 2.3, or in (1).

Example 2.2 Consider the case of cotangent bundles $M = T^*Z$ (to satisfy the general requirements above, we should impose real-analyticity conditions, but that is not actually necessary for the specific computations we are about to do). A typical example of a Lagrangian submanifold $L \subset M$ satisfying the conditions set out above is the conormal bundle $L = \nu^*W$ of a closed submanifold $W \subset Z$. If (L_0, L_1) are conormal bundles of transversally intersecting submanifolds (W_0, W_1), then

$$HF^*(L_0, L_1) \cong H^{*-codim(W_0)}(W_0 \cap W_1). \tag{5}$$

This is easy to see (except perhaps for the grading), since the only intersection points of the L_k lie in the zero-section. All of them have the same value of the action functional, and standard Morse–Bott techniques apply.

As a parallel but slightly different example, let $W \subset Z$ be an open subset with smooth boundary. Take a function $f : \overline{W} \to \mathbb{R}$ which is strictly positive in the interior, zero on the boundary, and has negative normal derivative at all boundary points. We can then consider the graph of $d(1/f)$, which is a Lagrangian submanifold of M, asymptotic to the positive part of the conormal bundle of ∂W. By a suitable isotopy, one can deform the graph so that it agrees at infinity with that conormal bundle. Denote the result by L. Given two such subsets W_k whose boundaries intersect transversally, one then has [23, 24, 31]

$$HF^*(L_0, L_1) \cong H^*(W_1 \cap \overline{W}_0, W_1 \cap \partial W_0). \tag{6}$$

Note that in both these cases, the Lagrangian submanifolds under consideration do admit natural gradings, so the isomorphisms are ones of $\mathbb{Z}$-graded groups.

We will need multiplicative structures on HF^*, realized on the chain level by an A_∞-category structure. The technical obstacle, in the first non-trivial case, is that the natural triangle product

$$\begin{aligned}
&CF^*(L_1, L_2; H_{12}) \otimes CF^*(L_0, L_1; H_{01}) \\
&= CF^*(\phi^1_{X_{12}}(L_1), L_2) \otimes CF^*(\phi^1_{X_{12}}\phi^1_{X_{01}}(L_0), \phi^1_{X_{12}}(L_1)) \\
&\longrightarrow CF^*(\phi^1_{X_{12}}\phi^1_{X_{01}}(L_0), L_2)
\end{aligned} \tag{7}$$

does not quite land in $CF^*(L_0, L_2; H_{02}) = CF^*(\phi^1_{X_{02}}(L_0), L_2)$. For instance, if one takes the same H for all pairs of Lagrangian submanifolds, the output of the product has $\phi_X^2 = \phi_{2X}^1$ instead of the desired ϕ_X^1. The solution adopted in [30] is (roughly speaking) to choose all functions h involved to be very small, in which case the deformation from X to $2X$ induces an actual isomorphism of Floer cochain groups. The downside is that this can only be done for a finite number of Lagrangian submanifolds and, more importantly, for a finite number of A_∞-products at a time. Hence, what one gets is a partially defined A_d-structure (for d arbitrarily large), from which one then has to produce a proper A_∞-structure; for some relevant algebraic results, see [13, Lemma 30.163]. As an alternative, one can take all functions h to satisfy $h(t) = \log t$, which means $H(r, y) = r$. Then, ϕ_X^2 is conjugate to a compactly supported perturbation of ϕ_X^1 (the conjugating diffeomorphism is the Liouville flow ϕ_Y^t for time $t = -\log(2)$). By making the other choices in a careful way, one can then arrange that (7) takes values in a Floer cochain group which is isomorphic to $CF^*(L_0, L_2; H_{02})$. In either way, one eventually ends up with an A_∞-category, which we denote by $\mathcal{F}(M)$.

2.2 Characteristic Cycles

Given a real-analytic manifold Z, one can consider sheaves of $\mathbb{K}$-vector spaces which are constructible (with respect to some real analytic stratification, which may depend on the sheaf). Denote by $D_c(Z)$ the full subcategory of the bounded-derived category of sheaves of $\mathbb{K}$-vector spaces comprising complexes with constructible cohomology. Kashiwara's characteristic cycle construction [22] associates to any object $\mathcal{G}$ in this category a Lagrangian cycle $CC(\mathcal{G})$ inside $M = T^*Z$, which is a cone (invariant under rescaling of cotangent fibres). If $\mathcal{G}$ is the structure sheaf of a closed submanifold, this cycle is just the conormal bundle, otherwise it tends to be singular. Nadler and Zaslow consider the structure sheaves of submanifolds $W \subset Z$ which are (real-analytic but) not necessarily closed. For each such "standard object", they construct a smoothing of $CC(\mathcal{G})$, which is a Lagrangian submanifold of M. In the special case where W is an open subset with smooth boundary, this is essentially equivalent to the construction indicated in Example 2.2. The singular boundary case is considerably more complicated and leads to Lagrangian submanifolds which are generally only asymptotically invariant under the Liouville flow (their limits at infinity are singular Legendrian cycles). Still, one can use them as objects of a Fukaya-type category, which is a variant of the previously described construction. We denote it by $\mathcal{A}(M)$, where $\mathcal{A}$ stands for "asymptotic". The main result of [30] is

Theorem 2.3 *The smoothed characteristic cycle construction gives rise to a full embedding of derived categories, $D_c(Z) \longrightarrow D\mathcal{A}(M)$.*

The proof relies on two ideas. One of them, namely that the standard objects generate $D_c(Z)$, is more elementary (it can be viewed as a fact about decompositions of real subanalytic sets). For purely algebraic reasons, this means that it is enough to define the embedding only on standard objects.

The other, more geometric, technique is the reduction of Floer cohomology to Morse theory, provided by the work of Fukaya and Oh [12].

2.3 Decomposing the Diagonal

In $D_c(Z)$, the structure sheaves of any two distinct points are algebraically disjoint (there are no morphisms between them). Of course, the same holds for cotangent fibres in the Fukaya category of M, which are the images of such structure sheaves under the embedding from Theorem 2.3. As a consequence, in $\mathcal{F}(M)$ (or $\mathcal{A}(M)$, the difference being irrelevant at this level) one cannot expect to have a finite resolution of a closed exact $L \subset M$ in terms of cotangent fibres. However, Nadler proves a modified version of this statement, where the fibres are replaced by standard objects associated with certain contractible subsets of Z.

Concretely, fix a real analytic triangulation of Z. Denote by x_i the vertices of the triangulation, by U_i their stars, and by $U_I = \bigcap_{i \in I} U_i$ the intersections of such stars, indexed by finite sets $I = \{i_0, \dots, i_d\}$. There is a standard Cech resolution of the constant sheaf $\mathbb{K}_Z$ in terms of the $\mathbb{K}_{U_I}$ (we will encounter a similar construction again later on, in Sect. 4.1). Rather than applying this to Z itself, we take the diagonal inclusion $\delta : Z \to Z \times Z$, and consider the induced resolution of $\delta_*(\mathbb{K}_Z)$ by the objects $\delta_*(\mathbb{K}_{U_I})$. Consider the embedding from Theorem 2.3 applied to $Z \times Z$. The image of $\delta_*(\mathbb{K}_Z)$ is the conormal bundle of the diagonal $\Delta = \delta(Z)$, and each $\delta_*(\mathbb{K}_{U_I})$ maps to the standard object associated to $\delta(U_I) \subset Z \times Z$. Since each U_I is contractible, one can deform $\delta(U_I)$ to $U_I \times \{x_{i_d}\}$ (as locally closed submanifolds of $Z \times Z$), and this induces an isotopy of the associated smoothed characteristic cycles. A priori, this isotopy is not compactly supported, hence not well-behaved in our category (it does not preserve the isomorphism type of objects). However, this is not a problem if one is only interested in morphisms from or to a given closed Lagrangian submanifold.

To formalize this, take $\mathcal{A}^{cpt}(M)$ to be the subcategory of $\mathcal{A}(M)$ consisting of closed Lagrangian submanifolds (this is in fact the Fukaya category in the classical sense). Dually, let $mod(\mathcal{A}(M))$, $mod(\mathcal{A}^{cpt}(M))$ be the associated categories of A_∞-modules. There is a chain of A_∞-functors

$$\mathcal{A}^{cpt}(M) \longrightarrow \mathcal{A}(M) \longrightarrow mod(\mathcal{A}(M)) \longrightarrow mod(\mathcal{A}^{cpt}(M)). \qquad (8)$$

The first and second one, which are inclusion and the Yoneda embedding, are full and faithful. The last one, restriction of A_∞-modules, will not generally have that property. However, the composition of all three is just the Yoneda embedding for $\mathcal{A}^{cpt}$, which is again full and faithful. In view of [40], and its chain-level analogue [27], each object C in $\mathcal{A}(M \times M)$ (and more generally, twisted complex built out of such objects) induces a convolution functor

$$\Phi_C : \mathcal{A}^{cpt}(M) \longrightarrow mod(\mathcal{A}^{cpt}(M)) \qquad (9)$$

(usually, one reverses the sign of the symplectic form on one of the two factors in $M \times M$, but for cotangent bundles, this can be compensated by a fibrewise reflection $\sigma : M \to M$). First, take C to be the conormal bundle of the diagonal, which is the same as the graph of σ. Then, convolution with C is isomorphic to the embedding (8). On the other hand, if C is the smoothed characteristic cycle of some product $U \times \{x\}$, then Φ_C maps each object to a direct sum of copies of T_x^* (the image of that fibre under the functor $\mathcal{A}(M) \to mod(\mathcal{A}^{cpt}(M))$, to be precise). Finally, if C is just a Lagrangian submanifold, Φ_C is invariant under Lagrangian isotopies which are not necessarily compactly supported. By combining those facts, one obtains the desired resolution of a closed exact $L \subset M$. Nadler actually pushes these ideas somewhat further, using a refined version of this argument, to show that:

Theorem 2.4 *The embedding $D_c(Z) \to D\mathcal{A}(M)$ from Theorem 2.3 is an equivalence.*

Remark 2.5 If one is only interested in the spectral sequence (1), there may be a potential simplification, which would bypass some of the categorical constructions above. First of all, rewrite $HF^*(L, L) \cong H^*(L; \mathbb{K})$ as $HF^*(L \times L, \nu^* \Delta)$. Then, using the resolution of $\nu^* \Delta$ in $\mathcal{A}(M \times M)$ described above, one gets a spectral sequence converging towards that group, whose E_1 page comprises the Floer cohomology groups between $L \times L$ and the smoothed characteristic cycles of $\delta(U_I)$. Since $L \times L$ is compact, one can deform $\delta(U_I)$ to $U_I \times \{x_{i_d}\}$, and from there construct an isotopy from its smoothed characteristic cycle to $T_{x_{i_d}}^* \otimes T_{x_{i_d}}^*$. In the terminology used in (1), this brings the terms in the E_1 page into the form $End(E_L)_{x_{i_d}}$. To get the desired E_2 term, one would further have to check that the differentials reproduce the ones in the Cech complex with twisted coefficients in $End(E_L)$. This of course follows from Theorem 2.4, but there ought to be a more direct geometric argument, just by looking at the relevant spaces of holomorphic triangles; this seems a worth while endeavour, but we have not attempted to study it in detail.

3 Lefschetz Thimbles

This section gives an overview of the paper [14], and an account – emphasizing geometric rather than algebraic aspects – of some of the underlying theory from the book [36].

3.1 Fukaya Categories of Lefschetz Fibrations

In principle, the notion of Lefschetz fibration can be defined in a purely symplectic way. However, we will limit ourselves to the more traditional algebro-geometric context. Let X be a smooth affine variety, and

$$\pi : X \longrightarrow \mathbb{C} \tag{10}$$

a polynomial, which has only non-degenerate critical points. For convenience, we assume that no two such points lie in the same fibre. Additionally, we impose a condition which excludes singularities at infinity, namely: let $\bar{X}$ be a projective completion of X, such that $D = \bar{X} \setminus X$ is a divisor with normal crossing. We then require that (for an approriate choice of $\bar{X}$) the closure of $\pi^{-1}(0)$ should be smooth in a neighbourhood of D, and intersect each component of D transversally. Finally, for Floer-theoretic reasons, we require X to be Calabi–Yau (have trivial canonical bundle).

Take any Kähler form on $\bar{X}$ which comes from a metric on $\mathcal{O}(D)$. Its restriction to X makes that variety into a Weinstein manifold (of finite type, but not complete; the latter deficiency can, however, be cured easily, by attaching the missing part of the conical end). Moreover, parallel transport for π is well-defined away from the singular fibres, in spite of its non-properness.

A vanishing path $\gamma : [0, \infty) \to \mathbb{C}$ is an embedding starting at a critical value $\gamma(0)$ of π, and such that for $t \gg 0$, $\gamma(t) = const. - it$ is a half-line going to $-i\infty$. To each such path one can associate a *Lefschetz thimble* $\Delta_\gamma \subset X$, which is a Lagrangian submanifold diffeomorphic to $\mathbb{R}^n$, projecting properly to $\gamma([0, \infty)) \subset \mathbb{C}$. More precisely, $\gamma^{-1} \circ \pi|\Delta_\gamma$ is the standard proper Morse function on $\mathbb{R}^n$ with a single minimum (placed at the unique critical point of π in the fibre over $\gamma(0)$). When defining the Floer cohomology between two Lefschetz thimbles, the convention is to rotate the semi-infinite part of the first path in anticlockwise direction for some small angle. Omitting certain technical points, this can be interpreted as adding a Hamiltonian term as in Sect. 2.1. In particular, one again has

$$HF^*(L_\gamma, L_\gamma) \cong H^*(L_\gamma; \mathbb{K}) = \mathbb{K}. \qquad (11)$$

Now suppose that $(\gamma_0, \ldots, \gamma_m)$ is a basis (sometimes also called a distinguished basis) of vanishing paths. We will not recall the definition here; for a sketch, see Fig. 1. In that situation, if one takes (γ_j, γ_k) with $j > k$ and applies the rotation described above to γ_j, the result remains disjoint from γ_k. Hence,

$$HF^*(L_{\gamma_j}, L_{\gamma_k}) = 0 \text{ for all } j > k. \qquad (12)$$

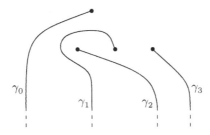

Fig. 1. A distinguished basis of vanishing paths

As usual, rather than working on the level of Floer cohomology, we want to have underlying A_∞-structures. There is a convenient shortcut, which eliminates non-compact Lagrangian submanifolds from the foundations of the theory (but which unfortunately requires $char(\mathbb{K}) \neq 2$). Namely, let $\tilde{X} \to X$ be the double cover branched over some fibre $\pi^{-1}(-iC)$, $C \gg 0$. Roughly speaking, one takes the ordinary Fukaya category $\mathcal{F}(\tilde{X})$, which contains only compact Lagrangian submanifolds, and defines $\mathcal{F}(\pi)$ to be its $\mathbb{Z}/2$-invariant part (only invariant objects and morphisms; obviously, getting this to work on the cochain level requires a little care). This time, let us allow only vanishing paths which satisfy $\gamma(t) = -it$ for $t \geq C$ (which is no problem, since each path can be brought into this form by an isotopy). One then truncates the Lefschetz thimble associated to such a path, so that it becomes a Lagrangian disc with boundary in $\pi^{-1}(-iC)$, and takes its preimage in $\tilde{X}$, which is a closed $\mathbb{Z}/2$-invariant Lagrangian sphere $\tilde{S}_\gamma \subset \tilde{X}$. On the cohomological level, the $\mathbb{Z}/2$-invariant parts of the Floer cohomologies of these spheres still satisfy the same properties as before, in particular reproduce (11) and (12).

Theorem 3.1 *If $(\gamma_0, \ldots, \gamma_m)$ is a basis of vanishing paths, the associated $\tilde{S}_{\gamma_j}$ form a full exceptional collection in the derived category $D\mathcal{F}(\pi)$.*

The fact that we get an exceptional collection is elementary; it just reflects the two Eqs. (11) and (12) (or rather, their counterparts for the modified definition of Floer cohomology involving double covers). Fullness, which is the property that this collection generates the derived category, is rather more interesting. The proof given in [36] relies on the fact that the product of Dehn twists along the $\tilde{L}_{\gamma_k}$ is isotopic to the covering involution in $\tilde{X}$. Hence, if $L \subset X$ is a closed Lagrangian submanifold which lies in $\pi^{-1}(\{\text{im}(z) > -C\})$, this product of Dehn twists will exchange the two components of the preimage $\tilde{L} \subset \tilde{X}$. The rest of the argument essentially consists in applying the long exact sequence from [34].

3.2 Postnikov Decompositions

We will use some purely algebraic properties of exceptional collections, see for instance [15] (the subject has a long history in algebraic geometry; readers interested in this might find the collection [32] to be a good starting point). Namely, let C be a triangulated category, linear over a field $\mathbb{K}$, and let $(Y_0, \ldots, Y_m)$ be a full exceptional collection of objects in C. Then, for any object X, there is a collection of exact triangles

$$Z_k \otimes Y_k \to X_k \to X_{k-1} \xrightarrow{[1]} Z_k \otimes Y_k \tag{13}$$

where $X_m = X$, $X_{-1} = 0$, and $Z_k = Hom_C^*(Y_k, X_k)$ (morphisms of all degrees; by assumption, this is finite-dimensional, so $Z_k \otimes Y_k$ is the direct sum of finitely many shifted copies of Y_k). The map $Z_k \otimes Y_k \to X_k$ is the canonical evaluation map, and X_{k-1} is defined (by descending induction on k) to be its mapping cone. To get another description of Z_k, one can use the unique (right) Koszul dual exceptional collection $(Y_m^!, \ldots, Y_0^!)$, which satisfies

$$Hom_C^*(Y_j, Y_k^!) = \begin{cases} \mathbb{K} \text{ (concentrated in degree zero)} & j = k, \\ 0 & \text{otherwise.} \end{cases} \tag{14}$$

It then follows by repeatedly applying (13) that $Z_k^\vee \cong Hom_C^*(X, Y_k^!)$. Now, given any cohomological functor R on C, we get an induced spectral sequence converging to $R(X)$, whose starting page has columns $E_1^{rs} = (Z_{m-r}^\vee \otimes R(Y_{m-r}))^{r+s}$. In particular, taking $R = Hom_C^*(-, X)$ and using the expression for Z_k explained above, we get a spectral sequence converging to $Hom_C^*(X, X)$, which starts with

$$E_1^{rs} = \left(Hom_C^*(X, Y_{m-r}^!) \otimes Hom_C^*(Y_{m-r}, X) \right)^{r+s}. \tag{15}$$

We now return to the concrete setting where $C = D\mathcal{F}(\pi)$. In this case, the Koszul dual of an exceptional collection given by a basis of Lefschetz thimbles is another such basis $\{\gamma_0^!, \ldots, \gamma_m^!\}$. This is a consequence of the more general relation between mutations (algebra) and Hurwitz moves on vanishing paths (geometry). Applying (15), and going back to the original definition of Floer cohomology, we therefore get the following result: for every (exact, graded, spin) closed Lagrangian submanifold $L \subset M$, there is a spectral sequence converging to $HF^*(L, L) \cong H^*(L; \mathbb{K})$, which starts with

$$E_1^{rs} = (HF(L, \Delta_{\gamma_{m-r}^!}) \otimes HF(\Delta_{\gamma_{m-r}}, L))^{r+s}. \tag{16}$$

3.3 Real Algebraic Approximation

The existence of (16) is a general statement about Lefschetz fibrations. To make the connection with cotangent bundles, we use a form of the Nash–Tognoli theorem, see for instance [19], namely:

Lemma 3.2 *If Z is a closed manifold and $p: Z \to \mathbb{R}$ is a Morse function, there is a Lefschetz fibration $\pi : X \to \mathbb{C}$ with a compatible real structure, and a diffeomorphism $f: Z \to X_\mathbb{R}$, such that $\pi \circ f$ is C^2-close to p.*

The diffeomorphism f can be extended to a symplectic embedding ϕ of a neighbourhood of the zero-section of $Z \subset M = T^*Z$ into X. Hence (perhaps after a preliminary radial rescaling) we can transport closed exact Lagrangian submanifolds $L \subset M$ over to X. The critical points of π fall into two classes, namely real and purely complex ones, and the ones in the first class correspond canonically to critical points of p. By a suitable choice of vanishing paths, one can ensure that

$$HF^*(\Delta_{\gamma_k}, \phi(L))$$
$$\cong \begin{cases} HF^*(T_{x_k}^*, L) & \text{if } \gamma(0) \in X_\mathbb{R} \text{ corresponds to } x_k \in Crit(p), \\ 0 & \text{otherwise.} \end{cases} \tag{17}$$

Here, the Floer cohomology on the right-hand side is taken inside T^*Z. The same statement holds for the other groups in (16), up to a shift in the grading which depends on the Morse index of x_k. As a consequence, the starting page of that spectral sequence can be thought of as $C^*_{Morse}(Z; End^*(E_L))$, where the Morse complex is taken with respect to the function p, and using the (graded) local coefficient system E_L. This is one page earlier than our usual starting term (1), but is already good enough to derive Theorem 1.1 by appealing to some classical manifold topology (after taking the product with a sphere if necessary, one can assume that $dim(Z) > 5$, in which case simple connectivity of Z implies that one can choose a Morse function without critical points of index or co-index 1).

Remark 3.3 The differential on the E_1-page of (16) is given in [36, Corollary 18.27], in terms of holomorphic triangle products between adjacent Lefschetz thimbles in the exceptional collection. In the special situation of (17), identifying the E_1 page with $C^*_{Morse}(Z; End(E_L))$, there is also the Morse differential δ coming from parallel transport in the local system $E_L \to Z$ (compare Remark 2.5). For Lefschetz fibrations arising from real algebraic approximation, rather than some more canonical construction, there seems to be no reason for these to agree in general; but one does expect the parts of the differential leading out of the first column, and into the last column, to agree. For instance, by deforming the final vanishing path to lie along the real axis, one can ensure that the entire thimble leading out of the maximum x_{max} of the Morse function is contained in the real locus, after which the intersection points between this thimble and one coming from a critical point x of index one less correspond bijectively to the gradient lines of the Morse function between x_{max} and x (the situation at the minimum x_{min} is analogous). If it was known that this part of the E_1-differential did reproduce the corresponding piece of δ, one could hope to study non-simply-connected cotangent bundles in this approach.

4 Family Floer Cohomology

This section covers the third point of view on Theorem 1.1. This time, the presentation is less linear, and occasionally several ways of reaching a particular goal are sketched. The reader should keep in mind the preliminary nature of this discussion. In some parts, this means that there are complete but unpublished constructions. For others, only outlines or strategies of proof exist, in which case we will be careful to formulate the relevant statements as conjectures.

4.1 Cech Complexes

At the start of the chapter, we mentioned that to an $L \subset M = T^*Z$ one can associate the bundle E_L of Floer cohomologies $E_x = HF^*(T^*_x, L)$. One naturally wants to replace E_L by an underlying cochain level object $\mathcal{E}_L$, which should be a "sheaf of complexes" in a suitable sense. In our interpretation, this will be a dg module over a dg algebra of Cech cochains (there are several other possibilities, with varying degrees of technical difficulty; see [10, Sect. 5] and [29, Sect. 4] for sketches of two of these, and compare also to [18]).

Fix a smooth triangulation of Z, with vertices x_i, and let U_I be the intersections of sets in the associated open cover, just as in Sect. 2.3 (but omitting the real analyticity condition). This time, we want to write down the associated Cech complex explicitly, hence fix an ordering of the i's. Let $\Gamma(U_I)$ be the space of locally constant $\mathbb{K}$-valued functions on U_I, which in our case is $\mathbb{K}$ if $U_I \neq \emptyset$, and 0 otherwise. The Cech complex is

$$\mathcal{C} = \bigoplus_I \Gamma(U_I)[-d] \tag{18}$$

where $d = |I| - 1$. This carries the usual differential, and also a natural associative product making it into a dg algebra. Namely, for every possible splitting of $I'' = \{i_0 < \cdots < i_d\}$ into $I' = \{i_0 < \cdots < i_k\}$ and $I = \{i_k < \cdots < i_d\}$, one takes

$$\Gamma(U_I) \otimes \Gamma(U_{I'}) \xrightarrow{\text{restriction}} \Gamma(U_{I''}) \otimes \Gamma(U_{I''}) \xrightarrow{\text{multiplication}} \Gamma(U_{I''}). \tag{19}$$

We want to consider (unital right) dg modules over $\mathcal{C}$. Denote the dg category of such modules by $\mathcal{M} = mod(\mathcal{C})$. This is *not* a dg model for the derived category: there are acyclic modules which are non-trivial in $H(\mathcal{M})$, and as a consequence, quasi-isomorphism does not imply isomorphism in that category.

All objects we will consider actually belong to a more restricted class, distinguished by a suitable "locality" property; we call these dg modules of *presheaf type*. The definition is that such a dg module $\mathcal{E}$ needs to admit a splitting

$$\mathcal{E} = \bigoplus_I \mathcal{E}_I[-d], \tag{20}$$

where the sum is over all $I = \{i_0 < \cdots < i_d\}$ such that $U_I \neq \emptyset$. This splitting is required to be compatible with the differential and module structure. This means that the differential maps $\mathcal{E}_I$ to the direct sum of $\mathcal{E}_{I'}$ over all $I' \supset I$; that $1 \in \Gamma(U_{i'})$ acts as the identity on $\mathcal{E}_I[-d]$ for all $I = \{i_0 < \cdots < i_d\}$ with $i_d = i'$, and as zero otherwise; and that the component $\mathcal{E}_I \otimes \Gamma(U_{I'}) \longrightarrow \mathcal{E}_{I''}$ of the module structure can only be non-zero if $I'' \supset I \cup I'$. In particular, the "stalks" $\mathcal{E}_I$ themselves are subquotients of $\mathcal{E}$, and inherit a dg module structure from that. The stalks associated to the smallest subsets of Z (i.e. to maximal index sets I) are actually chain complexes, with all chain homomorphisms being module endomorphisms, from which one can show:

Lemma 4.1 *Let $\mathcal{E}$ be a dg module of presheaf type. If each $\mathcal{E}_I$ is acyclic, $\mathcal{E}$ itself is isomorphic to the zero object in $H(\mathcal{M})$.*

Here is a first example. Let $P \to Z$ be a flat $\mathbb{K}$-vector bundle (or local coefficient system). For each $I = \{i_0 < \cdots < i_d\}$ such that $U_I \neq \emptyset$, define $(\mathcal{E}_P)_I = P_{x_{i_d}}$. Form the direct sum $\mathcal{E}_P$ as in (20). Equivalently, one can think of this as the sum of $P_{x_{i_d}} \otimes \Gamma(U_I)$ over all I, including empty ones. The differential consists of terms $(\mathcal{E}_P)_I \to (\mathcal{E}_P)_{I'}$ for $I = \{i_0 < \cdots < i_d\}$, $I' = I \cup \{i'\}$. If $i' < i_d$, these are given (up to sign) by restriction maps $P_{x_{i_d}} \otimes \Gamma(U_I) \to P_{x_{i_d}} \otimes \Gamma(U_{I'})$. In the remaining non-trivial case where $i' > i_d$ and $U_{I'} \neq \emptyset$, there is a unique edge of our triangulation going from x_{i_d} to $x_{i'}$, and one combines restriction with parallel transport along that edge. The module structure on $\mathcal{E}_P$ is defined in the obvious way, following the model (19); compatibility of the differential and the right $\mathcal{C}$-module structure is ensured by our choice $(\mathcal{E}_P)_I = P_{x_{i_d}}$, taking the fibre of P over the vertex corresponding to the last index i_d of I. Clearly, $H^*(\mathcal{E}_P) \cong H^*(Z; P)$ is ordinary cohomology with P-coefficients. Moreover, using the fact that every $\mathcal{E}_P$ is free as a module (ignoring the differential), one sees that

$$H^*(hom_{\mathcal{M}}(\mathcal{E}_{P_0}, \mathcal{E}_{P_1})) \cong H^*(Z; Hom(P_0, P_1)). \tag{21}$$

In particular, referring back to the remark made above, this leads to quite concrete examples of acyclic dg modules which are nevertheless non-trivial objects (since they have non-trivial endomorphism rings).

Still within classical topology, but getting somewhat closer to the intended construction, suppose now that $Q \to Z$ is a differentiable fibre bundle, equipped with a connection. For $I = i_0 < \cdots < i_d$ with $U_I \neq \emptyset$, define

$$(\mathcal{E}_Q)_I = C_{-*}(Q_{x_{i_d}}), \tag{22}$$

where C_{-*} stands for cubical cochains, with the grading reversed (to go with our general cohomological convention). As before, we want to turn the (shifted) direct sum of these into a dg module over $\mathcal{C}$. The module structure is straightforward, but the differential is a little more interesting, being the sum of three terms. The first of these is the ordinary boundary operator on each $(\mathcal{E}_Q)_I$. The second one is the Cech differential $(\mathcal{E}_Q)_I \to (\mathcal{E}_Q)_{I'}$, where $I = \{i_0 < \cdots < i_d\}$ and $I' = I \cup \{i'\}$ with $i' < i_d$. The final term consists of maps

$$C_{-*}(Q_{x_{i_d}}) \longrightarrow C_{-*}(Q_{x_{i'_e}})[1 - e] \tag{23}$$

where $I = \{i_0 < \cdots < i_d\}$, $I' = I \cup \{i'_1 < \cdots < i'_e\}$ with $i_d < i'_1$. Take the standard e-dimensional simplex Δ^e. It is a classical observation [3] that there is a natural family of piecewise smooth paths in Δ^e, parametrized by an $(e - 1)$-dimensional cube, which join the first to the last vertex. In our situation, the triangulation of Z contains a unique simplex with vertices $\{i_d, i'_1, \ldots, i'_e\}$, and we therefore get a family of paths joining x_{i_d} to $x_{i'_e}$. The resulting parametrized parallel transport map

$$[0; 1]^{e-1} \times Q_{x_{i_d}} \longrightarrow Q_{x_{i'_e}} \tag{24}$$

induces (23) (up to sign; the appearance of a cube here motivated our use of cubical chains). It is not difficult to show that the resulting total differential on $\mathcal{E}_Q$ indeed squares to zero, and is compatible with the module structure.

Finally, let us turn to the symplectic analogue of this, in which one starts with a closed exact Lagrangian submanifold $L \subset M = T^*Z$ (subject to the usual conditions: if one wants $\mathbb{Z}$-graded modules, L should be graded; and if one wants to use coefficient fields $char(\mathbb{K}) \neq 2$, it should be relatively spin). The appropriate version of (22) is

$$(\mathcal{E}_L)_I = CF^*(T^*_{x_{i_d}}, L) \tag{25}$$

and again, $\mathcal{E}_L$ is the sum of these. In the differential, we now use the Floer differential on each CF^* summand (replacing the differential on cubical chains), and continuation maps or their parametrized analogues (rather than parallel transport maps), which govern moving one cotangent fibre into the other. (Related continuation maps appeared in [24].) Of course, the details are somewhat different from the previous case. To get a chain map $CF^*(T^*_{x_{i_d}}, L) \to CF^*(T^*_{x_{i_1'}}, L)$, one needs a path

$$\gamma : \mathbb{R} \longrightarrow Z, \quad \gamma(s) = x_{i_d} \text{ for } s \ll 0, \ \gamma(s) = x_{i_1'} \text{ for } s \gg 0. \tag{26}$$

More generally, families of such paths parametrized by $[0;1)^{e-1}$ appear. In the limit as one (or more) of the parameters go to 1, the path breaks up into two (or more) pieces separated by increasingly long constant stretches. This ensures that the usual composition laws for continuation maps apply, compare [33]. Still, with these technical modifications taken into account, the argument remains essentially the same as before.

4.2 Wrapped Floer Cohomology

One naturally wants to extend the previous construction to allow L to be non-compact (for instance, a cotangent fibre). This would be impossible using the version of Floer cohomology from Sect. 2.1, since that does not have sufficiently strong isotopy invariance properties: $HF^*(T^*_x, L)$ generally depends strongly on x. Instead, we use a modified version called "wrapped Floer cohomology". This is not fundamentally new: it appears in [1] for the case of cotangent bundles, and is actually the open string analogue of Viterbo's symplectic cohomology [38].

Take a Weinstein manifold M (complete and of finite type) and consider exact Lagrangian submanifolds inside it which are Legendrian at infinity; this time, no real analyticity conditions will be necessary. We again use Hamiltonian functions of the form $H(r, y) = h(e^r)$ at infinity, but now require that $\lim_{r \to \infty} h'(e^r) = +\infty$. This means that as one goes to infinity, the associated Hamiltonian flow is an unboundedly accelerating version of the Reeb flow. Denote by $CW^*(L_0, L_1) = CF^*(L_0, L_1; H) = CF^*(\phi_X^1(L_0), L_1)$ the resulting Floer complex, and by

$$HW^*(L_0, L_1) = H(CW^*(L_0, L_1)) \qquad (27)$$

its cohomology, which we call *wrapped Floer cohomology*. This is independent of the choice of H. Moreover, it remains invariant under isotopies of either L_0 or L_1 inside the relevant class of submanifolds. Such isotopies need no longer be compactly supported; the Legendrian submanifolds at infinity may move (which is exactly the property we wanted to have). By exploiting this, it is easy to define a triangle product on wrapped Floer cohomology. In the case where $L_0 = L_1 = L$, we have a natural map from the ordinary cohomology $H^*(L)$ to $HW^*(L, L)$, which however is generally neither injective nor surjective. For instance, for $L = \mathbb{R}^n$ inside $M = \mathbb{C}^n$, $HW^*(L, L)$ vanishes. Another, far less trivial, example is the following one:

Theorem 4.2 (Abbondandolo–Schwarz) *Let $M = T^*Z$ be the cotangent bundle of a closed oriented manifold, and take two cotangent fibres $L_0 = T^*_{x_0}Z$, $L_1 = T^*_{x_1}Z$. Then $HW^*(L_0, L_1) \cong H_{-*}(\mathcal{P}_{x_0, x_1})$ is the (negatively graded) homology of the space of paths in Z going from x_0 to x_1.*

This is proved in [1]; the follow-up paper [2] shows that this isomorphism sends the triangle product on HW^* to the Pontryagin product (induced by composition of paths) on path space homology.

We want our wrapped Floer cochain groups to carry an A_∞-structure, refining the cohomology level product. When defining this, one encounters the same difficulty as in (7). Again, there is a solution based on a rescaling trick: one takes $h(t) = \frac{1}{2}t^2$, and uses the fact that ϕ_X^2 differs from $\phi_Y^{-\log(2)}\phi_X^1\phi_Y^{\log(2)}$ (ϕ_Y being the Liouville flow) only by a compactly supported isotopy. This is particularly intuitive for cotangent bundles, where the radial coordinate at infinity is $e^r = |p|$; then, $H = h(e^r) = \frac{1}{2}|p|^2$ gives rise to the standard geodesic flow (ϕ_X^t), and $Y = p\partial_p$ is the rescaling vector field, meaning that $\phi_Y^{\log(2)}$ doubles the length of cotangent vectors. An alternative (not identical, but ultimately quasi-isomorphic) approach to the same problem is to define wrapped Floer cohomology as a direct limit over functions H_k with more moderate growth $H_k(r, y) = ke^r$ at infinity. However, the details of this are quite intricate, and we will not describe them here. In either way, one gets an A_∞-category $\mathcal{W}(M)$, which we call the *wrapped Fukaya category*. For cotangent bundles $M = T^*Z$, a plausible cochain level refinement of Theorem 4.2 is the following one:

Conjecture 4.3 Let $L = T^*_x$ be a cotangent fibre. Then, the A_∞-structure on $CW^*(L, L)$ should be quasi-isomorphic to the dg algebra structure on $C_{-*}(\Omega_x)$, where Ω_x is the (Moore) based loop space of (Z, x).

Returning to the general case, we recall that a fundamental property of symplectic cohomology, established in [38], is *Viterbo functoriality* with respect to embeddings of Weinstein manifolds. One naturally expects a corresponding property to hold for wrapped Fukaya categories. Namely, take a bounded open subset $U \subset M$ with smooth boundary, such that Y points out-wards along ∂U. One can then attach an infinite cone to ∂U, to form another Weinstein manifold $M' = U \cup_{\partial U} ([0; \infty) \times \partial U)$. Suppose that $(L_1, \ldots, L_d)$ is a finite family of exact Lagrangian submanifolds in M, which are Legendrian at infinity, and with the following additional property: $\theta | L_k = dR_k$ for some compactly supported function R_k, which in addition vanishes in a neighbour-hood of $L_k \cap \partial U$. This implies that $L_k \cap \partial U$ is a Legendrian submanifold of ∂U. Again, by attaching infinite cones to $L_k \cap U$, one gets exact Lagrangian submanifolds $L'_k \subset M'$, which are Legendrian at infinity. Let $\mathcal{A} \subset \mathcal{W}(M)$ be the full A_∞-subcategory with objects L_k, and similarly $\mathcal{A}' \subset \mathcal{W}(M')$ for L'_k. Then,

Conjecture 4.4 (Abouzaid-Seidel) There is a natural (up to isomorphism) A_∞-functor $\mathcal{R} : \mathcal{A} \to \mathcal{A}'$.

Note that, even though we have not mentioned this explicitly so far, all A_∞-categories under consideration have units (on the cohomological level), and $\mathcal{R}$ is supposed to be a (cohomologically) unital functor. Hence, the conjecture implies that various relations between objects, such as isomorphism or exact triangles, pass from $\mathcal{A}$ to $\mathcal{A}'$, which is a non-trivial statement in itself.

4.3 Family Floer Cohomology Revisited

For $M = T^*Z$, there is a straightforward variation of the construction from Sect. 4.1, using wrapped Floer cohomology instead of ordinary Floer coho-mology. This associates to any exact Lagrangian submanifold $L \subset M$, which is Legendrian at infinity, a dg module $\mathcal{E}_L$ over $\mathcal{C}$. In fact, it gives rise to an A_∞-functor

$$\mathcal{G} : \mathcal{W}(M) \longrightarrow \mathcal{M} = mod(\mathcal{C}). \tag{28}$$

While little has been rigorously proved so far, there are plausible expectations for how this functor should behave, which we will now formulate precisely. Take $Q \to Z$ to be the path fibration (whose total space is contractible). Even though this is not strictly a fibre bundle, the construction from Sect. 4.1 applies, and yields a dg module $\mathcal{E}_Q$, which has $H^*(\mathcal{E}_Q) \cong \mathbb{K}$ (this can be viewed as a resolution of the simple $\mathcal{C}$-module).

Conjecture 4.5 Let $L = T^*_x$ be a cotangent fibre. Then, $\mathcal{E}_L$ is isomorphic to $\mathcal{E}_Q$ in $H(\mathcal{M})$. Moreover, if Z is simply connected, $\mathcal{G}$ gives rise to a quasi-isomorphism $C_{-*}(\Omega_x) \cong CW^*(L, L) \to hom_\mathcal{M}(\mathcal{E}_L, \mathcal{E}_L)$.

The first statement can be seen as a parametrized extension of Theorem 4.2. A possible proof would be to consist of checking that the chain level maps constructed in [1] can be made compatible with parallel transport (respectively continuation) maps, up to a suitable hierarchy of chain homotopies. This would yield a map of dg modules; to prove that it is an isomorphism, one would then apply Lemma 4.1 to its mapping cone. The second part of the conjecture is less intuitive. The assumption of simple connectivity is necessary, otherwise the endomorphisms of $\mathcal{E}_Q$ may not reproduce the homology of the based loop space (see Sect. 4.4 for further discussion of this); however, it is not entirely clear how that would enter into a proof.

Conjecture 4.6 Any one fibre $L = T_x^*$ generates the derived category (taken, as usual, to be the homotopy category of twisted complexes) of $\mathcal{W}(M)$.

There are two apparently quite viable approaches to this, arising from the contexts of Sects. 2 and 3, respectively. To explain the first one, we go back to the general situation where M is a finite-type complete Weinstein manifold, whose end is modelled on the contact manifold N, and where real analyticity conditions are imposed on N and its Legendrian submanifolds. Then, if (L_0, L_1) are exact Lagrangian submanifolds which are Legendrian at infinity, we have a natural homomorphism

$$HF^*(L_0, L_1) \longrightarrow HW^*(L_0, L_1), \tag{29}$$

which generalizes the map $H^*(L) \to HW^*(L, L)$ mentioned in Sect. 4.2. These maps are compatible with triangle products, and even though the details have not been checked, it seems plausible that they can be lifted to an A_∞-functor $\mathcal{F}(M) \to \mathcal{W}(M)$. Actually, what one would like to use is a variant of this, where $\mathcal{F}(M)$ is replaced by the Nadler–Zaslow category $\mathcal{A}(M)$, or at least a sufficiently large full subcategory of it. Assuming that this can be done, one can take the generators of $\mathcal{A}(M)$ provided by the proof of Theorem 2.4, and then map them to $\mathcal{W}(M)$, where isotopy invariance ought to ensure that they all become isomorphic to cotangent fibres (note that in the wrapped Fukaya category, any two cotangent fibres are mutually isomorphic).

The second strategy for attacking Conjecture 4.6 is fundamentally similar, but based on Lefschetz fibrations. Recall that with our definition, the total space of a Lefschetz fibration $\pi : X \to \mathbb{C}$ is itself a finite type Weinstein manifold. One then expects to have a canonical functor $\mathcal{F}(\pi) \to \mathcal{W}(X)$. Given a Lefschetz fibration constructed as in Sect. 3.3 by complexifying a Morse function on Z, one would then combine $\mathcal{F}(\pi) \to \mathcal{W}(X)$ with the functor from Conjecture 4.4, applied to a small neighbourhood of Z embedded into X. The outcome would be that the restrictions of Lefschetz thimbles generate the wrapped Fukaya category of that neighbourhood. One can easily check that all such restrictions are isomorphic to cotangent fibres.

Suppose now that Z is simply connected. In that case, if one accepts Conjectures 4.5 and 4.6, it follows by purely algebraic means that (28) is full and faithful. Take the dg module $\mathcal{E}_L = \bigoplus_I (\mathcal{E}_L)_I[-d]$ obtained from a closed exact $L \subset M$, and equip it with the decreasing filtration by values of $d = |I| - 1$. The associated graded space is precisely the dg module $\mathcal{E}_P$ constructed from the local coefficient system $P_x = HW^*(T_x^*, L) = HF^*(T_x^*, L)$. Hence, one gets a spectral sequence which starts with $H^*(Z; End(P))$ according to (21), and converges to the group $H(hom_{\mathcal{M}}(\mathcal{E}_L, \mathcal{E}_L))$, which would be equal to $HF^*(L, L) \cong H^*(L; \mathbb{K})$ as a consequence of Conjecture 4.5. This explains how (1) arises from this particular approach.

4.4 The (Co)bar Construction

There is a more algebraic perspective, which provides a shortcut through part of the argument above. To explain this, it is helpful to recall the classical bar construction. Let $\mathcal{C}$ be a dg algebra over our coefficient field $\mathbb{K}$, with an augmentation $\varepsilon : \mathcal{C} \to \mathbb{K}$, whose kernel we denote by $\mathcal{I}$. One can then equip the free tensor coalgebra $T(\mathcal{I}[1])$ with a differential, and then dualize it to a dg algebra $\mathcal{B} = T(\mathcal{I}[1])^\vee$. Consider the standard resolution $\mathcal{R} = T(\mathcal{I}[1]) \otimes \mathcal{C}$ of the simple $\mathcal{C}$-module $\mathcal{C}/\mathcal{I}$. One can prove that the endomorphism dga of $\mathcal{R}$ as an object of $mod(\mathcal{C})$ is quasi-isomorphic to $\mathcal{B}$. For standard algebraic reasons, this induces a quasi-equivalence between the subcategory of $mod(\mathcal{C})$ generated by $\mathcal{R}$, and the triangulated subcategory of $mod(\mathcal{B})$ generated by the free module $\mathcal{B}$. Denote that category by $modf(\mathcal{B})$.

The relevance of this duality to our discussion is a basic connection to loop spaces, which goes back to Adams [3]. He observed that if Z is simply connected, and $\mathcal{C}$ is the dg algebra of Cech cochains, then $\mathcal{B}$ is quasi-isomorphic to $C_{-*}(\Omega Z)$. Hence, reversing the functor above, we get a full and faithful embedding

$$H(modf(C_{-*}(\Omega Z))) \longrightarrow H(\mathcal{M}), \tag{30}$$

where $\mathcal{M} = mod(\mathcal{C})$ as in (28). If one moreover assumes that Conjectures 4.3 and 4.6 hold, it follows that $\mathcal{W}(M)$ itself is derived equivalent to $H(modf(C_{-*}(\Omega Z)))$. Hence, one would get a full embedding of the wrapped Fukaya category into $H(\mathcal{M})$ for algebraic reasons, avoiding the use of Cech complexes altogether.

5 The Non-simply-Connected Case

In this final section, we discuss exact Lagrangian submanifolds in non-simply-connected cotangent bundles. Specifically, we prove Corollary 1.3, and then make a few more observations about the wrapped Fukaya category.

5.1 A Finite Covering Trick

We start by recalling the setup from Sect. 2. Fix a real analytic structure on Z, and the associated category $\mathcal{A}(M)$. For any closed exact Lagrangian submanifold $L \subset M = T^*Z$ which is spin and has zero Maslov class, we have the spectral sequence

$$E_2^{pq} = H^p(Z; End^q(E_L)) \Rightarrow H^*(L; \mathbb{K})$$

arising from the resolution of L by "standard objects" in $\mathcal{A}(M)$. From now on, we assume that $char(\mathbb{K}) = p > 0$. In that case, one can certainly find a finite covering $b\colon \tilde{Z} \to Z$ such that b^*E_L is trivial (as a graded bundle of $\mathbb{K}$-vector spaces). In fact, E_L gives rise to a representation $\rho\colon \pi_1(Z) \to GL_r(\mathbb{K})$, and one takes $\tilde{Z}$ to be the Galois covering associated to $ker(\rho) \subset \pi_1(Z)$. Set $\tilde{M} = T^*\tilde{Z}$, denote by $\beta\colon \tilde{M} \to M$ the induced covering, and by $\tilde{L} = \beta^{-1}(L)$ the pre-image of our Lagrangian submanifold. This inherits all the properties of L, hence we have an analogous spectral sequence for $\tilde{L}$. Note also that for obvious reasons, the associated bundle of Floer cohomologies satisfies

$$E_{\tilde{L}} \cong b^*E_L, \tag{31}$$

hence is trivial by definition. Now, the discussion after (1) carries over with no problems to the non-simply-connected case if one assumes that the Lagrangian submanifold is connected, and its bundle of Floer cohomologies is trivial. In particular, one gets that $\tilde{L}$ is in fact connected: if it were not, the analogue of Theorem 1.1(iii) would apply to its connected components (since their Floer cohomology bundles are also trivial), implying that any two would have to intersect each other, which is a contradiction. With connectedness at hand, it then follows by the same argument that (31) has one-dimensional fibres. Hence, $End(E_L)$ is trivial, which means that the spectral sequence for the original L degenerates, yielding $H^*(L; \mathbb{K}) \cong H^*(Z; \mathbb{K})$. In fact, by borrowing arguments from [29] or from [14], one sees that in the Fukaya category of M, L is isomorphic to the zero-section equipped with a suitable *spin* structure (the difference between that and the a priori chosen spin structure on Z is precisely described by the bundle E_L, which, note, has structure group ± 1). Using that, one also gets the analogue of part (iii) of Theorem 1.1. Finally, the cohomological restrictions (i–iii) of Theorem 1.1 for arbitrary fields of positive characteristic imply them for $\mathbb{K}$ with $char(\mathbb{K}) = 0$, which completes the proof of Corollary 1.3.

5.2 The Eilenberg–MacLane Case

For general algebraic reasons, there is an A_∞-functor

$$\mathcal{W}(M) \longrightarrow mod(CW^*(T^*_x, T^*_x)). \tag{32}$$

Here, the right-hand side is the dg category of A_∞-modules over the endomorphism A_∞-algebra of the object T^*_x. Now suppose that $Z = K(\Gamma, 1)$ is an Eilenberg–MacLane space, so that the cohomology of $CW^*(T^*_x, T^*_x)$ is concentrated in degree zero. By the Homological Perturbation Lemma, this implies that the A_∞-structure is formal, hence by Theorem 4.2 quasi-isomorphic to the group algebra $\mathbb{K}[\Gamma]$ (this argument allows us to avoid Conjecture 4.3). However, in order to make (32) useful, we do want to appeal to (the currently unproven) Conjecture 4.6. Assuming from now on that it is true, one finds that (32) is a full embedding and actually lands (up to functor isomorphism) in the subcategory $modf(\mathbb{K}[\Gamma])$ generated by the one-dimensional free module. As far as that subcategory is concerned, one could actually replace A_∞-modules by ordinary chain complexes of $\mathbb{K}[\Gamma]$-modules. This gives a picture of $\mathcal{W}(M)$ in classical algebraic terms.

To see how this might be useful, let us drop the assumption that the Maslov class is zero and consider general closed, exact, and spin Lagrangian submanifolds $L \subset M$. Neither of the two arguments in favour of Conjecture 4.6 sketched in Sect. 4.3 actually uses the Maslov index. It is therefore plausible to assume that the description of $\mathcal{W}(M)$ explained above still applies. Denote by $\mathcal{N}_L$ the $\mathbb{Z}/2$-graded A_∞-module over $\mathbb{K}[\Gamma]$ corresponding to L, and by N_L its cohomology module. In fact, N_L is nothing other than our previous family Floer cohomology bundle E_L, now considered as a module over $\Gamma = \pi_1(Z)$. There is a purely algebraic obstruction theory, which determines when an A_∞-module is formal (isomorphic to its cohomology). The obstructions lie in

$$Ext^r_{\mathbb{K}[\Gamma]}(N_L, N_L[1-r]), \quad r \geq 2 \qquad (33)$$

where the $[1-r]$ now has to be interpreted mod 2. In particular, if all those groups vanish, it would follow directly that $\mathcal{N}_L$ is formal. However, there is no particular a priori reason why that should happen.

Returning to the trick from Sect. 5.1, assume now that $char(\mathbb{K}) = p > 0$. Then, after passing to a finite cover, one can assume that N_L is a direct sum of trivial representations, hence (33) is a direct sum of copies of $H^r(\Gamma; \mathbb{K})$. One can try to kill the relevant obstructions by passing to further finite covers (with respect to which the obstructions are natural). Generally, this is unlikely to be successful (there are examples of mod p cohomology classes which survive pullback to any finite cover, see for instance [9, Theorem 6.1]). However, in the special case where $H^*(\Gamma; \mathbb{K})$ is generated by degree 1 classes, such as for surfaces or tori, it is obviously possible. In those cases, one could then find a finite cover $b \colon \tilde{Z} \to Z$, inducing $\beta \colon \tilde{M} \to M$, such that the A_∞-module $\mathcal{N}_{\tilde{L}}$ associated with $\tilde{L} = \beta^{-1}(L)$ is isomorphic to a direct sum of ordinary trivial modules. Just by looking at the endomorphism ring of this object, it becomes clear that there can actually be only one summand, so $\tilde{L}$ is isomorphic to the zero-section. One can then return to M by the same argument as before, and obtain the same consequences as in the original Theorem 1.1. This would be a potential application of the machinery from Sect. 4 which has no obvious counterpart in the other approaches.

Finally, note that the appeal to Conjecture 4.6 above can be sidestepped, at least working over fields of characteristic not equal to two, in the special case where T^*Z admits a Lefschetz pencil for which, for suitable vanishing paths, the Lefschetz thimbles are *globally* cotangent fibres. This is, of course, an exceptional situation, but it can be achieved for suitable Lefschetz pencils on the affine variety $T^*T^n = (\mathbb{C}^*)^n$ (complexify a Morse function which is a sum of height functions on the distinct S^1-factors of T^n).

Acknowledgements

I.S. is grateful to Kevin Costello for encouragement and many helpful conversations. Some of this material was presented at the "Dusafest" (Stony Brook, Oct 2006). K.F. was partially supported by JSPS grant 18104001. P.S. was partially supported by NSF grant DMS-0405516. I.S. acknowledges EPSRC grant EP/C535995/1.

References

1. C. Abbondandalo and M. Schwarz, *On the Floer homology of cotangent bundles*, Commun. Pure Appl. Math. 59:254–316, 2006.
2. C. Abbondandolo and M. Schwarz, *Notes on Floer homology and loop space homology*, in "Morse-theoretic methods in nonlinear analysis and in symplectic topology", p. 75–108, Springer, 2006.
3. J. F. Adams, *On the cobar construction*, Proc. Natl. Acad. Sci. USA 42:409–412, 1956.
4. V. Arnol'd, *First steps in symplectic topology*, Russian Math. Surv. 41:1–21, 1986.
5. D. Austin and P. Braam, *Morse-Bott theory and equivariant cohomology*, in "The Floer Memorial Volume", Birkhäuser, 1995.
6. C. Barraud and O. Cornea, Lagrangian intersections and the Serre spectral sequence. Annals of Math. 166:657–722, 2007.
7. L. Buhovsky, *Homology of Lagrangian submanifolds of cotangent bundles*, Israel J. Math. 193:181–187, 2004.
8. Y. Eliashberg and L. Polterovich, *Local Lagrangian 2-knots are trivial*, Ann. of Math. 144:61–76, 1996.
9. B. Farb and S. Weinberger, *Isometries, rigidity, and universal covers*, Ann. Math. (to appear).
10. K. Fukaya, *Floer homology for families, I.* In "Integrable systems, topology and physics", Contemp. Math. 309:33–68, 2000.
11. K. Fukaya, *Mirror symmetry for abelian varieties and multi-theta functions*, J. Algebraic Geom. 11:393–512, 2002.
12. K. Fukaya and Y.-G. Oh, *Zero-loop open strings in the cotangent bundle and Morse homotopy*, Asian J. Math. 1:96–180, 1997.
13. K. Fukaya, Y.-G. Oh, K. Ono, and H. Ohta, *Lagrangian intersection Floer theory: anomaly and obstruction*, Book manuscript (2nd ed), 2007.
14. K. Fukaya, P. Seidel and I. Smith, *Exact Lagrangian submanifolds in simply-connected cotangent bundles*. Invent. Math. 172:1–27, 2008.
15. A. Gorodentsev and S. Kuleshov, *Helix theory*, Moscow Math. J. 4:377–440, 2004.
16. R. Hind, *Lagrangian isotopies in Stein manifolds*, Preprint, math.SG/0311093.
17. R. Hind and A. Ivrii, *Isotopies of high genus Lagrangian surfaces*, Preprint, math.SG/0602475.
18. M. Hutchings, *Floer homology of families, I.* Preprint, math/0308115.
19. N. V. Ivanov, *Approximation of smooth manifolds by real algebraic sets*, Russian Math. Surv. 37:3–52, 1982.

20. J. Johns, Ph. D. thesis, University of Chicago, 2006.
21. M. Kapranov, *Mutations and Serre functors in constructible sheaves*, Funct. Anal. Appl. 24:155–156, 1990.
22. M. Kashiwara and P. Schapira, *Sheaves on manifolds*. Springer, 1994.
23. R. Kasturirangan and Y.-G. Oh, *Floer homology of open subsets and a relative version of Arnold's conjecture*, Math. Z. 236:151–189, 2001.
24. R. Kasturirangan and Y.-G. Oh, *Quantization of Eilenberg-Steenrod axioms via Fary functors*, RIMS Preprint, 1999.
25. M. Kontsevich and Y. Soibelman, *Homological mirror symmetry and torus fibrations*, In "Symplectic geometry and mirror symmetry", World Scientific, 2001.
26. F. Lalonde and J.-C. Sikorav, *Sous-variétés lagrangiennes et lagrangiennes exactes des fibrés cotangents*, Comment. Math. Helv. 66:18-33, 1991.
27. S. Mau, K. Wehrheim and C. Woodward, *In preparation*.
28. J. Milnor, *Singular points of complex hypersurfaces*. Princeton Univ. Press, 1969.
29. D. Nadler, *Microlocal branes are constructible sheaves*. Preprint, math.SG/0612399.
30. D. Nadler and E. Zaslow, *Constructible sheaves and the Fukaya category*. J. Amer. Math. Soc, (to appear) math.SG/0406379.
31. Y.-G. Oh, *Naturality of Floer homology of open subsets in Lagrangian intersection theory*, in "Proc. of Pacific Rim Geometry Conference 1996", p. 261–280, International Press, 1998.
32. *Seminaire Rudakov*. London Math. Soc. Lecture Note Series, vol. 148. Cambridge Univ. Press, 1990.
33. D. Salamon and E. Zehnder, *Morse theory for periodic solutions of Hamiltonian systems and the Maslov index*, Comm. Pure Appl. Math. 45:1303–1360, 1992.
34. P. Seidel, *A long exact sequence for symplectic Floer cohomology*, Topology 42:1003–1063, 2003.
35. P. Seidel, *Exact Lagrangian submanifolds of T^*S^n and the graded Kronecker quiver*, In "Different faces of geometry", p. 349–364, Kluwer/Plenum, 2004.
36. P. Seidel, *Fukaya categories and Picard-Lefschetz theory*. Book Manuscript, to appear in ETH Lecture Notes series.
37. C. Viterbo, *Exact Lagrange submanifolds, periodic orbits and the cohomology of loop spaces*, J. Diff. Geom. 47: 420–468, 1997.
38. C. Viterbo, *Functors and computations in Floer homology, I.* Geom. Funct. Anal. 9:985–1033, 1999.
39. C. Viterbo, *Functors and computations in Floer homology, II.* Unpublished manuscript.
40. K. Wehrheim and C. Woodward, *Functoriality for Floer theory in Lagrangian correspondences*, Preprint, 2006.

B-Type D-Branes in Toric Calabi–Yau Varieties

M. Herbst[1], K. Hori[2], and D. Page[2]

[1] DESY Theory Group, 22603 Hamburg, Germany
   manfred.herbst@desy.de
[2] Department of Physics, University of Toronto, Toronto, Ontario, Canada
   hori@physics.utoronto.ca, d.c.page@gmail.com

Abstract. We define D-branes in linear sigma models corresponding to toric Calabi–Yau varieties. This enables us to study the transportation of D-branes along general paths in the Kähler moduli space. The most important finding is the grade restriction rule. It classifies Chan–Paton representations of the gauge group that are allowed when we cross the phase boundaries. Upon reduction to the chiral sector, we find equivalences of the derived categories on different toric varieties on the same Kähler moduli space.

1 Introduction

An important problem of string theory is to understand D-branes in Calabi–Yau varieties, in particular their dependence on complex structure as well as on Kähler moduli. In [1] a progress in describing B-type D-branes in non-geometric regimes is made purely from the worldsheet point of view. This note is a report on that work and attempts to present the basic idea of half of the story — D-branes in (non-compact) toric Calabi–Yau varieties. For complete details and for the other half of the story — D-branes in compact Calabi–Yau varieties — the reader is encouraged to see the original paper [1].

By B-type D-branes we mean holomorphic branes such as branes wrapped on complex submanifolds and those supporting holomorphic vector bundles. Upon truncation to the chiral sector, they are described through the derived category of coherent sheaves [2, 3, 4]. When investigating the dependence on the Kähler moduli space, its subdivision into phases (or Kähler cones) has to be taken into account. Each phase is associated with a toric variety and moving between them corresponds to generalized flop transitions — a sequence of blow-downs and blow-ups. When we consider B-type D-branes, a natural question comes up: How do they behave under such flops?

At this stage we realize that the above question can be asked only when we have a definition of D-branes deep inside the Kähler moduli space, in

Herbst, M., et al.: *B-Type D-Branes in Toric Calabi–Yau Varieties*. Lect. Notes Phys. **757**, 27–44 (2009)
DOI 10.1007/978-3-540-68030-7_2

particular near the phase boundaries. The main purpose of the work [1] is to provide such a definition. We do this by using linear sigma models [5] to realize the bulk theory.

On physical grounds the chiral sector of B-type D-branes does not change under Kähler deformations, which suggests that flops do not affect the derived category of coherent sheaves, i.e. the derived categories of the different phases of the same Kähler moduli space are indeed equivalent. The correct definition of B-type D-branes must respect this property and our construction passes this criterion. From the mathematical perspective such equivalences have been studied before. Widely known examples are the McKay correspondence [6] as well as the elementary flop transition of the resolved conifold [7]. A general treatment of transitions was given by Kawamata [8] and the one-parameter case was treated by van den Bergh [9]. In fact, the equivalences that appear in these works come out naturally in the context of our work and furthermore we are able to provide a new picture for multi-parameter cases. We also obtain a new perspective of monodromies around special loci in the Kähler moduli space and reproduce known results obtained through mirror symmetry. This provides further support of our contruction.

In this note we outline the definition of B-type D-branes in linear sigma models and their relation to the low energy D-branes on the toric varieties. Studying the moduli dependence leads us to an effective method of mapping D-branes between phases. The key point is a certain condition on D-branes that follows from the physics of the linear sigma model close to the boundary of two adjacent phases. We shall call it the *grade restriction rule*.

After summarizing some generalities on linear sigma models, we introduce B-type D-brane in Sect. 2. In order to keep the presentation simple we explain the main points in an example throughout the note. The maps to low energy D-branes on the toric varieties are explained in Sect. 3. Our key result is discussed thereafter in Sect. 4, where D-branes are transported across phase boundaries. We close this note with some comments on relations to earlier work in the literature in Sect. 5.

1.1 The Bulk Theory

Let us consider an $\mathcal{N} = (2,2)$ supersymmetric linear sigma model in $1+1$ dimensions with n complex scalar fields x_i for $i = 1, \ldots, n$ and an Abelian gauge group $T = U(1)^k$, where the action of the ath component $U(1)_a$, $a = 1, \ldots, k$, is specified by the integral, coprime gauge charges $(Q_1^a, \ldots, Q_n^a)$:

$$(x_1, \ldots, x_n) \mapsto (e^{iQ_1^a \varphi_a} x_1, \ldots, e^{iQ_n^a \varphi_a} x_n) .$$

Each x_i has one Dirac fermion and a complex auxiliary scalar as its superpartner, while the partners of the $U(1)_a$ gauge field v_a are one complex scalar σ_a, one Dirac fermion and a real auxilary scalar field. For simplicity, we turn off the superpotential in this report.

The scalar potential of the theory is

$$U = \sum_{a=1}^{k} \frac{e_a^2}{2} \left(\sum_{i=1}^{n} Q_i^a |x_i|^2 - r^a \right)^2 + \sum_{i=1}^{n} \left| \sum_{a=1}^{k} Q_i^a \sigma_a x_i \right|^2 . \tag{1}$$

Here, e_a are the gauge coupling constants. The real parameters r^a for $a = 1, \ldots, k$ are referred to as Fayet–Iliopoulos parameters and we denote the space spanned by them by $\mathbb{R}_{\mathrm{FI}}^k$. The theory also depends on the theta angles θ^a that appear in the topological term,

$$\frac{1}{2\pi} \sum_{a=1}^{k} \int_{\Sigma} \theta^a dv_a . \tag{2}$$

They are angular parameters, $\theta^a \equiv \theta^a + 2\pi$. We combine r^a and θ^a into a complex parameter $t^a = r^a - i\theta^a$.

We focus on models with *axial* and *vector* R-symmetries at all energies, that is, models obeying the Calabi-Yau condition,

$$\sum_{i=1}^{n} Q_i^a = 0, \qquad \text{for} \quad a = 1, \ldots, k . \tag{3}$$

The absence of the axial anomaly ensures that the theory does depend on the k theta angles. The vector R-symmetry that becomes a part of the superconformal symmetry in the infra-red limit will play an important rôle when we consider D-branes later on. In the model without superpotential, the correct R-symmetry is the one with all the fields $x_1, \ldots, x_n$ having trivial R-charges.

1.2 The Phases

We are interested in the low-energy theory. Its character is determined by the locus of classical vacua, $U = 0$. At a generic point of $\mathbb{R}_{\mathrm{FI}}^k$, the first term in U leads to the vacuum equation (the D-term equation), $\sum_{i=1}^{n} Q_i^a |x_i|^2 = r^a$, which requires that x_i are 'maximal rank', so that the gauge group is completely broken or broken to a discrete subgroup. Then, the second term in the potential sets $\sigma_a = 0$ for $a = 1, \ldots, k$.

However, at particular $(k-1)$-dimensional walls in $\mathbb{R}_{\mathrm{FI}}^k$ the D-term equation admits a solution with an unbroken $U(1)$ subgroup of T so that σ is allowed to have large values in the corresponding direction. This signals a singularity of the theory. The walls divide the FI-parameter space into cone domains, which are called *phases* or *Kähler cones*. Accordingly, the walls themselves are called *phase boundaries* [5, 14].

Taking into account the gauge group T the (classical) vacuum configuration is given by the symplectic quotient $\{\sum_{i=1}^{n} Q_i^a |x_i|^2 = r^a\}_{\forall a}/T$, which is a *toric variety* in each phase. It will be more convenient for us to express the toric variety as an algebraic quotient,

$$X_r = \frac{\mathbb{C}^n - \Delta_r}{T_{\mathbb{C}}} \ ,$$

with the complexified torus $T_{\mathbb{C}} = (\mathbb{C}^\times)^k$. Here, the *deleted set* $\Delta_r \subset \mathbb{C}^n$ is the set of points whose $T_{\mathbb{C}}$-orbits do not pass through $\{\sum_{i=1}^n Q_i^a |x_i|^2 = r^a\}_{\forall a}$. Δ_r is constant within each phase and jumps at phase boundaries, so that every phase, $I, II, \ldots$, is equipped with its particular deleted set, $\Delta_I, \Delta_{II}, \ldots$.

In order to illustrate how the deleted sets and thus the toric varieties depend on the Kähler parameters, let us consider the following model with gauge group $T = U(1)$:

fields	x	y	z	p
$U(1)$	1	1	1	-3

Studying the potential U readily shows that the deleted sets are

$$\begin{aligned}
\Delta_+ &= \{x = y = z = 0\} &&\text{for} \quad r > 0 \ , \\
\Delta_- &= \{p = 0\} &&\text{for} \quad r < 0 \ .
\end{aligned} \tag{4}$$

Thus, the toric varieties, X_+ resp. X_-, are the line bundle $\mathcal{O}(-3)$ over $\mathbb{CP}^2$ for positive r and the orbifold $\mathbb{C}^3/\mathbb{Z}_3$ (set $p = 1$) for negative r.

The classical theory is divided into disconnected domains by the phase boundaries. In the quantum theory, this picture is completely altered by the theta angles [5]: The singular locus is a complex hypersurface $\mathfrak{S}$ in the space $(\mathbb{C}^\times)^k$ of complexified parameters $t^a = r^a - i\theta^a$. The Kähler moduli space of the low energy theory is thus the complement

$$\mathfrak{M}_K = (\mathbb{C}^\times)^k \backslash \mathfrak{S} \ ,$$

which is connected. Let us have a closer look at $\mathfrak{S}$ in one-parameter models. The singularity occurs when there is a non-compact σ branch and that is where the quantum effective potential vanishes. For large values of σ the second term in (1) gives masses to the fields x_i and we can integrate out the latter. This results in the effective potential [5] (under condition (3)):

$$U_{eff} = \frac{e^2}{2} |r_{eff} - i\theta_{min}|^2 \quad \text{for } |\sigma| \gg e. \tag{5}$$

Here, $r_{eff} = r + \sum_i Q_i \log|Q_i|$ and $\theta_{min}^2 = \min_{q \in \mathbb{Z}}(\theta + S\pi + 2\pi q)^2$. The constant S is the sum of positive charges Q_i. Thus, the singular locus $\mathfrak{S}$ is at $r = -\sum_i Q_i \log|Q_i|$ and $\theta = S\pi \mod 2\pi$. The moduli space $\mathfrak{M}_K$ for a one-parameter model is depicted in Fig. 1.

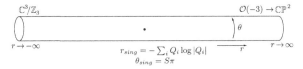

Fig. 1. The moduli space of a one-parameter model. In our example we have $r_{sing} = 3\log 3$ and $\theta_{sing} = 3\pi \mod 2\pi$

2 D-branes in the Linear Sigma Model

Studying D-branes in a two-dimensional quantum field theory means that we have to put the model on a world sheet Σ *with boundary*. In this section it will suffice to consider the half-plane with coordinates $(x^0, x^1) \in \mathbb{R} \times (-\infty, 0)$. We will mainly discuss one-parameter models.

The presence of a boundary has two consequences: First, at least half of the $\mathcal{N} = (2, 2)$ supersymmetry algebra is broken — we consider boundary conditions that preserve the $\mathcal{N} = 2_B$ subalgebra, cf. [10]. Second, we have to introduce some additional boundary interactions in order to ensure invariance of the action S under $\mathcal{N} = 2_B$ supersymmetry. There is some freedom in doing so, which comes from a part in the boundary action, let us call it S_{bdry}, which is supersymmetric by itself. It will suffice to concentrate on this part for the present purposes.

The term S_{bdry} in fact gives rise to a Wilson line with connection $\mathcal{A}$, i.e., we may write $S_{bdry} = - \int_{\partial \Sigma} \mathcal{A}_0 dx^0$. The simplest choice for a $\mathcal{N} = 2_B$ invariant connection is

$$\mathcal{A}_0 = q \left[v_0 - \mathrm{Re}(\sigma) \right] , \tag{6}$$

which yields a gauge covariant Wilson line provided q is an integral charge. Let us refer to the D-brane defined through (6) as *Wilson line brane*

$$\mathcal{W}(q) .$$

Note that the theta angle term (2) is not topological on a bounded world sheet with B-type boundary conditions [11]. This has the consequence that θ is not an angular variable anymore. Indeed, we can combine the theta angle and the charge q in a single boundary term

$$S_{bdry} = - \int_{\partial \Sigma} \left(\frac{\theta}{2\pi} + q \right) \left[v_0 - \mathrm{Re}(\sigma) \right] . \tag{7}$$

The action is then invariant under combined shifts $\theta \to \theta - 2\pi$ and $q \to q + 1$.

We may think of the Wilson line brane $\mathcal{W}(q)$ as one-dimensional vector space $\mathcal{V}$, the *Chan–Paton space*, which carries a representation $\rho(e^{i\varphi}) = e^{iq\varphi}$ of the gauge group T. In a similar way, we assign to $\mathcal{V}$ representations of the two global symmetries of the action functional S: an integral R-degree j coming from the representation $R(\lambda) = \lambda^j$ of the vector R-symmetry and a $\mathbb{Z}_2$-degree $\sigma = (-1)^F$ from the fermion number symmetry. The latter distinguishes branes and antibranes. We choose the R-degree so that it reduces modulo 2 to the $\mathbb{Z}_2$ grading, $\sigma = R(e^{i\pi})$. Therefore, the Wilson line brane is completely characterized by the $\mathbb{Z}$-valued (or more generally $\mathbb{Z}^k$-valued) gauge charge q and the $\mathbb{Z}$-degree j from the R-symmetry.

Given the basic B-type D-branes of the gauged linear sigma model we may form bound states, i.e. $\mathbb{Z}$-graded direct sums of Wilson line branes $\mathcal{W} = \oplus_{j \in \mathbb{Z}} \mathcal{W}^j$ with interaction terms in S_{bdry}, where the direct summand of R-degree j is

$$\mathcal{W}^j = \oplus_{i=1}^{n_j} \mathcal{W}(q_{ij}) \ .$$

We assume the finiteness condition $\sum_{j\in\mathbb{Z}} n_j < \infty$, i.e. we only consider bound states of finitely many Wilson line branes. Then the data for a general D-brane $\mathfrak{B}$ is a complex,

$$C(\mathfrak{B}) : \ \ldots \xrightarrow{d^{j-1}} \mathcal{W}^{j-1} \xrightarrow{d^j} \mathcal{W}^j \xrightarrow{d^{j+1}} \mathcal{W}^{j+1} \xrightarrow{d^{j+2}} \ldots \ ,$$

where the maps d^j are holomorphic in the fields x_i and compatible with the gauge charge assignment q_{ij}. Such a data determines the boundary interaction of the form

$$\mathcal{A}_0 = \sum_{a=1}^{k} \rho_* \big[v_0 - \mathrm{Re}(\sigma) \big] \ + \ \frac{1}{2} \{Q, Q^\dagger\} \ + \ \text{fermions} \ . \tag{8}$$

Here, $\rho_*(\cdot) \colon \mathcal{W} \to \mathcal{W}$ is the infinitesimal form of the gauge representation ρ and $Q \colon \mathcal{W} \to \mathcal{W}$ is the collection of the maps d^j. By the fact that the d^j's form a complex, $d^{j+1} \circ d^j = 0$, we have

$$Q^2 = 0 \ , \tag{9}$$

the condition for $\mathcal{N} = 2_B$ supersymmetry of $\mathcal{A}_0$. The compatibility of d^j with the gauge charge assignment ensures the gauge invariance condition

$$\rho(e^{i\varphi})^{-1} \ Q(e^{iQ_1\varphi} x_1, \ldots, e^{iQ_n\varphi} x_n) \ \rho(e^{i\varphi}) = Q(x_1, \ldots, x_n) \ .$$

Q has R-degree 1 as d^j's do

$$R(\lambda) Q(x_1, \ldots, x_n) R(\lambda)^{-1} = \lambda Q(x_1, \ldots, x_n) \ ,$$

which ensures invariance of the interaction $\mathcal{A}_0$ under R-symmetry. We denote the *set of D-branes* in the gauged linear sigma model by $\mathfrak{D}(\mathbb{C}^n, T)$.

So far we have only considered the supersymmetric action including new terms at the boundary of the worldsheet. It remains to specify the $\mathcal{N} = 2_B$ supersymmetric boundary conditions on the bulk fields. They must ensure that we acquire a well-defined action for the non-linear sigma model on the toric variety at low energies. In fact, supersymmetry requires us to pick a (real) one-dimensional subspace in the σ plane and the particular choice of the latter has severe influences on the low-energy theory. As it turns out, the following conditions on the bosons together with their supersymmetry transforms lead to a well-defined low-energy theory for a Wilson line brane $\mathcal{W}(q)$:

$$\begin{aligned}
\partial_1 x_i &= 0 \ , \\
v_1 &= \mathrm{Im}(\sigma) = 0 \ , \\
\partial_1 (v_0 - \mathrm{Re}(\sigma)) &= 0 \ , \\
v_{01} &= -e^2(\theta + 2\pi q) \ .
\end{aligned} \tag{10}$$

The last condition is the Gauss law constraint for the gauge field strength $v_{01} = \partial_0 v_1 - \partial_1 v_0$. We will comment on the consequences of the boundary conditions (10) in the course of the subsequent discussions.

Now that we have fully specified the gauged linear sigma model in the presence of a boundary, let us see how D-branes in $\mathfrak{D}(\mathbb{C}^n, T)$ descend to D-branes on the toric varieties $X_I, X_{II}, \ldots$ at low energies.

3 D-Branes on Toric Varieties

Assume that we are deep inside a Kähler cone. Since the gauge coupling e sets the energy scale of the problem, the low-energy limit is equivalent to the $e \to \infty$ limit. Consequently, the kinetic term for the gauge multiplet vanishes and the gauge field v as well as the scalars, $\mathrm{Re}(\sigma)$ and $\mathrm{Im}(\sigma)$, acquire algebraic equations of motion.

As an example consider v_0. After the $e \to \infty$ limit, it enters the action S as follows:

$$\frac{1}{2\pi} \int_\Sigma \sum_i Q_i^2 |x_i|^2 \left(v_0^2 - \frac{\sum_j i Q_j (\bar{x}_j \partial_0 x_j - \partial_0 \bar{x}_j x_j)}{\sum_i Q_i^2 |x_i|^2} v_0 + \text{ferm.} \right) - \int_{\partial\Sigma} \left(\frac{\theta}{2\pi} + q \right) v_0.$$

$$(11)$$

The first part comes from the kinetic term for the fields x_i. The last is due to the boundary term (7) for a Wilson line brane $\mathcal{W}(q)$. Completing the square has the effect, familiar from the case without boundary, that the gauge field v becomes the pull-back, $x^* A$, of the connection on the holomorphic line bundle $\mathcal{O}(1)$ over X_r. In particular, the Wilson line (6) then corresponds to the line bundle $\mathcal{O}(q)$ with its divisor class specified by q.[3]

The last term in (11) causes some problem when completing the square, it gives rise to a singularity $[\delta_{\partial\Sigma}(x^1)]^2$, where $\delta_{\partial\Sigma}(x^1)$ denotes the delta distribution localized at the boundary. Fortunately, when we integrate out $\mathrm{Re}(\sigma)$, another singularity with opposite sign precisely cancels the one from v_0. This cancelation is in fact a consequence of the boundary conditions (10). For the imaginary part of σ the conditions (10) ensure that such problems with singularities do not occur in the first place.

In summary, a D-brane $\mathfrak{B}$ of the gauged linear sigma model is mapped to a D-brane $\mathcal{B}$ on the toric variety X_r that corresponds to a complex of holomorphic vector bundles,

$$\mathcal{C}(\mathcal{B}) : \quad \ldots \xrightarrow{d^{j-1}} \mathcal{E}^{j-1} \xrightarrow{d^j} \mathcal{E}^j \xrightarrow{d^{j+1}} \mathcal{E}^{j+1} \xrightarrow{d^{j+2}} \ldots \, ,$$

---

[3] If the toric variety X_r has orbifold singularities, the gauge charge q turns into the data for the representation of the discrete orbifold group on the Chan–Paton space. In order to stick to a uniform terminology we, somewhat loosely, refer to D-branes on orbifolds as 'line bundles' or 'vector bundles' as well.

with $\mathcal{E}^j = \oplus_{i=1}^{n_j} \mathcal{O}(q_{ij})$. All we have to do is to apply the map

$$\mathcal{W}(q) \mapsto \mathcal{O}(q)$$

on all component branes. The Neumann boundary condition, $\partial_1 x_i = 0$, in (10) ensures that all the entries in the complex correspond to space-filling D-branes.

3.1 Trivial D-branes and D-isomorphisms

We are interested in the low-energy behaviour of the bulk and the boundary theory. Just as in the bulk, the boundary theory reduces at low energies to the bottom of the boundary potential. In fact, in order to find a non-empty superconformal interaction in the deep infra-red limit, we need a *zero* of the potential. Note that the matrix $\{Q, Q^\dagger\}$ in (8) serves as a boundary potential for the D-brane $\mathfrak{B}$. If $\{Q, Q^\dagger\}$ is everywhere positive on X_r there is no chance of obtaining a vanishing potential and complete brane–antibrane annihilation will take place [12], thus such a D-brane is empty at low energies.

The simplest example is the trivial brane antibrane annihilation

$$\mathcal{O}(q) \xrightarrow{\;1\;} \mathcal{O}(q) \qquad \text{with} \quad \{Q, Q^\dagger\} = \mathrm{id}_\mathcal{V} \ .$$

Another class of examples which play a crucial role in our study are D-branes such that $\{Q, Q^\dagger\}$ has zero eigenvalues only in the deleted set Δ_r. Since the latter is not part of the vacuum variety, such a D-brane will be infra-red empty on X_r. The fact that the deleted sets differ in the low-energy phases then implies that the set of trivial D-branes in $\mathfrak{D}(\mathbb{C}^n, T)$ changes across phase boundaries.

There is a simple (sufficient) condition for two D-branes to flow to the same infra-red fixed point. In order to formulate it, let us borrow the notion of a quasi-isomorphism from the theory of derived categories. Suppose there are two D-branes, $\mathcal{B}_1$ and $\mathcal{B}_2$, and a chain map $\varphi = \varphi(x_1, \ldots, x_n)$ from $\mathcal{C}(\mathcal{B}_1)$ to $\mathcal{C}(\mathcal{B}_2)$, i.e. it satiesfies $Q_2 \varphi - \varphi Q_1 = 0$. We define the bound state D-brane $\mathcal{B}_{C(\varphi)}$ (known as the *cone*) by $\mathcal{E}^j_{C(\varphi)} = \mathcal{E}^{j+1}_1 \oplus \mathcal{E}^j_2$ and

$$Q_{C(\varphi)} = \begin{pmatrix} -Q_1 & 0 \\ \varphi & Q_2 \end{pmatrix} \ .$$

Then the chain map φ is called a *quasi-isomorphism*, if $\mathcal{B}_{C(\varphi)}$ is trivial over X_r, i.e. the matrix $\{Q_{C(\varphi)}, Q^\dagger_{C(\varphi)}\}$ is invertible. Two D-branes, $\mathcal{B}_1$ and $\mathcal{B}_2$, that are related by a chain of quasi-isomorphisms are called quasi-isomorphic. This definition of a quasi-isomorphism φ is equivalent to the one commonly used in the literature, which is based on the propety of φ at the cohomology level. Now we can state the condition: *two D-branes in $\mathfrak{D}(\mathbb{C}^n, T)$ flow to the same D-brane at low energies if they are quasi-isomorphic.* (The converse is not true.)

This follows from the following *non-trivial* fact: two quasi-isomorphic D-branes are related through a combination of *(i)* brane–antibrane annihilations, as described above and *(ii)* certain deformations of the connection $\mathcal{A}$, called *D-term deformations*, that do not change the low-energy behaviour. Since we are not dealing with the derived category, but rather with boundary interactions in a quantum field theory, we prefer to refer to the combination of brane–antibrane annihilations and D-term deformations as *D-isomorphisms*. In the chiral sector of the theory the two notions, quasi-isomorphism and D-isomorphism, are interchangeable.

We summarize our findings by identifying the *set of low-energy D-branes* with D-branes in $\mathfrak{D}(\mathbb{C}^n, T)$ up to D-isomorphisms and we refer to this set by $D(X_r)$. Let us denote the map from linear sigma model D-branes to geometric D-branes by

$$\pi_r : \mathfrak{D}(\mathbb{C}^n, T) \longrightarrow D(X_r) \ .$$

Then, taking into account the phase structure of the Kähler moduli space, we obtain a pyramid of maps:

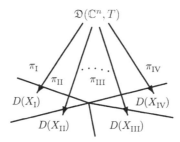

Every Kähler cone comes with its own deleted set, thus inducing specific D-isomorphisms in $D(X_I), D(X_{II}), \ldots$.

Let us illustrate some D-isomorphisms in our example. Recall the deleted sets, $\Delta_+ = \{x = y = z = 0\}$ and $\Delta_- = \{p = 0\}$. Let us consider the D-brane, $\mathfrak{B}_+$, given by the complex[4]

$$\mathcal{C}(\mathfrak{B}_+): \quad \mathcal{W}(-1) \xrightarrow{\begin{pmatrix} x \\ y \\ z \end{pmatrix}} \mathcal{W}(0)^{\oplus 3} \xrightarrow{\begin{pmatrix} 0 & z & -y \\ -z & 0 & x \\ y & -x & 0 \end{pmatrix}} \mathcal{W}(1)^{\oplus 3} \xrightarrow{(x,y,z)} \underline{\mathcal{W}(2)} \ . \quad (12)$$

The R-degree 0 part of the complex is underlined. The potential term in the boundary interaction is

$$\{Q_+, Q_+^\dagger\} \stackrel{\cdot}{=} \left(|x|^2 + |y|^2 + |z|^2\right) \cdot \mathrm{id}_{\mathcal{V}_+} \ , \quad (13)$$

---

[4] In the following we use the short-hand notation $\mathcal{W}(-1) \xrightarrow{X} \mathcal{W}(0)^{\oplus 3} \xrightarrow{X} \mathcal{W}(1)^{\oplus 3} \xrightarrow{X} \mathcal{W}(2)$ for Koszul-like complexes (12).

In the low-energy theory of the positive volume phase, $r \gg 0$, this boundary potential is strictly non-vanishing, since $x = y = z = 0$ is deleted. As a consequence complete brane–antibrane annihilation takes place, $\pi_+(\mathfrak{B}_+) \cong 0$, where 0 denotes the trivial D-brane. On the other hand, we may view $\mathfrak{B}_+$ as a result of binding two D-branes, say

$$\mathcal{C}(\mathfrak{B}_1): \quad \mathcal{W}(-1) \xrightarrow{X} \mathcal{W}(0)^{\oplus 3} \xrightarrow{X} \underline{\mathcal{W}(1)^{\oplus 3}} \quad \text{and} \quad \mathcal{C}(\mathfrak{B}_2) = \mathcal{W}(2) \ .$$

The map that binds them is the right-most one in (12). Then, invertibility of $\{Q_+, Q_+^\dagger\}$ tells us that $\mathfrak{B}_1$ and $\mathfrak{B}_2$ determine D-isomorphic low-energy D-branes in $D(X_+)$:

$$\pi_+(\mathfrak{B}_1) \ \cong \ \pi_+(\mathfrak{B}_2) \ .$$

Let us next study these branes in the negative volume phase $r \ll 0$, where the low-energy theory is the free orbifold $\mathbb{C}^3/\mathbb{Z}_3$. In this phase the fields x, y, z are allowed to vanish at the same time, so that the boundary potential $\{Q_+, Q_+^\dagger\}$ is not everywhere invertible anymore. Consequently, the D-brane $\mathfrak{B}_+$ localizes at the orbifold fixed point $\mathfrak{p} = \{x = y = z = 0\}$ and becomes the fractional D-brane, $\mathcal{O}_{\mathfrak{p}}(2)$. The gauge charge is now a $\mathbb{Z}_3$ representation label. We also find that $\pi_-(\mathfrak{B}_1)$ and $\pi_-(\mathfrak{B}_2)$ are not D-isomorphic in $D(X_-)$.

A complementary example is provided by the complex

$$\mathcal{C}(\mathfrak{B}_-): \quad \mathcal{W}(q+3) \xrightarrow{p} \underline{\mathcal{W}(q)} \ ,$$

for some $q \in \mathbb{Z}$. The associated boundary potential is $\{Q_-, Q_-^\dagger\} = |p|^2 \cdot \mathrm{id}_{\mathcal{V}_-}$. In the orbifold phase the latter is positive everywhere and hence the image $\pi_-(\mathfrak{B}_-)$ is trivial in the low-energy limit. This can likewise be interpreted as the D-isomorphism,

$$\pi_-(\mathcal{W}(q+3)) \ \cong \ \pi_-(\mathcal{W}(q)) \ , \tag{14}$$

in $D(X_-)$, which reflects the breaking of the gauge group $T = U(1)$ to the discrete subgroup $\mathbb{Z}_3$.

At large volume, $r \gg 0$, we find that $\pi_+(\mathfrak{B}_-)$ is D-isomorphic to $\mathcal{O}_E(q)$, a line bundle supported on the exceptional divisor $E \cong \mathbb{CP}^2$ (or more accurately the structure sheaf on E tensored with $\mathcal{O}(q)$). Of course, the D-branes $\pi_+(\mathcal{W}(q+3)) = \mathcal{O}(q+3)$ and $\pi_+(\mathcal{W}(q)) = \mathcal{O}(q)$ are not D-isomorphic in $D(X_+)$.

4 Changing the Phase — The Grade Restriction Rule

We now discuss the key result of our work. We relate D-branes on toric varieties, say X_+ and X_-, of two adjacent phases.

Let us pick an arbitrary Wilson line brane $\mathcal{W}(q) \in \mathfrak{D}(\mathbb{C}^n, T)$ and a path from $r \gg 0$ to $r \ll 0$ as depicted in Fig. 2. Naively, when we follow the path

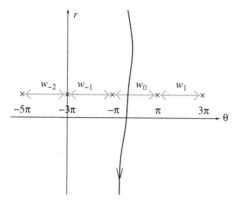

Fig. 2. A path in the moduli space for a one-parameter model with S odd. The window w_0 determines the set of grade restricted branes $\mathcal{T}^{w_0}$

we start with the D-brane $\mathcal{O}(q) \in D(X_+)$ and end up with $\mathcal{O}(q) \in D(X_-)$. But we know already from the discussion in the previous section that this cannot be true in general. We have to be careful when we come close to the phase boundary at $r = -\sum_i Q_i \log |Q_i|$, where the scalar field σ could become large and give rise to a singularity.

In order to observe possible singularities we have to redo the computation of the effective potential (5) at large σ in the presence of a boundary. To this end we formulate the model on a strip, so that the finite width L enables us to compare the energies from the bulk and the boundary of the interval $[0, L]$. We work in the NS sector in which the left and the right boundaries preserve the opposite $\mathcal{N} = 2_B$ supersymmetries.

We first notice that the boundary interaction (7) includes the term $(\theta/2\pi + q)\mathrm{Re}(\sigma)$ which plays the role of the scalar potential term in the large σ region. (There is no such term for $\mathrm{Im}(\sigma)$ because of the boundary condition (10).) In the NS sector with a constant σ, the contributions from the left and the right boundaries add up, yielding

$$V_{classical} = L\frac{e^2}{2}r^2 - 2\left(\frac{\theta}{2\pi} + q\right)\mathrm{Re}(\sigma)$$

as the classical potential. We find a problem here: this potential for $\mathrm{Re}(\sigma)$ is unbounded from below. However, in order to see what really happens at large values of σ we have to take into account quantum corrections from integrating out the heavy fields x_i, just as in the bulk. The effective potential turns out to be,

$$V_{eff} = L\frac{e^2}{2}|r_{eff} - i\theta_{eff}|^2 + 2\left[-\left(\frac{\theta}{2\pi} + q\right)\mathrm{Re}(\sigma) + \frac{S}{2}|\mathrm{Re}(\sigma)|\right] , \qquad (15)$$

where $\theta_{eff} = \theta - \mathrm{sgn}(\mathrm{Re}(\sigma))S\pi + 2\pi q$ and $S = \sum_{Q_i > 0} Q_i$.

Since the bulk part is constant and the boundary part is linear, the latter will always dominate for large enough $|\mathrm{Re}(\sigma)|$. The problem of an unbounded potential could be resolved by the quantum correction $S|\mathrm{Re}(\sigma)|$ in V_{eff}. Indeed, the boundary potential is bounded from below for Wilson line branes $\mathcal{W}(q)$ that satisfy

$$-\frac{S}{2} < \frac{\theta}{2\pi} + q < \frac{S}{2} . \tag{16}$$

We call this condition the *grade restriction rule*.

The grade restriction rule (16) provides us with a simple condition on D-branes in $\mathfrak{D}(\mathbb{C}^n, T)$, which can safely be transported along a path from large to small r. For a given window w between two neighboring singular points at $\theta = S\pi$ mod 2π, as shown in Fig. 2, we define the set of charges

$$N^w = \{q \in \mathbb{Z} \mid \forall \theta \in w : -S\pi < \theta + 2\pi q < S\pi\} .$$

We call a D-branes $\mathcal{B}$ grade restricted with respect to the window w if all its Wilson line components $\mathcal{W}(q)$ carry charges $q \in N^w$. Let us denote the set of grade restricted D-branes by $\mathcal{T}^w \subset \mathfrak{D}(\mathbb{C}^n, T)$.

4.1 Transport of D-Branes

We are interested in transporting D-branes from one phase to another through a given window at the phase boundary. Inside either phase, we are free to use the D-isomorphisms to replace the representatives in $\mathfrak{D}(\mathbb{C}^n, T)$. When we cross the phase boundary, we must use the one obeying the grade restriction rule. In this way we can transport D-branes unambiguously. Here is the procedure.

Start with an arbitrary D-brane $\mathcal{B} \in D(X_+)$ at $r \gg 0$ and choose a convenient lift $\mathfrak{B} \in \mathfrak{D}(\mathbb{C}^n, T)$, so that $\pi_+(\mathfrak{B}) \cong \mathcal{B}$. One may try to find a D-brane, which is D-isomorphic to $\mathfrak{B}$ and, at the same time, is an element of the subset $\mathcal{T}^w$. This is indeed always possible, i.e. we can lift any D-brane in $D(X_+)$ to a D-brane in the grade restricted set $\mathcal{T}^w$. Let us denote the lifting map by $\omega_{+,-}^w : D(X_+) \to \mathcal{T}^w$. In order for $\omega_{+,-}^w$ to be a well-defined map we have to check whether the lift to $\mathcal{T}^w$ is unique in an appropriate sense. As it turns out this is ensured by the fact that there are no D-isomorphisms between grade restricted D-branes. We will sketch the construction of $\omega_{+,-}^w$ in our example below. From there the generalization to arbitrary models with $T = U(1)$ will be clear.

Once we have lifted a D-brane from $D(X_+)$ to a grade restricted one, we can transport it through a window w to $r \ll 0$ and project it down to $D(X_-)$ via the map π_-. For the path in opposite direction this works similarly and we can subsume the transport of low-energy D-branes between adjacent phases in the following hat diagram:

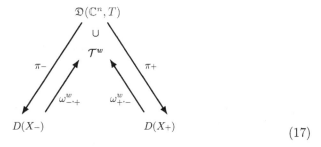

$$(17)$$

This diagram defines the composite maps

$$
\begin{aligned}
F^w_{-,+} &: D(X_-) \xrightarrow{\omega^w_{-,+}} \mathcal{T}^w \xrightarrow{\pi_+} D(X_+), \\
F^w_{+,-} &: D(X_+) \xrightarrow{\omega^w_{+,-}} \mathcal{T}^w \xrightarrow{\pi_-} D(X_-),
\end{aligned}
\tag{18}
$$

which, by the uniqueness of the lifts to $\mathcal{T}^w$, are inverses of each other:

$$
F^w_{-,+} \circ F^w_{+,-} \cong \mathrm{id}_{D(X_+)}, \qquad F^w_{+,-} \circ F^w_{-,+} \cong \mathrm{id}_{D(X_-)}.
$$

In our example the singular locus $\mathfrak{S}$ of the Kähler moduli space was at $\theta \in 3\pi + 2\pi\mathbb{Z}$ and $r = 3\log 3$. Let us choose the window $w = \{-\pi < \theta < \pi\}$. The corresponding grade restriction rule gives $N^w = \{-1, 0, 1\}$ and hence the D-branes in $\mathcal{T}^w$ are made of

$$
\mathcal{W}(-1), \ \mathcal{W}(0), \ \mathcal{W}(1).
$$

We start at $r \gg 0$ with the holomorphic line bundle $\mathcal{O}(2)$ over X_+. The most naive lift to the linear sigma model is the D-brane $\mathcal{C}(\mathfrak{B}_2) = \mathcal{W}(2)$; but, as we have seen previously, the D-brane

$$
\mathcal{C}(\mathfrak{B}_1): \quad \mathcal{W}(-1) \xrightarrow{X} \mathcal{W}(0)^{\oplus 3} \xrightarrow{X} \underline{\mathcal{W}(1)^{\oplus 3}}
$$

also satisfies $\pi_+(\mathfrak{B}_1) \cong \mathcal{O}(2)$. In fact, there are infinitely many D-branes $\mathfrak{B}$ with $\pi_+(\mathfrak{B}) \cong \mathcal{O}(2)$. However, among those the D-brane $\mathfrak{B}_1$ is special in that it is an object in the grade restricted subset $\mathcal{T}^w$. Thus, $\mathfrak{B}_1$ is the right representative to cross the phase boundary through the window w. After arriving at the orbifold phase, $r \ll 0$, we apply the projection π_- to $\mathfrak{B}_1$ to obtain the low-energy D-brane in $D(X_-)$:

$$
\mathcal{O}(-1) \xrightarrow{X} \mathcal{O}(0)^{\oplus 3} \xrightarrow{X} \underline{\mathcal{O}(1)^{\oplus 3}} \ .
$$

This is the image of $\mathcal{O}(2) \in D(X_+)$ under the transportation $F^w_{+,-}$ through the window w.

We next start at $r \ll 0$ with the $\mathbb{Z}_3$-equivariant line bundle $\mathcal{O}(2)$. Again this can be lifted into infinitely many D-branes $\mathcal{W}(2+3n)$, $n \in \mathbb{Z}$, but only one of them, $\mathcal{W}(-1)$, is in the grade restriction range. Thus, the transportation of $\mathcal{O}(2) \in D(X_-)$ through the window w yields $\mathcal{O}(-1) \in D(X_+)$.

Finally, let us consider the fractional D-brane $\mathcal{O}_{\mathfrak{p}}(2) \in D(X_-)$. Its naive lift is the D-brane $\mathfrak{B}_+$ given in (12). However, the rightmost entry, $\mathcal{W}(2)$, is not in the grade restriction range. We can replace it by

$$\mathcal{W}(-1) \xrightarrow{X} \mathcal{W}(0)^{\oplus 3} \xrightarrow{X} \mathcal{W}(1)^{\oplus 3} \xrightarrow{pX} \underline{\mathcal{W}(-1)}$$

using the D-isomorphism relation $\mathcal{O}(2) \cong \mathcal{O}(-1)$ that comes from $\mathcal{W}(2) \xrightarrow{p} \mathcal{W}(-1)$. This D-brane can be transported safely to $r \gg 0$ through the window w and we obtain the complex $\mathcal{O}(-1) \xrightarrow{X} \mathcal{O}(0)^{\oplus 3} \xrightarrow{X} \mathcal{O}(1)^{\oplus 3} \xrightarrow{pX} \underline{\mathcal{O}(-1)}$. Another D-isomorphism tells us that this D-brane is $\mathcal{O}(2) \xrightarrow{p} \underline{\mathcal{O}(-1)}$. The fractional D-brane $\mathcal{O}_{\mathfrak{p}}(2)$ of the orbifold phase is, therefore, mapped by $F^w_{-,+}$ to the line bundle $\mathcal{O}_E(-1)$, which is compactly supported on the exceptional divisor $E = \{p = 0\} \cong \mathbb{CP}^2$ in X_+.

From our examples we see that the key point is to find, in the phase of the starting point, the lift of a given brane $\mathcal{B}$ to the grade restricted collection $\mathcal{T}^w$. In order to obtain this particular D-brane, we start with a convenient lift $\mathfrak{B} \in \mathfrak{D}(\mathbb{C}^n, T)$, so that $\pi_-(\mathfrak{B}) \cong \mathcal{B}$. Then the complex $\mathcal{C}(\mathfrak{B})$ can always be changed into a complex made of grade restricted Wilson line branes without changing the D-isomorphism class in the low-energy theory. For $r \gg 0$ one can do so using the D-brane associated to the deleted set Δ_+:

$$\mathcal{W}(q-3) \xrightarrow{X} \mathcal{W}(q-2)^{\oplus 3} \xrightarrow{X} \mathcal{W}(q-1)^{\oplus 3} \xrightarrow{X} \mathcal{W}(q).$$

It is a trivial D-brane in the low-energy theory and can be used to eliminate a Wilson line component $\mathcal{W}(q)$, whose charge is above the grade restriction range N^w, in favour of component branes $\mathcal{W}(q-3), \mathcal{W}(q-2), \mathcal{W}(q-1)$, i.e. we can decrease the charges. Likewise, we may use this trivial D-brane to increase charges, which lie below the grade restriction range. Applying this procedure iteratively allows us to bring any D-brane $\mathcal{B}$ into the grade restriction range. The trivial complex has just the right length in the gauge charges, so that one can make sure that the process of decreasing or increasing the charges does not overshoot. Also, the trivial D-brane is too long to give rise to non-trivial D-isomorphisms within the grade restricted set $\mathcal{T}^w$, thus making the grade restricted D-brane unique up to trivial branes of the form $\mathcal{W}(q) \xrightarrow{1} \mathcal{W}(q)$. In the phase $r \ll 0$ a similar rôle is played by the complex associated with the deleted set Δ_-:

$$\mathcal{W}(q) \xrightarrow{p} \mathcal{W}(q-3).$$

4.2 Some Remarks on Multi-parameter Models

In order to generalize our result to models with higher-rank gauge group, let us recall that at the phase boundary between two phases, say Phase I and II, there is a Coulomb branch where a $U(1)$ subgroup of the gauge group is unbroken. We denote it $U(1)_{I,II}$. It is this unbroken subgroup that plays the rôle of the $U(1)$ gauge group in the one-parameter model.

Nearly everything that we said for one-parameter models carries over to the multi-parameter case. The grade restriction rule (16) applies only to the charges with respect to the unbroken subgroup $U(1)_{I,II}$ and is, therefore, more appropriately called a *band restriction rule*. If we choose $U(1)_{I,II}$ to be at the first component of the gauge group $T \cong U(1)^k$, the band of admissible charges for a given window w is given by

$$N^w = \{(q^1, \ldots, q^k) \in \mathbb{Z}^k \mid \forall \theta^1 \in w : -S^1 \pi < \theta^1 + 2\pi q^1 < S^1 \pi\} \, ,$$

where S^1 is the sum over all positive Q_i^1's. The subset $\mathcal{T}^w \subset \mathfrak{D}(\mathbb{C}^n, T)$ is now generated from Wilson line branes $\mathcal{W}(q^1, \ldots, q^k)$ with charges in N^w.

When we transport a low-energy D-brane from one phase to another, we have to band restrict using D-isomorphisms, just as in the one-parameter case. There is one difference form the $k = 1$ case: there are non-trivial D-isomorphisms between branes in $\mathcal{T}^w$ and the lifting maps $\omega^w_{*,**}$ are not unique. However, such D-isomorphisms are common to both phases and are projected out under the maps π_*. The key to this statement is the relation of the deleted sets

$$\Delta_I = \Delta_+ \cup (\Delta_I \cap \Delta_{II}), \qquad \Delta_{II} = \Delta_- \cup (\Delta_I \cap \Delta_{II}),$$

where Δ_+ (*resp.* Δ_-) is the common zeroes of the fields x_i with positive (*resp.* negative) charge with respect to $U(1)_{I,II}$, $Q_i^1 > 0$ (*resp.* $Q_i^1 < 0$). The parts $\Delta_\pm$ are relevant for the band restriction rule and the part $\Delta_I \cap \Delta_{II}$ is responsible for the common D-isomorphism relations.

Thus, we again have maps

$$F^w_{I,II} : D(X_I) \xrightarrow{\omega^w_{I,II}} \mathcal{T}^w \xrightarrow{\pi_{II}} D(X_{II}),$$
$$F^w_{II,I} : D(X_{II}) \xrightarrow{\omega^w_{II,I}} \mathcal{T}^w \xrightarrow{\pi_I} D(X_I),$$

which are inverses of each other. An immediate consequence of this result is the one-to-one correspondence of the sets, $D(X_r)$, of D-isomorphic D-branes in all the Kähler cones.

5 Some Relations to Earlier Work

Let us briefly comment on relations of our results to the literature.

5.1 McKay Correspondence

Consider a gauged linear sigma model, which reduces in a particular phase to an orbifold $X_{orb} \cong \mathbb{C}^n / \Gamma$, where Γ is a discrete subgroup of the gauge group T and is, therefore, always *abelian*. The representation ρ of the gauge group on the Chan–Paton space reduces to a representation of the orbifold group Γ, so that the low-energy D-branes in this phase are given by complexes of Γ-equivariant vector bundles on $\mathbb{C}^n$, $D(X_{orb}) \cong D_\Gamma(\mathbb{C}^n)$. Other phases in such a model are partial or complete crepant resolutions, X_{res}, of the orbifold singularity.

The maps $F^w_{+,-}$ of the previous subsection define a one-to-one correspondence, Φ, between D-isomorphism classes of D-branes on X_{orb} and any of its resolutions X_{res}, i.e.

$$D_\Gamma(\mathbb{C}^n) \xrightarrow{\Phi} D(X_{res}) .$$

In the chiral sector of the theory this reduces to McKay correspondence:

Given a finite group $\Gamma \subset SL(n, \mathbb{C})$ and a crepant resolution, X_{res}, of the quotient $\mathbb{C}^n/\Gamma$, there exists an equivalence of derived categories, $\mathbf{D}_\Gamma(\mathbb{C}^n) \xrightarrow{\cong} \mathbf{D}(X_{res})$.

For abelian finite groups and arbitrary n, this was first proven as a special case in [8]. For $n \leq 3$ and non-abelian finite groups, it was shown in [6].

5.2 Flop Transition

Let us discuss the flop transition of the resolved conifold. It is realized by the following model:

fields	x	y	u	v
$U(1)$	$+1$	$+1$	-1	-1

The flop is realized in terms of the linear sigma model coordinates in the following way:

$$X_- = \begin{bmatrix} \mathbb{CP}^1_{[u:v]} \\ \uparrow \\ \mathcal{O}(-1)^{\oplus 2}_{x,y} \end{bmatrix} \xleftarrow{\;\text{flop}\;} X_+ = \begin{bmatrix} \mathcal{O}(-1)^{\oplus 2}_{u,v} \\ \downarrow \\ \mathbb{CP}^1_{[x:y]} \end{bmatrix}$$

The trivial D-branes are

$$\mathcal{K}_+ : \quad \mathcal{W}(0) \xrightarrow{\binom{y}{-x}} \mathcal{W}(1)^{\oplus 2} \xrightarrow{(x,y)} \mathcal{W}(2) \qquad \text{for} \quad r \gg 0 ,$$

$$\mathcal{K}_- : \quad \mathcal{W}(2) \xrightarrow{\binom{v}{-u}} \mathcal{W}(1)^{\oplus 2} \xrightarrow{(u,v)} \mathcal{W}(0) \qquad \text{for} \quad r \ll 0 ,$$

and shifts thereof. Note that $S = 2$. At the phase boundary we choose the window $w = \{-2\pi < \theta < 0\}$, so that

$$N^w = \{0, 1\} .$$

A particularly interesting D-brane in this context, say at $r \gg 0$, is the D0-brane, $\mathcal{O}_\mathfrak{p}$, located at $\mathfrak{p} = \{u = v = P = 0\}$, where $P = \alpha x + \beta y$. The parameters $[\alpha : \beta] \in \mathbb{CP}^1$ parametrize the location of the D0-brane. The most naive lift to the linear sigma model is the Koszul complex

$$\mathcal{W}(1) \xrightarrow{\begin{pmatrix} p \\ u \\ v \end{pmatrix}} \begin{matrix} \mathcal{W}(2) \\ \oplus \\ \mathcal{W}(0)^{\oplus 2} \end{matrix} \xrightarrow{\begin{pmatrix} -v & 0 & p \\ u & -p & 0 \\ 0 & v & -u \end{pmatrix}} \begin{matrix} \mathcal{W}(1)^{\oplus 2} \\ \oplus \\ \mathcal{W}(-1) \end{matrix} \xrightarrow{(u,v,p)} \underline{\mathcal{W}(0)} \ .$$

Clearly, $\mathcal{W}(-1)$ and $\mathcal{W}(2)$ have to be eliminated using the complexes $\mathcal{K}_+$ resp. $\mathcal{K}_+(-1)$. One can check that the grade restricted representative in the linear sigma model is given by the complex

$$\mathcal{C}(\mathfrak{B}_p) : \ \mathcal{W}(0) \xrightarrow{\begin{pmatrix} P \\ uQ \\ vQ \end{pmatrix}} \begin{matrix} \mathcal{W}(1) \\ \oplus \\ \mathcal{W}(0)^{\oplus 2} \end{matrix} \xrightarrow{\begin{pmatrix} -vQ & 0 & P \\ uQ & -P & 0 \\ 0 & Qv & -Qu \end{pmatrix}} \begin{matrix} \mathcal{W}(1)^{\oplus 2} \\ \oplus \\ \mathcal{W}(0) \end{matrix} \xrightarrow{(Qu,Qv,P)} \underline{\mathcal{W}(1)} \ ,$$

where Q is a linear combination of x and y, so that $[P : Q]$ provide new coorindates on $\mathbb{CP}^1_{[x:y]}$.

The D-brane $\mathfrak{B}_p$ can now be transported through the window w to small r and projected down to $D(X_-)$ via π_-. There, one might suspect that there exist D-isomorphisms in order to eliminate brane–antibrane pairs, so that $\pi_-(\mathfrak{B}_p)$ can be expressed in terms of compactly supported D-branes on $E_- = \mathbb{CP}^1_{[u:v]}$. This is, however, not possible. The best we can say about $\pi_-(\mathfrak{B}_p)$ is that as a complex of coherent sheaves its cohomology is $H^{-1}(\pi_-(\mathfrak{B}_p)) \cong \mathcal{O}_{E_-}(0)$ and $H^0(\pi_-(\mathfrak{B}_p)) \cong \mathcal{O}_{E_-}(1)$, which is indeed compactly supported. In fact, the D-brane $\pi_-(\mathfrak{B}_p)$ is an example for perverse point sheaves as they were discussed in [7] and later in [13]. It has the same moduli space, namely $\mathbb{CP}^1$, and the same charges as a D0-brane on X_-, but it is not D-isomorphic to a D0-brane.

References

1. M. Herbst, K. Hori, and D. Page. Phases of $N = 2$ Theories in $1+1$ Dimensions With Boundary, arXiv:0803.2045[hep-th].
2. M. Kontsevich. Homological Algebra of Mirror Symmetry. Proceedings of the International Congress of Mathematicians (Zurich, 1994) 120–139, Birkhauser, Basel, 1995. [alg-geom/9411018].
3. E. R. Sharpe. D-branes, derived categories, and Grothendieck groups. *Nucl. Phys.*, B561:433–450, 1999.
4. M. R. Douglas. D-branes, Categories and $N = 1$ Supersymmetry. *J. Math. Phys.*, 42:2818–2843, 2001. [hep-th/0011017].
5. E. Witten. Phases of $N = 2$ Theories in two Dimensions. *Nucl. Phys.*, B403: 159–222, 1993. [hep-th/9301042].
6. T. Bridgeland, A. King, and M. Reid. The McKay correspondence as an equivalence of derived categories. *J. Am. Math. Soc.*, 14:535–554, 2001.
7. T. Bridgeland. Flops and derived categories. *Invent. Math.*, 147:613–632, 2002.
8. Y. Kawamata. Log crepant birational maps and derived categories. *J. Math. Sci. Univ. Tokyo*, 12:211–231, 2005.

9. M. van den Bergh, Non-commutative crepant resolutions, in The legacy of Niels Henrik Abel. 749–770, Springer, Berlin, 2004, [arXiv:math/0211064].
10. K. Hori, S. Katz, A. Klemm, R. Pandharipande, R. Thomas, C. Vafa, R. Vakil, and E. Zaslow. Mirror Symmetry. Providence, USA:AMS(2003) 929 p. (Clay mathematics monographs. 1).
11. K. Hori, A. Iqbal, and C. Vafa. D-branes and Mirror Symmetry. 2000. [hep-th/0005247].
12. A. Sen. Tachyon Condensation on the Brane Antibrane System. J. High. Energy Phys., 08:012, 1998. [hep-th/9805170].
13. P. S. Aspinwall. D-branes on Calabi–Yau Manifolds. Published in Boulder 2003, Progress in string theory, 1–152. [hep-th/0403166].
14. D. R. Morrison and M. Ronen Plesser, Summing the instantons: Quantum cohomology and mirror symmetry in toric varieties. *Nucl. Phys.*, B440:279, 1995 [arXiv:hep-th/9412236].

Topological String Theory on Compact Calabi–Yau: Modularity and Boundary Conditions

M.-x. Huang[1], A. Klemm[1,2], and S. Quackenbush[1]

[1] Department of Physics, University of Wisconsin, Madison, WI 53706, USA
`minxin@physics.wisc.edu`, `aklemm@physics.wisc.edu`,
`squackenbush@wisc.edu`
[2] Department of Mathematics, University of Wisconsin, Madison, WI 53706, USA

Abstract. The topological string partition function $Z(\lambda, t, \bar{t}) = \exp(\lambda^{2g-2} F_g(t, \bar{t}))$ is calculated on a compact Calabi–Yau M. The $F_g(t, \bar{t})$ fulfil the holomorphic anomaly equations, which imply that $\Psi = Z$ transforms as a wave function on the symplectic space $H^3(M, \mathbb{Z})$. This defines it everywhere in the moduli space $\mathcal{M}(M)$ along with preferred local coordinates. Modular properties of the sections F_g as well as local constraints from the 4d effective action allow us to fix Z to a large extent. Currently with a newly found gap condition at the conifold, regularity at the orbifold and the most naive bounds from Castelnuovo's theory, we can provide the boundary data, which specify Z, e.g. up to genus 51 for the quintic.

1 Outline

Coupling topological matter to topological gravity is a key problem in string theory. Conceptually most relevant is the topological matter sector of the critical string as it arises, e.g. in Calabi–Yau compactifications. Topological string theory on non-compact Calabi–Yau manifolds such as $\mathcal{O}(-3) \to \mathbb{P}^2$ is essentially solved by either localization – [1] or large N-techniques [2] and has intriguing connections to Chern–Simons theory [3], open–closed string duality [4], matrix models [5], integrable hierarchies of non-critical string theory [6] and 2d Yang–Mills theory [7].

However, while local Calabi–Yau manifolds are suitable to study gauge theories and more exotic field theories in 4d and specific couplings to gravity, none of the techniques above extends to compact Calabi–Yau spaces, which are relevant for important questions in 4d quantum gravity concerning, e.g. the properties of 4d black holes [8] and the wave function in mini superspace [9].

Moreover, while the genus dependence is encoded in the Chern–Simons and matrix model approaches in a superior fashion by the $1/N^2$-expansion, the moduli dependence on the parameter t is reconstructed locally and in

Huang, M.-x., et al.: *Topological String Theory on Compact Calabi–Yau: Modularity and Boundary Conditions*. Lect. Notes Phys. **757**, 45–102 (2009)
DOI 10.1007/978-3-540-68030-7_3   © Springer-Verlag Berlin Heidelberg 2009

a holomorphic limit, typically by sums over partitions. This yields an algorithm, which grows exponentially in the world–sheet degree or the space–time instanton number.

As the total $F_g(t, \bar{t})$ are modular invariant sections over the moduli space $\mathcal{M}(M)$, they must be generated by a ring of almost holomorphic modular forms. This solves the dependence on the moduli in the most effective way. In the following we will show that space–time modularity, the holomorphic anomaly equations of Bershadsky, Cecotti, Ooguri and Vafa (BCOV), as well as boundary conditions at various boundary components of the moduli space, solve the theory very efficiently.

For compact (and non-compact) Calabi–Yau spaces mirror symmetry is proven at genus zero. The modular properties that we need are also established at genus zero. Moreover it has been argued recently that the holomorphic anomaly recursions follow from categorical mirror symmetry [10, 11]. To establish mirror symmetry at higher genus, one needs merely to prove that the same boundary data fix the $F_g(t, \bar{t})$ in the A- and the B-model.

1.1 Extending the Seiberg–Witten Approach to Gravity

Seiberg–Witten reconstructed the non-perturbative N=2 gauge coupling from meromorphic sections over $\mathcal{M}(M)$ using their modular properties and certain local data from the effective action at singular divisors of $\mathcal{M}(M)$. In [12] we reconsidered the problem of topological string on local Calabi–Yau from the modular point of view and found that the singular behaviour of the gravitational couplings is restrictive enough to reconstruct them globally. This can be viewed as the most straightforward extension of the Seiberg–Witten approach to gravitational couplings.

Note that the problem of instanton counting in these cases is solved either by geometric engineering, one of the techniques mentioned above, or more directly by the localization techniques in the moduli space of gauge theory instantons by Nekrasov, Nakajima et al. It is nevertheless instructive to outline the general idea in this simple setting.[3] We focused on the N=2 $SU(2)$ Seiberg–Witten case, but the features hold for any local Calabi–Yau whose mirror is an elliptic curve with a meromorphic differential [13, 14][4] and are as follows

- The genus g topological string partition functions are given by

$$F^{(g)}(\tau, \bar{\tau}) = \xi^{2g-2} \sum_{k=0}^{3(g-1)} \hat{E}_2^k(\tau, \bar{\tau}) c_k^{(g)}(\tau) . \tag{1}$$

---

[3] Maybe the simplest example of the relation between modularity and the holomorphic anomaly equations is provided by Hurwitz theory on elliptic curves [15].

[4] With fairly obvious generalizations for the cases where the mirror is a higher genus curve. In this case the traditional modular forms of subgroups Γ of $\Gamma_0 := SL(2, \mathbb{Z})$ have to be replaced by Siegel modular forms of subgroups of $Sp(2g, \mathbb{Z})$ [14].

Here $\hat{E}_2(\tau, \bar{\tau}) := E_2(\tau) + 6i/\pi(\bar{\tau} - \tau)$ is the modular invariant anholomorphic extension of the second Eisenstein series $E_2(\tau)$ and the holomorphic 'Yukawa coupling' $\xi := C_{ttt}^{(0)} = \partial \tau/\partial t$ is an object of weight -3 under the modular $\Gamma \in \Gamma_0 = SL(2, \mathbb{Z})$. For example for pure $N = 2$ $SU(2)$ gauge theory $\Gamma = \Gamma(2)$ [16]. Modular invariance implies then that $c_k^{(g)}(\tau)$ are modular forms of Γ of weight $6(g-1) - 2k$.

- The simple anti-holomorphic dependence of (1) implies that the only part in $F^{(g)}(\tau, \bar{\tau})$ not fixed by the recursive anomaly equations is the weight $6(g-1)$ holomorphic forms $c_0^{(k)}(\tau)$, which are finitely generated as a weighted polynomial $c_0^{(g)}(\tau) = p_{6(g-2)}(k_1, \ldots, k_m)$ in the holomorphic generators $G_{k_1}, \ldots, G_{k_m}$ of forms of Γ.

- The finite data needed to fix the coefficients in $p_{6(g-2)}(k_1, \ldots, k_m)$ are provided in part by the specific leading behaviour of the $F^{(g)}$ at the conifold divisor

$$F_{\text{conifold}}^{(g)} = \frac{(-1)^{g-1} B_{2g}}{2g(2g-2)t_D^{2g-2}} + \mathcal{O}(t_D^0), \tag{2}$$

in special local coordinates t_D. The order of the leading term was established in [17], the coefficient of the leading term in [18], and the "gap condition", i.e. the vanishing of the following $2g - 3$ negative powers in t_D in [12]. This property in particular carries over to the compact case and we can give indeed a string theoretic explanation of the finding in [12].

- Further conditions are provided by the regularity of the $F^{(g)}$ at orbifold points in $\mathcal{M}(M)$. These conditions unfortunately turn out to be somewhat weaker in the global case than in the local case.

Similar forms as (1) for the $F^{(g)}$ appear in the context of Hurwitz theory on elliptic curves [15], of mirror symmetry in $K3$ fibre limits[19] and on rational complex surfaces [20, 21]. In the local cases, which have elliptic curves as mirror geometry, we found [12, 13], that the above conditions (over)determine the unknowns in $p_{6(g-2)}(k_1, \ldots, k_m)$ and solve the theory. This holds also for the gauge theories with matter, which from geometric engineering point of view correspond to local Calabi–Yau manifolds with several (Kähler) moduli [13]. Using the precise anholomorphic dependence and restrictions from space–time modularity one can iterate the holomorphic anomaly equation with an algorithm which is exact in the moduli dependence and grows polynomially in complexity with the genus.

Here we extend this approach further to compact Calabi–Yau spaces and focus on the class of one Kähler moduli Calabi–Yau spaces M such as the quintic. More precisely we treat the class of one modulus cases whose mirror W has, parameterized by a suitable single cover variable, a Picard–Fuchs system with exactly three regular singular points: The point of maximal monodromy, a conifold point, and a point with rational branching. The latter can be simply a $\mathbb{Z}_d$ orbifold point. This example is the case for the hypersurfaces where the string theory has an exact conformal field theory description at this point in terms of an orbifold of a tensor product of minimal $(2, 2)$ SCFT field theories, the so called the Gepner-model. For some complete intersections there are massless BPS particles at the branch locus, which lead in addition to logarithmic singularities.

We find a natural family of coordinates in which the conifold expansion as well as the rational branched logarithmic singularities exhibit the gap condition (2). Despite the fact that the modular group, in this case a subgroup of $\mathrm{Sp}(h^3, \mathbb{Z})$, is poorly understood,[5] we will see that the essential feature carry over to the compact case. Modular properties, the "gap condition", together with regularity at the orbifold, the leading behaviour of the F_g at large radius, and Castelenovo's Bound determine topological string on one modulus Calabi–Yau to a large extent.

2 The Topological B-Model

In this section we give a quick summary of the approach of [17, 22] to the topological B-model, focusing as fast as possible on the key problems that need to be overcome: namely the problem of integrating the anomaly equation efficiently and the problem of fixing the boundary conditions.

2.1 The Holomorphic Anomaly Equations

The definition of $F^{(g)}$ is $F^{(g)} = \int_{\mathcal{M}_g} \mu_g$ with measure on $\mathcal{M}_g$

$$\mu_g = \prod_{i=1}^{3g-3} \mathrm{d}m_i \mathrm{d}\bar{m}_{\bar{\imath}} \left\langle \prod_{i,\bar{\imath}} \int_{\Sigma} G_{zz} \mu_{\bar{z}}^{(i)\,z} \mathrm{d}^2 z \int_{\Sigma} G_{\bar{z}\bar{z}} \mu_z^{(i)\,\bar{z}} \mathrm{d}^2 z \right\rangle . \tag{3}$$

Here the Beltrami differentials $\mu_{\bar{z}}^{(i)\,z} \mathrm{d}\bar{z}$ span $H^1(\Sigma, T\Sigma)$, the tangent space to $\mathcal{M}_g$. The construction of the measure μ_g is strikingly similar to the one for the bosonic string, once the BRST partner of the energy–momentum tensor is identified with the superconformal current $G_{zz} \mathrm{d}z$ and the ghost number with the $U(1)$ charge [23]. $\langle\rangle$ is to be evaluated in the internal $(2, 2)$ SCFT, but it is easy to see that it gets only contributions from the topological (c, c) sector.

The holomorphic anomaly equation reads for $g = 1$ [22]

$$\bar{\partial}_{\bar{k}} \partial_m F^{(1)} = \frac{1}{2} \bar{C}_{\bar{k}}^{ij} C_{mij}^{(0)} + \left(\frac{\chi}{24} - 1 \right) G_{\bar{k}m} , \tag{4}$$

where χ is the Euler number of the target space M, and for $g > 1$ [17]

$$\bar{\partial}_{\bar{k}} F^{(g)} = \frac{1}{2} \bar{C}_{\bar{k}}^{ij} \left(D_i D_j F^{(g-1)} + \sum_{r=1}^{g-1} D_i F^{(r)} D_j F^{(g-r)} \right) . \tag{5}$$

The right-hand side of the equations comes from the complex codimension one boundary of the moduli space of the worldsheet $\mathcal{M}_g$, which corresponds to pinching of handles. The key idea is that $\bar{\partial}_{\bar{k}} F^{(g)} = \int_{\mathcal{M}_g} \bar{\partial} \partial \lambda_g$, where $\bar{\partial} \partial$ are derivatives on $\mathcal{M}_g$ so that $\bar{\partial}_{\bar{k}} F^{(g)} = \int_{\partial \mathcal{M}_g} \lambda_g$. The contribution to the latter integral is from the codimension one boundary $\partial \mathcal{M}_g$.

The first Eq. (4) can be integrated using special geometry up to a holomorphic function [22], which is fixed by the consideration in Sect. 2.2.

The Eq. (5) is solved in BCOV using the fact that due to

$$\bar{D}_{\bar{i}} \bar{C}_{\bar{j} \bar{k} \bar{l}} = \bar{D}_{\bar{j}} \bar{C}_{\bar{i} \bar{k} \bar{l}} \tag{6}$$

one can integrate

$$\bar{C}_{\bar{j} \bar{k} \bar{l}} = e^{-2K} \bar{D}_{\bar{i}} \bar{D}_{\bar{j}} \bar{\partial}_{\bar{k}} S \tag{7}$$

as

$$S_{\bar{i}} = \bar{\partial}_{\bar{i}} S, \quad S_{\bar{i}}^j = \bar{\partial}_{\bar{i}} S^j, \quad \bar{C}_{\bar{k}}^{ij} = \bar{\partial}_{\bar{k}} S^{ij} . \tag{8}$$

The idea is to write the right-hand side of (5) as a derivative w.r.t. $\bar{\partial}_{\bar{k}}$. In the first step one writes

$$\begin{aligned}
\bar{\partial}_{\bar{k}} F^{(g)} = \bar{\partial}_{\bar{k}} & \left(\frac{1}{2} S^{ij} \left(D_i D_j F^{(g-1)} + \sum_{r=1}^{g-1} D_i F^{(r)} D_j F^{(g-r)} \right) \right) \\
& - \frac{1}{2} S^{ij} \bar{\partial}_{\bar{k}} \left(D_i D_j F^{(g-1)} + \sum_{r=1}^{g-1} D_i F^{(r)} D_j F^{(g-r)} \right) .
\end{aligned} \tag{9}$$

With the commutator $R_{i\bar{k}j}^l = -\bar{\partial}_{\bar{k}} \Gamma_{ij}^l = [D_i, \partial_{\bar{k}}]_j^l = G_{i\bar{k}} \delta_j^l + G_{j\bar{k}} \delta_i^l - C_{ijm}^{(0)} \bar{C}_{\bar{k}}^{ml}$ the second term can be rewritten so that the $\bar{\partial}_{\bar{k}}$ derivative acts in all terms directly on $F^{(g)}$. Then using (4, 5) with $g' < g$ one can iterate the procedure, which produces an equation of the form

$$\bar{\partial}_{\bar{k}} F^{(g)} = \bar{\partial}_{\bar{k}} \Gamma^{(g)} (S^{ij}, S^i, S, C_{i_1,\dots,i_n}^{(<g)}) , \tag{10}$$

where $\Gamma^{(g)}$ is a functional of S^{ij}, S^i, S and $C_{i_1,\dots,i_n}^{(<g)}$. This implies that

$$F^{(g)} = \Gamma^{(g)} (S^{ij}, S^i, S, C_{i_1,\dots,i_n}^{(<g)}) + c_0^{(g)}(t) , \tag{11}$$

is a solution. Here $c_0^{(g)}(t)$ is the holomorphic ambiguity, which is not fixed by the recursive procedure. It is holomorphic in t as well as modular invariant. The *major conceptual problem* of topological string theory on compact Calabi–Yau is to find the *boundary conditions* which fix $c_0^{(g)}(t)$. Note that the problem is not well defined without the constraints from modular invariance.

Using the generalization of the gap condition in Sect. 2.2, the behaviour of
the orbifold singularities in Sect. 3.4 and Castelnuovo's bound in Sect. 5.2 we
can achieve this goal to a large extent.

Properties of the $\Gamma^{(g)}(S^{ij}, S^i, S, C^{(<g)}_{i_1,\ldots,i_n})$ are established using the auxiliary action

$$Z = \int \mathrm{d}x \mathrm{d}\phi \exp(Y + \tilde{W}) \tag{12}$$

where

$$\tilde{W}(\lambda, x, \phi, t, \bar{t}) = \sum_{g=0}^{\infty} \sum_{m=0}^{\infty} \sum_{n=0}^{\infty} \frac{1}{m!n!} \tilde{C}^{(g)}_{i_1,\ldots,i_n,\phi^m} x_{i_1} \cdots x_{i_n} \phi^m$$

$$= \sum_{g=0}^{\infty} \sum_{n=0}^{\infty} \frac{\lambda^{2g-2}}{n!} C^{(g)}_{i_1,\ldots,i_n} x_{i_1} \cdots x_{i_n} (1-\phi)^{2-2g-n} \tag{13}$$

$$+ \left(\tfrac{\chi}{24} - 1\right) \log\left(\tfrac{1}{1-\phi}\right) \, ,$$

with $C^{(g)}_{i_1,\ldots,i_n} = D_{i_1} \ldots D_{i_n} F^{(g)}$ and the "kinetic term" is given by

$$Y(\lambda, x, \phi; t, \bar{t}) = -\frac{1}{2\lambda^2} (\Delta_{ij} x^i x^j + 2\Delta_{i\phi} x^i \phi + \Delta_{\phi\phi} \phi^2) + \frac{1}{2} \log\left(\frac{\det \Delta}{\lambda^2}\right) \, . \tag{14}$$

In [17] it was shown that $\exp(\tilde{W})$ fulfills an equation

$$\frac{\partial}{\partial \bar{t}_i} \exp(\tilde{W}) = \left[\frac{\lambda^2}{2} \bar{C}^{jk}_{\bar{i}} \frac{\partial^2}{\partial x^j \partial x^k} - G_{\bar{i}j} x^j \frac{\partial}{\partial \phi}\right] \exp(\tilde{W}) \tag{15}$$

that is equivalent to the holomorphic anomaly equations, by checking the
coefficients of the λ powers, and $\exp(Y)$ fulfills

$$\frac{\partial}{\partial \bar{t}_{\bar{i}}} \exp(Y) = \left[-\frac{\lambda^2}{2} \bar{C}^{jk}_{\bar{i}} \frac{\partial^2}{\partial x^j \partial x^k} - G_{\bar{i}j} x^j \frac{\partial}{\partial \phi}\right] \exp(Y) \tag{16}$$

implying that Δ_{ij}, $\Delta_{i\phi}$ and $\Delta_{\phi\phi}$ are the inverses to the propagators $K^{ij} = -S^{ij}$, $K^{i\phi} := -S^i$ and $K^{\phi\phi} := -2S$. A saddle point expansion of Z gives

$$\log(Z) = \sum_{g=2}^{\infty} \lambda^{2g-2} \left[F^{(g)} - \Gamma^{(g)}(S^{ij}, S^i, S, C^{(<g)}_{i_1,\ldots,i_n})\right] \, , \tag{17}$$

where $\Gamma^{(g)}(S^{ij}, S^i, S, C^{(<g)}_{i_1,\ldots,i_n})$ is simply the Feynman graph expansion of the
action (12) with the vertices $\tilde{C}^{(g)}_{i_1,\ldots,i_n,\phi^m}$ and the propagators above. Moreover it can be easily shown that $\partial/\partial \bar{t}_i Z = 0$, which implies to all orders that
$F^{(g)}$ can be written as (11). This establishes the reduction of the whole calculation to the determination of the holomorphic modular invariant sections
$c_0^{(g)}(t) \in \mathcal{L}^{2g-2}$. However it also reflects the *major technical problem* in the

approach of BCOV, namely that the procedure to determine the recursive anholomorphic part *grows exponentially with the genus*. It has been observed in [25] that in concrete cases the terms appearing in the Feynman graph expansion are not functionally independent. This is a hint for finitely generated rings of anholomorphic modular forms over $\mathcal{M}(M)$. Using the modular constraints systematically in each integration step Yamaguchi and Yau developed a recursive procedure for the quintic whose complexity grows asymptotically only polynomially, see (53).

Since the B-model is 2d gravity coupled to 2d matter, let us compare the situation with pure 2d gravity, where the objects of interest are correlation functions of $\tau_{d_i} = (2d_i + 1)!! c_1(L_i)^{d_i}$ which are forms on $\overline{\mathcal{M}}_g$ constructed from the descendent fields

$$F_g(t_0, t_1, \ldots) = \sum_{\{d_i\}} \langle \prod \tau_{d_i} \rangle_g \prod_{r>0} \frac{t_r^{n_r}}{n_r!} . \tag{18}$$

Here $\{d_i\}$ is the set of all non-negative integers and $n_r := \mathrm{Card}(i : d_i = r)$.

The linear second order differential equations (15) is the small phase space analogue of the *Virasoro constraints*

$$L_n Z = 0, \qquad n \geq -1 \tag{19}$$

on $Z = e^F$ with $F = \sum_{g=0}^{\infty} \lambda^{2g-2} F_g$ the free energy of 2d topological gravity [26]. Indeed the L_n with $[L_n, L_m] = (n - m)L_{n+m}$ are second order linear differential operators in the t_i. The *non-linear KdV Hierarchy*, which together with dilaton and string equation are equivalent to (19) [26] and correspond in the small phase space of the B-model to the holomorphic anomaly equations (4, 5). In the A-model approaches to topological string on Calabi–Yau manifolds, such as relative GW-theory, localization or attempts to solve the theory via massive (2, 2) models, the descendents are introduced according to the details of the geometrical construction and then "summed away".

The combinatorial cumbersome information in the descendent sums is replaced in the B-model by the contraints from the modular group, holomorphicity and boundary information from the effective 4d action. As a consequence of this beautiful interplay between space–time and world–sheet properties one needs only the small phase space equations (4, 5, 15).

This approach requires the ability to relate various local expansions of $F^{(g)}$ near the boundary of the moduli space. Sensible local expansions (of terms in the effective action) are in locally monodromy invariant coordinates. As explained in [14] these coordinates in various patches are related by symplectic transformations on the phase space $H^3(M)$. The latter extend as metaplectic transformations to the wave function $\psi = Z$ of the topological string on the Calabi–Yau [27], which defines the transformation on the $F^{(g)}$. It will be important for us that the real polarization [14, 28, 29] defines an unique splitting (11, 54) of local expansions of the $F^{(g)}$ in the anholomorphic modular part determined by the anomaly equations and the holomorphic modular part

$c_0^{(g)}(t)$. Aspects of the wave functions properties and the various polarizations have been further discussed in [14, 28, 29].

2.2 Boundary Conditions from Light BPS States

Boundaries in the moduli space $\mathcal{M}(M)$ correspond to degenerations of the manifold M and general properties of the effective action can be inferred from the physics of the lightest states. More precisely the light states relevant to the $F^{(g)}$ terms in the $N = 2$ actions are the Bogomol'nyi-Prasad-Sommerfield (BPS) states. Let us first discuss the boundary conditions for $F^{(1)}$ at the singular points in the moduli space.

- At the point of maximal unipotent monodromy in the mirror manifold W, the Kähler areas, four and six volumes of the original manifold M are all large. Therefore the lightest string states are the constant maps $\Sigma_g \to pt \in M$. For these Kaluza–Klein reduction, i.e. a zero mode analysis of the A-twisted non-linear σ-model is sufficient to calculate the leading behaviour[6] of $F^{(1)}$ as [22]

$$F^{(1)} = \frac{t_i}{24} \int c_2 \wedge J_i + \mathcal{O}(e^{2\pi it}) \ . \tag{20}$$

Here $2\pi i\, t_i = \frac{X^i}{X^0}$ are the canonical Kähler parameters, c_2 is the second Chern class and J_i is the basis for the Kähler cone dual to two-cycles C_i defining the $t_i := \int_{C_i} \hat{J} = \int_{C_i} \sum_i t_i J_i$.

- At the conifold divisor in the moduli space $\mathcal{M}(W)$, W develops a nodal singularity, i.e. a collapsing cycle with S^3 topology. As discussed in Sect. 3.5 this corresponds to the vanishing of the total volume of M. The leading behaviour at this point is universally [30]

$$F^{(1)} = -\frac{1}{12} \log(t_D) + O(t_D) \ . \tag{21}$$

This leading behaviour has been physically explained as the effect of integrating out a non-perturbative hypermultiplet, namely the extremal black hole of [31]. Its mass $\sim t_D$, see (26), goes to zero at the conifold and it couples to the $U(1)$ vector in the $N = 2$ vectormultiplet, whose lowest component is the modulus t_D. The factor $1/12$ comes from the gravitational one-loop β-function, which describes the running of the $U(1)$ coupling [32]. A closely-related situation is the one of a shrinking lens space S/G. As explained in [33] one gets in this case several BPS hyper multiplets as the bound states of wrapped D-branes, which modifies the factor $1/12 \to |G|/12$ in the one loop β-function (21).

- The gravitational β-function argument extends also to non-perturbative spectra arising at more complicated singularities, e.g. with gauge symmetry enhancement and adjoint matter [34].

---

[6] The leading of $F^{(0)}$ at this point is similarly calculated and given in (37).

For the case of the one-parameter families the above boundary information and the fact is sufficient to fix the holopmorphic ambiguity in $F^{(1)}$.

To learn from the effective action point of view about the higher genus boundary behaviour, let us recall that the $F^{(g)}$ as in $F(\lambda, t) = \sum_{g=1}^{\infty} \lambda^{2g-2} F^{(g)}(t)$ give rise to the following term:

$$S_{1-loop}^{N=2} = \int d^4 x R_+^2 F(\lambda, t) , \qquad (22)$$

where R_+ is the self-dual part of the curvature and we identify λ with F_+, the self-dual part of the graviphoton field strength. As explained in [35, 36], see [23] for a review, the term is computed by a one-loop integral in a constant graviphoton background, which depends only on the left $(SO(4) = SU(2)_L \otimes SU(2)_R)$ Lorentz quantum numbers of BPS particles P in the loop. The calculation is very similar to the normal Schwinger-loop calculation. The latter computes the one-loop effective action in an $U(1)$ gauge theory, which comes from integrating out massive particles P coupling to a constant background $U(1)$ gauge field. For a self-dual background field $F_{12} = F_{34} = F$ it leads to the following one-loop determinant evaluation:

$$\begin{aligned} S_{1-loop}^S &= \log \det \left(\nabla + m^2 + 2e\, \sigma_L F \right) \\ &= \int_\epsilon^\infty \frac{ds}{s} \frac{\text{Tr}(-1)^f \exp(-sm^2) \exp(-2se\sigma_L F)}{4 \sin^2 (seF/2)} . \end{aligned} \qquad (23)$$

Here the $(-1)^f$ takes care of the sign of the log of the determinant depending on whether P is a boson or a fermion and σ_L is the Cartan element in the left Lorentz representation of P. To apply this calculation to the $N = 2$ supergravity case one notes, that the graviphoton field couples to the mass, i.e. we have to identify $e = m$. The loop has two R_+ insertions and an arbitrary, (for the closed string action even) number, of graviphoton insertions. It turns out [36] that the only supersymmetric BPS states with the Lorentz quantum numbers

$$\left[\left(\frac{1}{2}, 0 \right) + 2(0, 0) \right] \otimes \mathcal{R} \qquad (24)$$

contribute to the loop. Here $\mathcal{R}$ is an arbitrary Lorentz representation of $SO(4)$. Moreover the two R_+ insertions are absorbed by the first factor in the Lorentz representation (24) and the coupling of the particles in the loop to F_+ insertions in the $N = 2$ evaluation works exactly as in the non-supersymmetric Schwinger-loop calculation above for P in the representation $\mathcal{R}$.

What are the microscopic BPS states that run in the loop? They are related to non-perturbative RR states, which are the only charged states in the Type II compactification. They come from branes wrapping cycles in the Calabi–Yau and as BPS states their masses are proportional to their central charge (38). For example, in the large radius in the type IIA string on M, the mass is determined by integrals of complexified volume forms over even cycles.

Example the mass of a 2 brane wrapping a holomorphic curve $\mathcal{C}_\beta \in H^2(M, \mathbb{Z})$ is given by

$$m_\beta = \frac{1}{\lambda} \int_{\mathcal{C}_\beta \in H^2(M, \mathbb{Z})} (iJ + B) = \frac{1}{\lambda} 2\pi i t \cdot \beta =: \frac{1}{\lambda} t_\beta . \tag{25}$$

We note that $H^2(M, \mathbb{Z})$ plays here the role of the charge lattice. In the type IIB picture the charge is given by integrals of the normalized holomorphic $(3,0)$-form Ω. In particular the mass of the extremal black hole that vanishes at the conifold is given by

$$m_{BH} = \frac{1}{\lambda \int_{A_D} \Omega} \int_{S^3} \Omega =: \frac{1}{\lambda} t_D , \tag{26}$$

where A_D is a suitable non-vanishing cycle at the conifold. It follows from the discussion in previous paragraph that with the identification $e = m$ and after a rescaling $s \to s\lambda/e$ in (23), as well as absorbing F into λ, one gets a result for (22)

$$F(\lambda, t) = \int_\epsilon^\infty \frac{ds}{s} \frac{\mathrm{Tr}(-1)^f \exp(-st) \exp(-2s\sigma_L \lambda)}{4\sin^2(s\lambda/2)} . \tag{27}$$

Here t is the regularized mass, c.f. (25,26) of the light particles P that are integrated out, f is their spin in $\mathcal{R}$ and σ_L is the Cartan element in the representation $\mathcal{R}$.

At the large volume point one can relate the relevant BPS states actually to bound states of $D2$ with and infinite tower of $D0$ branes with quantized momenta along the M-theory circle. Moreover the left spin content of the bound state can be related uniquely to the genus of $\mathcal{C}_\beta$. This beautiful story leads, as explained in [35, 36], after summing over the momenta of the $D0$ states to (105), which together with Castelnuovo's bound for smooth curves leads to very detailed and valuable boundary information as explained in Sect. 5. It is important to note that all states are massive so that there are no poles in $F^{(g)}$ for $g > 1$. Hence the leading contribution is regular and can be extracted from the constant $\beta = 0$ contribution in (105) as

$$\lim_{t \to \infty} F^{(g)}_{\text{A-model}} = \frac{(-1)^{g-1} B_{2g} B_{2g-2}}{2g(2g-2)(2g-2)!} \cdot \frac{\chi}{2} . \tag{28}$$

The moduli space of constant maps factors into tow components: the moduli space of the worldsheet curve Σ_g and the location of the image point, i.e. M. The measures on the components are λ^{3g-3} and $c_3(T_M)$ respectively, explaining the above formula.

Let us turn to type IIB compactifications near the conifold. As it was checked with the β-function in [32] there is precisely one BPS hypermultiplet with the Lorentz representation of the first factor in (24) becoming massless at the conifold. In this case the Schwinger-loop calculation (27) simply becomes

$$F(\lambda, t_D) = \int_\epsilon^\infty \frac{ds}{s} \frac{\exp(-st_D)}{4\sin^2(s\lambda/2)} + \mathcal{O}(t_D^0) = \sum_{g=2}^\infty \left(\frac{\lambda}{t_D}\right)^{2g-2} \frac{(-1)^{g-1} B_{2g}}{2g(2g-2)} + \mathcal{O}(t_D^0) \ .$$

(29)

Since there are no other light particles, the above Eq. (29) encodes all singular terms in the effective action. There will be regular terms coming from other massive states. This is precisely the gap condition.

3 Quintic

We consider the familiar case of the quintic hypersurface in $\mathbb{P}^4$. The topological string amplitudes $F^{(g)}$ were computed up to genus 4 in [17, 25] using the holomorphic anomaly equation and fixing the holomorphic ambiguity by various geometric data. It was also observed a long time ago [18] that the leading terms in $F^{(g)}$ around the conifold point are the same as $c = 1$ strings at the self-dual radius, thus providing useful information for the holomorphic ambiguity. We want to explore whether we can find coordinates in which the $F^{(g)}$ on the compact Calabi–Yau exhibit the gap structure around the conifold point that was recently found for local Calabi–Yau geometries [12]. In order to do this, it is useful to rewrite the topological string amplitudes as polynomials [37]. We briefly review the formalism in [37].

The quintic manifold M has one Kähler modulus t and its mirror W has one complex modulus ψ and is given by the equation [7]

$$W = x_1^5 + x_2^5 + x_3^5 + x_4^5 + x_5^5 - 5\psi^{\frac{1}{5}} x_1 x_2 x_3 x_4 x_5 = 0 \ . \tag{30}$$

There is a relation between t and ψ known as the mirror map $t(\psi)$. The mirror map and the genus zero amplitude can be obtained using the Picard–Fuchs equation on W

$$\{(\psi\partial_\psi)^4 - \psi^{-1}(\psi\partial_\psi - \frac{1}{5})(\psi\partial_\psi - \frac{2}{5})(\psi\partial_\psi - \frac{3}{5})(\psi\partial_\psi - \frac{4}{5})\}\omega = 0 \tag{31}$$

We can solve the equation as an asymptotic series around $\psi \to \infty$.

3.1 $\psi = \infty$ Expansion and Integer Symplectic Basis

Here we set the notation for the periods in the integer symplectic basis on W and the relation of this basis to the D-brane charges on M. Equations (34, 37) and following, apply to all one-parameter models.

The point $\psi = \infty$ has maximal unipotent monodromy and corresponds to the large radius expansion of the mirror M [38]. In the variable $z = \frac{1}{5^5\psi}$ and using the definitions

---

[7] Here for later convenience we use a slightly different notation from that of [37]. Their notation is simply related to ours by a change of variable $\psi^5 \to \psi$.

$$\omega(z,\rho) := \sum_{n=0}^{\infty} \frac{\Gamma(5(n+\rho)+1)}{\Gamma^5(n+\rho+1)} z^{n+\rho} \qquad D_\rho^k \omega := \frac{1}{(2\pi i)^k k!} \frac{\partial^k}{\partial^k \rho} \omega \bigg|_{\rho=0} \qquad (32)$$

one can write the solutions [30]

$$\omega_0 = \omega(z,0) = \sum_{n=0}^{\infty} \frac{(5n)!}{(n!)^5 (5^5 \psi)^n}$$

$$\omega_1 = D_\rho \omega(z,0) = \frac{1}{2\pi i} (\omega_0 \ln(z) + \sigma_1)$$

$$\omega_2 = \kappa D_\rho^2 \omega(z,\rho) - c\omega_0 = \frac{\kappa}{2 \cdot (2\pi i)^2} \left(\omega_0 \ln^2(z) + 2\sigma_1 \ln(z) + \sigma_2 \right)$$

$$\omega_3 = \kappa D_\rho^3 \omega(z,\rho) - c\omega_1 + e\omega_0 = \frac{\kappa}{6 \cdot (2\pi i)^3} \left(\omega_0 \ln^3(z) \right.$$

$$\left. + 3\sigma_1 \ln^2(z) + 3\sigma_2 \ln(z) + \sigma_3 \right) \qquad (33)$$

The constants κ, c, e are topological intersection numbers, see below. The σ_k are also determined by (31). To the first few orders, $\omega_0 = 1 + 120z + 113400z^2 + \mathcal{O}(z^3)$, $\sigma_1 = 770z + 810225z^2 + \mathcal{O}(z^3)$, $\sigma_2 = 1150z + 4208175/2z^2 + \mathcal{O}(z^3)$ and $\sigma_3 = -6900z - 9895125/2z^2 + \mathcal{O}(z^3)$.

The solutions (33) can be combined into the period vector $\mathbf{\Pi}$ with respect to an integer symplectic basis[8] (A^i, B_j) of $H^3(W, \mathbb{Z})$ as follows [38]:

$$\mathbf{\Pi} = \begin{pmatrix} \int_{B_1} \Omega \\ \int_{B_2} \Omega \\ \int_{A^1} \Omega \\ \int_{A^2} \Omega \end{pmatrix} = \begin{pmatrix} F_0 \\ F_1 \\ X_0 \\ X_1 \end{pmatrix} = \omega_0 \begin{pmatrix} 2\mathcal{F}^{(0)} - t\partial_t \mathcal{F}^{(0)} \\ \partial_t \mathcal{F}^{(0)} \\ 1 \\ t \end{pmatrix} = \begin{pmatrix} \omega_3 + c\,\omega_1 + e\,\omega_0 \\ -\omega_2 - a\,\omega_1 + c\,\omega_0 \\ \omega_0 \\ \omega_1 \end{pmatrix}. \qquad (34)$$

Physically, t is the complexified area of a degree one curve and is related by the mirror map

$$2\pi i t(\psi) = \int_C (iJ + B) = \frac{\omega_1}{\omega_0} = -\log(5^5 \psi) + \frac{154}{625\psi} + \frac{28713}{390625\psi^2} + \ldots \quad (35)$$

$$\frac{1}{z} = 5^5 \psi = \frac{1}{q} + 770 + 421375\,q + 274007500\,q^2 + \ldots \qquad (36)$$

to ψ. In (36), we inverted (35) with $q = e^{2\pi i t}$. The prepotential is given by

$$\mathcal{F}^{(0)} = -\frac{\kappa}{3!} t^3 - \frac{a}{2} t^2 + ct + \frac{e}{2} + f_{inst}(q) \qquad (37)$$

where the instanton expansion $f_{inst}(q)$ vanishes in the large radius $q \to 0$ limit. The constants in (34, 37) can be related to the classical geometry of

---

[8] With $A^i \cap B_j = \delta_j^i$ and zero intersections otherwise.

the mirror manifold [30, 38]. Denote by $J \in H_2(M, \mathbb{Z})$ the Kähler form which spans the one-dimensional Kähler cone. Then $\kappa = \int_M J \wedge J \wedge J$ is the classical triple intersection number, $c = 1/24 \int_M c_2(T_M) \wedge J$ is proportional to the evaluation of the second Chern class of the tangent bundle T_M against J, $e = \zeta(3)/(2\pi i)^3 \int_M c_3(T_M)$ is proportional to the Euler number and a is related to the topology of the divisor D dual to J. A^1 is topologically a $T_{\mathbb{R}}^3$, while analysis at the conifold shows that the dual cycle B_1 has the topology of an S^3.

The identification of the central charge formula for compactified type II supergravity [39] with the K-theory charge of D-branes as objects $A \in K(M)$ [40, 41],

$$\mathbf{Q} \cdot \mathbf{\Pi} = -\int_M e^{-\hat{J}} \mathrm{ch}(A) \sqrt{\mathrm{td}(M)} = Z(A), \tag{38}$$

with $\hat{J} = t(iJ + B)$ for the one-dimensional Kähler cone, checks the D-brane interpretation of (34) [42, 43, 44] in the $q \to 0$ limit on the classical terms κ, a, c, but misses the $\chi\zeta(3)/(2\pi i)^3$ term. Based on their scaling with the area parameter t the periods (F_0, F_1, X_0, X_1) are identified with the masses of (D_6, D_4, D_0, D_2) branes. For smooth intersections and D the restriction of the hyperplane class we can readily calculate κ, a, c, χ using the adjunction formula, see Appendix A.

3.2 Polynomial Expansion of $F^{(g)}$

From the periods, or equivalently the prepotential $\mathcal{F}^{(0)}$, we can compute the Kähler potential $K := -\log(i(\bar{X}^{\bar{a}} F_a - X^a \bar{F}_{\bar{a}}))$ and metric $G_{\psi\bar\psi} := \partial_\psi \bar\partial_{\bar\psi} K$ in the moduli space. The genus zero Gromov–Witten invariants are obtained by expanding $\mathcal{F}^{(0)}$ in large Kähler parameter in a power series in q, see Sect. 5.

We use the notation of [37] and introduce the following symbols:

$$A_p := \frac{(\psi\partial_\psi)^p G_{\psi\bar\psi}}{G_{\psi\bar\psi}}, \quad B_p := \frac{(\psi\partial_\psi)^p e^{-K}}{e^{-K}}, \quad (p = 1, 2, 3, \cdots)$$

$$C := C_{\psi\psi\psi} \psi^3, \quad X := \frac{1}{1-\psi} =: -\frac{1}{\delta} \tag{39}$$

Here $C_{\psi\psi\psi} \sim \frac{\psi^{-2}}{1-\psi}$ is the three-point Yukawa coupling, and $A := A_1 = -\psi \Gamma_{\psi\psi}^\psi$ and $B := B_1 = -\psi\partial_\psi K$ are the Christoffel and Kähler connections. In the holomorphic limit $\bar\psi \to \infty$, the holomorphic part of the Kähler potential and the metric go like

$$e^{-K} \sim \omega_0, \quad G_{\psi\bar\psi} \sim \partial_\psi t, \tag{40}$$

so in the holomorphic limit, the generators A_p and B_p are

$$A_p = \frac{(\psi\partial_\psi)^p(\partial_\psi t)}{\partial_\psi t}, \quad B_p = \frac{(\psi\partial_\psi)^p \omega_0}{\omega_0}. \tag{41}$$

It is straightforward to compute them in an asymptotic series using the Picard–Fuchs equation. There are also some derivative relations among the generators,

$$\psi\partial_\psi A_p = A_{p+1} - AA_p, \quad \psi\partial_\psi B_p = B_{p+1} - BB_p, \quad \psi\partial_\psi X = X(X-1).$$

The topological amplitudes in the "Yukawa coupling frame" are denoted as

$$P_g := C^{g-1}F^{(g)}, \quad P_g^{(n)} = C^{g-1}\psi^n C_{\psi^n}^{(g)}. \tag{42}$$

The A-model higher genus Gromov–Witten invariants are obtained in the holomorphic limit $\bar{t} \to \infty$ with a familiar factor of ω_0 as

$$F_{\text{A-model}}^{(g)} = \lim_{\bar{t}\to\infty} \omega_0^{2(g-1)} (\frac{1-\psi}{5\psi})^{g-1} P_g. \tag{43}$$

The $P_g^{(n)}$ are defined for $g = 0, n \geq 3$, $g = 1, n \geq 1$ and $g = 2, n \geq 0$. We have the initial data

$$P_{g=0}^{(3)} = 1$$
$$P_{g=1}^{(1)} = -\frac{31}{3}B + \frac{1}{12}(X-1) - \frac{1}{2}A + \frac{5}{3}, \tag{44}$$

and using the Christoffel and Kähler connections we have the following recursion relation in n,

$$P_g^{(n+1)} = \psi\partial_\psi P_g^{(n)} - [n(A+1) + (2-2g)(B - \frac{1}{2}X)]P_g^{(n)}. \tag{45}$$

One can also use the Picard–Fuchs equation and the special geometry relation to derive the following relations among generators:

$$B_4 = 2XB_3 - \frac{7}{5}XB_2 + \frac{2}{5}XB - \frac{24}{625}X$$
$$A_2 = -4B_2 - 2AB - 2B + 2B^2 - 2A + 2XB + XA + \frac{3}{5}X - 1. \tag{46}$$

By taking derivatives w.r.t. ψ one see that all higher A_p ($p \geq 2$) and B_p ($p \geq 4$) can be written as polynomials of A, B, B_2, B_3, X. It is convenient to change variables from (A, B, B_2, B_3, X) to (u, v_1, v_2, v_3, X) as

$$B = u, \quad A = v_1 - 1 - 2u, \quad B_2 = v_2 + uv_1,$$
$$B_3 = v_3 - uv_2 + uv_1 X - \frac{2}{5}uX. \tag{47}$$

Then the main result of [37] is the following proposition.

Proposition: Each P_g ($g \geq 2$) is a degree $3g - 3$ inhomogeneous polynomial of v_1, v_2, v_3, X, where one assigns the degree $1, 2, 3, 1$ for v_1, v_2, v_3, X, respectively.

For example, at genus two the topological string amplitude is

$$P_2 = \frac{25}{144} - \frac{625}{288}v_1 + \frac{25}{24}v_1^2 - \frac{5}{24}v_1^3 - \frac{625}{36}v_2 + \frac{25}{6}v_1v_2 + \frac{350}{9}v_3 - \frac{5759}{3600}X$$
$$- \frac{167}{720}v_1X + \frac{1}{6}v_1^2X - \frac{475}{12}v_2X + \frac{41}{3600}X^2 - \frac{13}{288}v_1X^2 + \frac{X^3}{240}. \qquad (48)$$

The expression for the P_g, $g = 1, \ldots, 12$ for all models discussed in this chapter can be obtained in a Mathematica readable form on [45], see "Pgfile.txt".

3.3 Integration of the Holomorphic Anomaly Equation

The anti-holomorphic derivative $\partial_{\bar\psi} B_p$ of $p \geq 2$ can be related to $\partial_{\bar\psi} B$ [37]. Assuming $\partial_{\bar\psi} A$ and $\partial_{\bar\psi} B$ are independent, one obtains two relations for $P_g^{(n)}$ from the BCOV holomorphic anomaly equation as the following:

$$\frac{\partial P_g}{\partial u} = 0 \qquad (49)$$

$$\left(\frac{\partial}{\partial v_1} - u\frac{\partial}{\partial v_2} - u(u+X)\frac{\partial}{\partial v_3}\right)P_g = -\frac{1}{2}\left(P_{g-1}^{(2)} + \sum_{r=1}^{g-1} P_r^{(1)}P_{g-r}^{(1)}\right) \qquad (50)$$

The first Eq. (49) implies there is no u dependence in P_g, as already taken into account in the main proposition of [37]. The second Eq. (50) provides a very efficient way to solve P_g recursively up to a holomorphic ambiguity. To do this, one writes down an ansatz for P_g as a degree $3g - 3$ inhomogeneous polynomial of v_1, v_2, v_3, X, plugs in Eq. (50), and uses the lower genus results $P_r^{(1)}$, $P_r^{(2)}$ ($r \leq g-1$) to fix the coefficients of the polynomials. The number of terms in general inhomogeneous weighted polynomials in (v_1, v_2, v_3, X) with weights $(1, 2, 3, 1)$ is given by the generating function

$$\frac{1}{(1-x)^3(1-x^2)(1-x^3)} = \sum_{n=0}^{\infty} p(n)x^n . \qquad (51)$$

It is easy to see from the Eq. (50) that the terms $v_1, \ldots v_1^{2g-4}$ vanish, which implies that the number of terms in P_g is

$$n_g = p(3g - 3) - (2g - 4), \qquad (52)$$

example $n_g = 14, 62, 185, 435, 877, 1590, 2666, 4211, 6344$ for $g = 1, \ldots, 10$. Comparing with $\sum_{n=1}^{\infty} \tilde{p}(n)x^n = \frac{1}{(1-x)^5}$ it follows in particular that asymptotically

$$n_g \preceq (3g - 3)^4 . \qquad (53)$$

Note that Eq. (50) determines the term $\hat{P}_g(v_1, v_2, v_3, X)$ in

$$P_g =: \hat{P}_g(v_1, v_2, v_3, X) + f^{(g)}(X) \qquad (54)$$

completely. We can easily understand why the terms in the modular as well as holomorphic ambiguity $f^{(g)}(X)$ are not fixed, by noting that the Eq. (50) does not change when we add a term proportional to X^i to P_g. This ambiguity must have the form [25, 37]

$$f^{(g)} = \sum_{i=0}^{3g-3} a_i X^i \, . \tag{55}$$

The maximal power of X is determined by (29). This follows from (42) and the universal behaviour of $t_D \sim \delta + \mathcal{O}(\delta^2)$ at the conifold (73). Note in particular from (75) that $\hat{P}_g$ is less singular at that point. The minimal power in (55) follows from (28) for $\psi \sim \infty$ and the leading behaviour of the solutions in Sect. 3.1.

We will try to fix these $3g - 2$ unknown constants a_i $(i = 0, 1, 2 \cdots 3g - 3)$ by special structure of expansions of $F^{(g)}$ around the orbifold point $\psi \sim 0$ and the conifold point $\psi \sim 1$. Before proceeding to this, we note the constant term is fixed by the known leading coefficients in large complex structure modulus limit $\psi \sim \infty$ in [19, 36, 46].

The leading constant terms in the A-model expansion $\psi \sim \infty$ come from the constant map from the worldsheet to the Calabi–Yau (28). The large complex structure modulus behaviour of X is

$$X \sim \frac{1}{\psi} \sim q = e^{2\pi i t}. \tag{56}$$

So only the constant term a_0 in the holomorphic ambiguity contributes to leading term in A-model expansions (28) and is thus fixed. We still have $3g - 3$ coefficients a_i $(i = 1, 2, \cdots, 3g - 3)$ to be fixed.

3.4 Expansions Around the Orbifold Point $\psi = 0$

To analyse the $F^{(g)}$ in a new region of the moduli space we have to find the right choice of polarization. To do this we analytically continue the periods to determine the symplectic pairing in the new region and pick the new choice of conjugated varaibles.

The solutions of the Picard–Fuchs equation around the orbifold point$\psi \sim 0$ are four power series solutions with the indices $1/5, 2/5, 3/5, 4/5$

$$\omega_k^{\mathrm{orb}} = \psi^{\frac{k}{5}} \sum_{n=0}^{\infty} \frac{\left(\left[\frac{k}{5}\right]_n\right)^5}{[k]_{5n}} \left(5^5 \psi\right)^n$$

$$= -\frac{\Gamma(k)}{\Gamma^5\left(\frac{k}{5}\right)} \int_{\mathcal{C}_0} \frac{ds}{e^{2\pi i s} - 1} \frac{\Gamma^5\left(s + \frac{k}{5}\right)}{\Gamma(5s + k)} \left(5^5 \psi\right)^{s + \frac{k}{5}} \, , \qquad k = 1, \ldots, 4 \, . \tag{57}$$

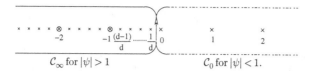

$\mathcal{C}_\infty$ for $|\psi| > 1$ $\mathcal{C}_0$ for $|\psi| < 1$.

The Pochhammer symbol are defined as $[a]_n := \Gamma(a+n)/\Gamma(a)$ and we normalized the first coefficient in $\omega_k^{\text{orb}} = \psi^{\frac{k}{5}} + \mathcal{O}(\psi^{5+k/5})$ to one. The expression in the first line is recovered from the integral representation by noting that the only poles inside $\mathcal{C}_0$ for which the integral converges for $|\psi| < 0$ are from $g(s) = 1/\exp(2\pi i s) - 1$, which behaves at $s_{-n}^\epsilon = n - \epsilon$, $n \in \mathbb{N}$ as $g(s_{-n}^\epsilon) \sim -1/2\pi i \epsilon$.

Up to normalization this basis of solutions is canonically distinguished, as it diagonalizes the $\mathbb{Z}_5$ monodromy at $\psi = 0$. Similar as for the $\mathbb{C}^3/\mathbb{Z}_3$ orbifold [14], it can be viewed as a twist field basis. Here this basis is induced from the twist field basis of $\mathbb{C}^5/\mathbb{Z}_5$. As it was argued in [14] for $\mathbb{C}^3/\mathbb{Z}_3$, this twist field basis provides the natural coordinates in which the $F^{(g)}$ near the orbifold point can be interpreted as generating functions for orbifold Gromov–Witten invariants. Following up on foundational work on orbifold Gromov–Witten theory [47] and examples in two complex dimensions [48] this prediction has been checked by direct computation of orbifold Gromov–Witten invariants [49] at genus zero. This provides a beautiful check on the global picture of mirror symmetry.

As explained in [14] the relation between the large radius generating function of Gromov–Witten invariants and generating function of orbifold Gromov–Witten invariants is provided by the metaplectic transformation of the wave function [27]. Since the change of the phase space variables from large radius to the orbifold the symplectic form is only invariant up to scaling one has to change the definition of the string coupling, which plays the role of $\hbar$ in the metaplectic transformation. These can be viewed as small phase space specialization of the metaplectic transformation on the large phase space [50].

Using the modular invariance of the anholomorphic $\hat{F}^g(t, \bar{t})$ it has been further shown in [14] that this procedure of obtaining the transformed holomorphic wave function is simply equivalent to taking the holomorphic limit on the $\hat{F}^g(t, \bar{t})$. This point was made [14] for the local case. But the only point one has to keep in mind for the global case is that $\hat{F}^g(t, \bar{t})$ are globally well defined sections of the Kähler line bundle, i.e. one has to perform a Kähler transformation along with the holomorphic limit.

We will now study the transformation from the basis (34) to the basis (57) to make B-model prediction along the lines of [14] with the additional Kähler transformation. Since the symplectic form ω on the moduli space is invariant under monodromy and ω_k^{orb} diagonalizes the $\mathbb{Z}_5$ monodromy, we must have in accordance with the expectation from the orbifold cohomology $H^*(\mathbb{C}^5/\mathbb{Z}_5) = \mathbb{C}1_0 \oplus \mathbb{C}1_1 \oplus \mathbb{C}1_2 \oplus \mathbb{C}1_3 \oplus \mathbb{C}1_4$

$$\omega = \mathrm{d}F_k \wedge \mathrm{d}X_k = -\frac{5^4 s_1}{6}\mathrm{d}\omega_4^{orb} \wedge \mathrm{d}\omega_1^{orb} + \frac{5^4 s_2}{2}\mathrm{d}\omega_3^{orb} \wedge \mathrm{d}\omega_2^{orb} . \tag{58}$$

The rational factors above have been chosen to match constraints from special geometric discussed below. Similarly the monodromy invariant Kähler potential must have the form

$$e^{-K} = \sum_{k=1}^{4} r_k \omega_k^{orb} \overline{\omega_k^{orb}} . \tag{59}$$

To obtain the s_i, r_i by analytic continuation to the basis (34) we follow [38] for the quintic and the generalization in [51] for other cases and note that the integral converges for $|\psi| > 1$ due to the asymptotics of the $f(s) = \Gamma^5(s + k/5)/\Gamma(5s + k)$ term, when the integral is closed along $\mathcal{C}_\infty$ [38]. At $s_n^\epsilon = -n - \epsilon$ the $g(s_n^\epsilon)$ pole is compensated by the $f(s_n^\epsilon)$ zero and at $s_{n,k}^\epsilon = -n - k/5 - \epsilon$ we note the expansions

$$g(s_{n,k}^\epsilon) = \frac{\alpha^k}{1-\alpha^k} + \frac{2\pi i \alpha^k}{(1-\alpha^k)^2}\epsilon + \frac{(2\pi i)^2 \alpha^k(1+\alpha^k)}{2(1-\alpha^k)^3}\epsilon^2 + \frac{(2\pi i)^3 \alpha^k(1+4\alpha^k+\alpha^{2k})}{6(1-\alpha^k)^4}\epsilon^3 + \mathcal{O}(\epsilon^4)$$

$$f(s_{n,k}^\epsilon) = \frac{\kappa\omega_0(n)}{\epsilon^4} + \frac{\kappa\sigma_1(n)}{\epsilon^3} + \frac{1}{\epsilon^2}\left(\frac{\kappa\sigma_2(n)}{2} + \frac{(2\pi i)^2 c_{2J}\omega_0(n)}{24}\right) +$$

$$\frac{1}{\epsilon}\left(\frac{\kappa\sigma_3(n)}{6} + \frac{(2\pi i)^2 c_{2J}\sigma_1(n)}{24} + \chi\zeta(3)\omega_0(n)\right) + \mathcal{O}(\epsilon^0)$$

$$(5^5\psi)^{s_n^\epsilon} = z^n(1 + \log(z)\epsilon + \tfrac{1}{2}\log(z)^2\epsilon^2 + \tfrac{1}{6}\log(z)^3\epsilon^3 + \mathcal{O}(\epsilon^4))$$
$$\tag{60}$$

Here $\alpha = \exp(2\pi i/5)$. $\kappa = \int_M J^3$, $c_{2J} = \int_M c_2 J$ and $\chi = \int_M c_3$ are calculated in Appendix A.1.[9] The $\omega_0(n)$, $\sigma_i(n)$ are coefficients of the series we encountered in Sect. 3.1. Performing the residue integration and comparing with (33, 34) we get

$$\omega_k^{orb} = \frac{(2\pi i)^4 \Gamma(k)}{\Gamma^5\left(\frac{k}{5}\right)}\left(\frac{\alpha^k F_0}{1-\alpha^k} - \frac{\alpha^k F_1}{(1-\alpha^k)^2} + \frac{5\alpha^k(\alpha^{2k}-\alpha^k+1)X_0}{(1-\alpha^k)^4} + \frac{\alpha^k(8\alpha^k-3)X_1}{(1-\alpha^k)^3}\right) \tag{61}$$

It follows with $r_i = \frac{\Gamma^{10}\left(\frac{k}{5}\right)}{\Gamma^2(k)}c_i$ that

$$c_1 = -c_4 = \alpha^2(1-\alpha)(2+\alpha^2+\alpha^3), \quad c_3 = -c_2 = \alpha(2+\alpha-\alpha^2-2\alpha^3)$$

$$s_1 = s_2 = -\frac{1}{5^5(2\pi i)^3} \tag{62}$$

$$\begin{pmatrix} F_0 \\ F_1 \\ X_0 \\ X_1 \end{pmatrix} = \psi^{1/5}\frac{\alpha\Gamma^5\left(\frac{1}{5}\right)}{(2\pi i)^4}\begin{pmatrix} (1-\alpha)(\alpha-1-\alpha^2) \\ \frac{1}{5}(8-3\alpha)(1-\alpha)^2 \\ (1-\alpha+\alpha^2) \\ \frac{1}{5}(1-\alpha)^3 \end{pmatrix} + \mathcal{O}(\psi^{2/5}) . \tag{63}$$

---

[9] In fact using the generalization of (57) in [51] it is easily shown that the combinatorics, which leads to (116), are the same as the ones leading to the occurrence of the classical intersections here.

Equation (62) implies that up to a rational rescaling of the orbifold periods the transformation of the wave functions from infinity to the orbifold is given by a metaplectic transformation with the same rescaling of the string coupling as for the $\mathbb{C}^3/\mathbb{Z}_3$ case in [2]. Equation (63) implies that there are no projective coordinates related to an $Sp(4,\mathbb{Z})$ basis, which would vanish at the orbifold. This means that there is no massless RR state in the K-theory charge lattice which vanishes at the orbifold point. We further note that after rescaling of the orbifold periods the transformation (61) can be chosen to lie in $Sp(4,\mathbb{Z}[\alpha,1/5])$.

We can define the analogue of mirror map at the orbifold point,

$$
s = \frac{\omega_2^{orb}}{\omega_1^{orb}} = \psi^{\frac{1}{5}}\left(1 + \frac{13\psi}{360} + \frac{110069\psi^2}{9979200} + \mathcal{O}(\psi^3)\right) \tag{64}
$$

where we use the notation s, as in [14], to avoid confusion with the mirror map in the large volume limit. We next calculate the genus zero prepotential at the orbifold point. For convenience let us rescale our periods $\hat{\omega}_{k-1} = 5^{3/2}\omega_k^{orb}$. The Yukawa–Coupling is transformed to the s variables as

$$
C_{sss} = \frac{1}{\hat{\omega}_0^2} \frac{5}{\psi^2(1-\psi)}\left(\frac{\partial\psi}{\partial s}\right)^3 = 5 + \frac{5}{3}s^5 + \frac{5975}{6048}s^{10} + \frac{34521785}{54486432}s^{15} + \mathcal{O}(s^{20}) .
$$
$$\tag{65}$$

A trivial consistency check of special geometry is that the genus zero prepotential $F^{(0)} = \int ds \int ds \int ds \, C_{sss}$ appears in the periods $\hat{\Pi}^{orb} = (\hat{\omega}_0, \hat{\omega}_1, 5/2!\hat{\omega}_2, -5/3!\hat{\omega}_3)^T$ as

$$
\hat{\Pi}^{orb} = \hat{\omega}_0 \begin{pmatrix} 1 \\ s \\ \partial_s F^{(0)}_{\text{A-orbf.}} \\ 2F^{(0)}_{\text{A-orbf.}} - s\partial_s F^{(0)}_{\text{A-orbf.}} \end{pmatrix} . \tag{66}
$$

This can be viewed also as a simple check on the lowest order meta-plectic transformation of Ψ which is just the Legendre transformation. Note that the Yukawa coupling is invariant under the $\mathbb{Z}_5$ which acts as $s \mapsto \alpha s$. $\mathbb{Z}_5$ implies further that there can be no integration constants, when passing from C_{sss} to F_0 and the coupling λ must transform with $\lambda \mapsto \alpha^{3/2}\lambda$ to render $F(\lambda, s, \bar{s})$ invariant.

The holomorphic limit $\bar{\psi} \to 0$ of Kähler potential and metric follows from (59) by extracting the leading anti-holomorphic behaviour. Denoting[10] by a_k the leading powers of ω_k^{orb} we find

$$
\lim_{\bar{\psi}\to 0} e^{-K} = r_1\bar{\psi}^{a_1}\omega_1^{orb}, \qquad \lim_{\bar{\psi}\to 0} G_{\psi\bar{\psi}} = \bar{\psi}^{a_2-a_1-1}\frac{r_2}{r_1}\left(\frac{a_2}{a_1} - 1\right)\frac{\partial s}{\partial\psi} . \tag{67}
$$

---

[10] This is to make contact with the other one modulus cases. Of course if $a_1 = a_2$ a log singularity appears and the formula does not apply.

Note that the constants and $\bar{\psi}$ and its leading power are irrelvant for the holomorphic limit of the generators (39)

$$X = \frac{1}{1 - \psi} = 1 + \psi + \psi^2 + \mathcal{O}(\psi^3)$$

$$A = -\frac{4}{5} + \frac{13}{60}\psi + \frac{3551}{18144}\psi^2 + \mathcal{O}(\psi^3)$$

$$B = \frac{1}{5} + \frac{1}{120}\psi + \frac{17}{4032}\psi^2 + \mathcal{O}(\psi^3)$$

$$B_2 = \frac{1}{25} + \frac{7}{600}\psi + \frac{1027}{100800}\psi^2 + \mathcal{O}(\psi^3)$$

$$B_3 = \frac{1}{125} + \frac{43}{3000}\psi + \frac{1633}{72000}\psi^2 + \mathcal{O}(\psi^3).$$

Using this information and integrating (44) we obtain the genus one free energy $F^{(1)}_{\text{A-orbf.}} = -s^5/9 + \ldots$. The regularity of $F^{(1)}$, i.e. the absence of log terms, is expected as there are no massless BPS states at the Gepner-point. Because of this, the considerations in Sect. 2.2 also imply that the higher genus amplitudes

$$F^{(g)}_{\text{A-orbf.}} = \lim_{\bar{t} \to 0} (\hat{\omega}_0)^{2(g-1)} \left(\frac{1 - \psi}{5\psi}\right)^{g-1} P_g \qquad (68)$$

have no singularity at the orbifold point $\psi \sim 0$. This is in accordance with the calculations in [25] and implies that $P_g/\psi^{\frac{3}{5}(g-1)}$ is regular at $\psi \sim 0$.

The situation for the higher genus amplitudes for the compact $\mathcal{O}(5)$ constraint in $\mathbb{P}^4$ is considerably different from the one for the resolution $\mathcal{O}(-5) \to \mathbb{P}^4$ of the $\mathbb{C}^5/\mathbb{Z}_5$. In normal Gromov–Witten theory for genus $g > 0$ on $\mathcal{O}(5)$ in $\mathbb{P}^4$ there is no bundle whose Euler class of its pullback from the ambient space $\mathbb{P}^4$ to the moduli space of maps gives rise to a suitable measure on $\mathcal{M}_{g,\beta}$ that counts the maps to the quintic. This is the same difficulties one has to face for higher genus calculation for the orbifold GW theory in $\mathcal{O}(5)$ in $\mathbb{P}^4$ and it is notably different[11] from the equivariant GW theory on $\mathbb{C}^5/\mathbb{Z}_5$.

However we claim that our $F^{(g)}_{\text{A-orbf.}}$ predictions from the B-model computation contain the information about the light even RR states at the orbifold point in useful variables and could in principle be checked in the A-model by some version of equivariant localization. Below we give the first few order results. They are available to genus 20 at [45].

---

[11] Since the brane-bound state cohomology at infinity can only be understood upon including higher genus information, see Sects. 5 and 5.2, the claims that one can learn essential properties about the D-branes of the quintic at small volume from the $\mathbb{C}^5/\mathbb{Z}_5$ orbifold might be overly optimistic.

$$F^{(0)}_{\text{A-orbf.}} = \frac{5\,s^3}{6} + \frac{5\,s^8}{1008} + \frac{5975\,s^{13}}{10378368} + \frac{34521785\,s^{18}}{266765571072} + \cdots$$

$$F^{(1)}_{\text{A-orbf.}} = -\frac{s^5}{9} - \frac{163\,s^{10}}{18144} - \frac{85031\,s^{15}}{46702656} - \frac{6909032915\,s^{20}}{20274183401472} + \cdots$$

$$F^{(2)}_{\text{A-orbf.}} = \frac{155\,s^2}{18} - \frac{5\,s^7}{864} + \frac{585295\,s^{12}}{14370048} + \frac{1710167735\,s^{17}}{177843714048} + \cdots$$

$$F^{(3)}_{\text{A-orbf.}} = \frac{488305\,s^4}{9072} - \frac{3634345\,s^9}{979776} - \frac{1612981445\,s^{14}}{7846046208} - \frac{2426211933305\,s^{19}}{116115777662976} + \cdots$$

$$F^{(4)}_{\text{A-orbf.}} = \frac{48550\,s}{567} + \frac{36705385\,s^6}{163296} + \frac{16986429665\,s^{11}}{603542016} + \frac{341329887875\,s^{16}}{70614415872} + \cdots$$

$$F^{(5)}_{\text{A-orbf.}} = \frac{1237460905\,s^3}{224532} + \frac{108607458385\,s^8}{28740096} - \frac{2079654832074515\,s^{13}}{1553517149184}$$
$$- \frac{50102421421803185\,s^{18}}{438808843984896} + \cdots$$

The holomorphic ambiguity (55) is a power series of ψ starting from a constant term, so requiring $P_g/\psi^{3/5(g-1)}$ to be regular imposes

$$\lceil \tfrac{3}{5}(g-1) \rceil \tag{69}$$

number of relations in a_i in (55), where $\lceil 3/5(g-1) \rceil$ is the ceiling, i.e. the smallest integer greater or equal to $\frac{3}{5}(g-1)$. We note that the leading behaviour $\omega_1^{orb} \sim \psi^{1/d}$ with $d = 5$ for the quintic, which is typical for an orbifold point in compact Calabi–Yau and which "shields" the singularity and diminishes the boundary conditions at the orbifold point from $g - 1$ to $\lceil d - 2/d(g-1) \rceil$. If this period is non-vanishing at $\psi = 0$ and it is indeed simply a constant for all local cases [13], one gets $g - 1$ conditions, which together with the gap condition at the conifold and the constant map information is already sufficient to completely solve the model. An example of this type is $\mathcal{O}(-3) \to \mathbb{P}^2$.

3.5 Expansions Around the Conifold Point $\psi = 1$

An new feature of the conifold region is that there is an choice in picking the polarization, but as we will show the gap property is independent of this choice.

A basis of solutions of the Picard–Fuchs equation around the conifold point $\psi - 1 = \delta \sim 0$ is the following

$$\mathbf{\Pi}_c = \begin{pmatrix} \omega_0^c \\ \omega_1^c \\ \omega_2^c \\ \omega_3^c \end{pmatrix} = \begin{pmatrix} 1 + \frac{2\delta^3}{625} - \frac{83\delta^4}{18750} + \frac{757\delta^5}{156250} + \mathcal{O}(\delta^6) \\ \delta - \frac{3\delta^2}{10} + \frac{11\delta^3}{25} - \frac{217\delta^4}{2500} + \frac{889\delta^5}{15625} + \mathcal{O}(\delta^6) \\ \delta^2 - \frac{23\delta^3}{30} + \frac{1049\delta^4}{1800} - \frac{34343\delta^5}{75000} + \mathcal{O}(\delta^6) \\ \omega_1^c \log(\delta) - \frac{9\,d^2}{20} - \frac{169\,d^3}{450} + \frac{27007\,d^4}{90000} - \frac{152517\,d^5}{625000} + \mathcal{O}(\delta^6) \end{pmatrix} \tag{70}$$

Here we use the superscript "c" in the periods to denote them as solutions around the conifold point. We see that one of the solutions ω_1^c is singled out as it multiplies the log in the solution ω_3^c. By a Lefshetz argument [52] it corresponds to the integral over the vanishing S^3 cycle B_1 and moreover a solution containing the log is the integral over dual cycle A_1. Comparing with (34, 38) shows in the Type IIA interpretation that the $D6$ brane becomes massless. To determine the symplectic basis we analytically continue the solutions (34) from $\psi = \infty$ and get

$$
\begin{pmatrix} F_0 \\ F_1 \\ X_0 \\ X_1 \end{pmatrix} = \begin{pmatrix} 0 & \frac{\sqrt{5}}{2\pi i} & 0 & 0 \\ a - \frac{11i}{2}g & b - \frac{11i}{2}h & c - \frac{11i}{2}r & 0 \\ d & e & f & -\frac{\sqrt{5}}{(2\pi i)^2} \\ ig & ih & ir & 0 \end{pmatrix} \begin{pmatrix} \omega_0^c \\ \omega_1^c \\ \omega_2^c \\ \omega_3^c \end{pmatrix}
\tag{71}
$$

Six of the real numbers $a, \ldots, r$ are only known numerically.[12] Nevertheless we can give the symplectic form exactly in the new basis

$$
dF_k \wedge dX_k = -\frac{1}{(2\pi i)^3} \left(\frac{5}{2} d\omega_2^c \wedge d\omega_0^c + (-5) d\omega_3^c \wedge d\omega_1^c \right) .
\tag{72}
$$

The mirror map should be invariant under the conifold monodromy and vanishing at the conifold. The vanishing period has $D6$ brane charge and is singled out to appear in the numerator of the mirror map. The numerator is not fixed up to the fact that ω_3^c should not appear. The simplest mirror map compatible with symplectic form (72) is

$$
t_D(\delta) := \frac{\omega_1^c}{\omega_0^c} = \delta - \frac{3\delta^2}{10} + \frac{11\delta^3}{75} - \frac{9\delta^4}{100} + \frac{5839 t_D^5}{93750} + \mathcal{O}(t_D^6)
\tag{73}
$$

$$
\delta(t_D) = t_D + \frac{3t_D^2}{10} + \frac{t_D^3}{30} + \frac{t_D^4}{200} + \frac{169 t_D^5}{375000} + \mathcal{O}(t_D^6)
\tag{74}
$$

We call this the dual mirror map and denote it t_D to distinguish from the large complex structure modulus case.

In the holomorphic limit $\bar{\delta} \to 0$, the Kahler potential and metric should behave as $e^{-K} \sim \omega_0^c$ and $G_{\delta\bar{\delta}} \sim \partial_\delta t_D$. We can find the asymptotic behaviour of various generators,

---

[12] $a = 6.19501627714957\ldots$, $b = 1.016604716702582\ldots$, $c = -0.140889979448831\ldots$, $d = 1.07072586843016\ldots$, $e = -0.0247076138044847\ldots$, $g = 1.29357398450411\ldots$, $h = \frac{2bg\pi - (\sqrt{5}d)}{2a\pi}$, $r = \frac{5+16cg\pi^3}{16a\pi^3}$, $f = \frac{\sqrt{5}b+8cd\pi^2}{8a\pi^2}$.

$$X = \frac{1}{1 - \psi} = -\frac{1}{\delta}$$

$$A = -\frac{3}{5} - \frac{2}{25}\delta + \frac{2}{125}\delta^2 - \frac{52}{9375}\delta^3 + \mathcal{O}(\delta^4)$$

$$B = \frac{6}{625}\delta^2 - \frac{76}{9375}\delta^3 + \frac{611}{93750}\delta^4 + \mathcal{O}(\delta^5)$$

$$B_2 = \frac{12}{625}\delta - \frac{16}{3125}\delta^2 + \frac{82}{46875}\delta^3 + \mathcal{O}(\delta^4)$$

$$B_3 = \frac{12}{625} + \frac{28}{3125}\delta - \frac{78}{15625}\delta^2 + \mathcal{O}(\delta^3). \tag{75}$$

Now we can expand

$$F^{(g)}_{\text{conifold}} = \lim_{\delta \to 0} (\omega_0^c)^{2(g-1)} \left(\frac{1-\psi}{\psi}\right)^{g-1} P_g \tag{76}$$

around the conifold point in terms of t_D using the dual mirror map (73).

Remarkably it turns out that shifts $\omega_0^c \to \omega_0^c + b_1\omega_1^c + b_2\omega_2^c$ does not affect the structure we are interested in. The fact that the b_1 shifts do not affect the amplitudes is reminiscent of the $SL(2,\mathbb{C})$ orbit theorem [53, 54] and is proven in Appendix A.3. It is therefore reasonable to state the results in the more general polarization and define $\hat{\omega}_0^c = \omega_0^c + b_1\omega_1^c + b_2\omega_2^c$. We first determine the genus 0 prepotential checking consistency of the solutions with special geometry. Defining $\hat{t}_D = \omega_1^c/\hat{\omega}_0^c$ and $\Pi_{con} = (\hat{\omega}_0^c, \omega_1^c, 5/2\omega_2^c, -5\omega_3^c)^T$ we get

$$\hat{\Pi}^{orb} = \hat{\omega}_0 \begin{pmatrix} 1 \\ \hat{t}_D \\ 2F^{(0)}_{\text{conif.}} - \hat{t}_D \partial_{\hat{t}_D} F^{(0)}_{\text{conif.}} \\ \partial_{\hat{t}_D} F^{(0)}_{\text{conif.}} \end{pmatrix}. \tag{77}$$

We can now calculate the $F^{(g)}$ in the generalized polarization. Since the normalization of $\hat{\omega}_0^c$ is not fixed by the Picard–Fuchs equation, we pick the normalization $b_0 = 1$ that will be convenient later on.[13] Using the known expression for the ambiguity at genus 2, 3 [17, 25, 37] for $g = 0 - 3$ and Castelnuovos bound for $g > 3$ we find the same interesting structure first observed in [12]

$$F^{(0)}_{\text{conif.}} = -\frac{5}{2} \log(\hat{t}_D)\hat{t}_D^2 + \frac{5}{12} (1 - 6b_1) \hat{t}_D^3$$

$$+ \left(\frac{5}{12} (b_1 - 3b_2) - \frac{89}{1440} - \frac{5}{4} b_1^2 \right) \hat{t}_D^4 + \mathcal{O}(\hat{t}_D^5)$$

---

[13] The b_0 dependence can be restored noting that each order in $\hat{t}_D$ is homogeneous in b_i.

$$F^{(1)}_{\text{conif.}} = -\frac{\log(\hat{t}_D)}{12} + \left(\frac{233}{120} - \frac{113\,b_1}{12}\right)\hat{t}_D$$

$$+ \left(\frac{233\,b_1}{120} - \frac{113\,b_1{}^2}{24} - \frac{107b_2}{12} - \frac{2681}{7200}\right)\hat{t}_D^2 + \mathcal{O}(\hat{t}_D^3)$$

$$F^{(2)}_{\text{conif.}} = \frac{1}{240\hat{t}_D^2} - \left(\frac{120373}{72000} + \frac{11413b_2}{144}\right)$$

$$+ \left(\frac{107369}{150000} - \frac{120373\,b_1}{36000} + \frac{23533\,b_2}{720} - \frac{11413b_1b_2}{72}\right)\hat{t}_D + \mathcal{O}(\hat{t}_D^2)$$

$$F^{(3)}_{\text{conif.}} = \frac{1}{1008\,\hat{t}_D^4} - \left(\frac{178778753}{324000000} + \frac{2287087\,b_2}{43200} + \frac{1084235\,b_2{}^2}{864}\right) + \mathcal{O}(\hat{t}_D)$$

$$F^{(4)}_{\text{conif.}} = \frac{1}{1440\,\hat{t}_D^6} - \left(\frac{977520873701}{3402000000000} + \frac{162178069379\,b_2}{3888000000} + \frac{5170381469\,b_2{}^2}{2592000}\right.$$

$$\left. + \frac{490222589\,b_2{}^3}{15552}\right) + \mathcal{O}(\hat{t}_D). \tag{78}$$

As explained in Sect. 2.2 we expect this gap structure to be present at higher genus as in (2), i.e.

$$F^{(g)}_{\text{conifold}} = \frac{(-1)^{g-1}B_{2g}}{2g(2g-2)(i\hat{t}_D)^{2g-2}} + \mathcal{O}(\hat{t}_D^0). \tag{79}$$

If this is true then it will impose $2g-2$ conditions on the holomorphic ambiguity (55). We can further note that because the prefactor in (76) goes like δ^{g-1} and the generator X goes like $X \sim 1/\delta$, the terms $a_i X^i$ in holomorphic ambiguity (55) with $i \le g-1$ do not affect the gap structure in (2). Therefore the gap structure fixes the coefficients a_i of $i = g, \cdots, 3g-3$ in (55), but not the coefficients a_i with $i \le g-1$. Note that the choice of b_i, $i = 0, 1, 2$ does not affect the gap structure at all. In Appendix A.3 it is proven that the generators of the modular forms in P_g do not change under the shift b_1. The effect of this shift is hence merely a Kähler gauge transformation. On the other hand we note that imposing the invariance of the gap structure under the b_2 shift does not give further conditions on the a_i, but it does affect the subleading expansion.

3.6 Fixing the Holomorphic Ambiguity: A Summary of Results

Putting all the information together, let us do a counting of number of unknown coefficients. Originally we have $3g-2$ coefficients in (55). The constant map calculation [19, 35, 46] in A-model large complex structure modulus limit $\psi \sim \infty$ fixes one constant a_0. The conifold expansion around $\psi \sim 1$ fixes $2g-2$

coefficients a_i of $i = g, \cdots, 3g - 3$ and the orbifold expansion around $\psi \sim 0$ further fixes $\lceil 3(g - 1)/5 \rceil$ coefficients. So the number of unknown coefficients at genus g is

$$3g - 2 - (1 + 2g - 2 + \lceil \frac{3(g - 1)}{5} \rceil) = \lceil \frac{2(g - 1)}{5} \rceil. \tag{80}$$

This number is zero for genus $g = 2, 3$. So we could have computed the $g = 2, 3$ topological strings using this information, although the answers were already known. At $g \geq 4$ there are still some unknown constants. However, in the A-model expansion when we rewrite the Gromov–Witten invariants in terms of Gopakumar–Vafa invariants, one can in principle use the Castelnuovos bound to fix the $F^{(g)}$ up to genus 51. This will be shown in Sect. 5.2.

4 One-Parameter Calabi–Yau Spaces with Three Regular Singular Points

We generalize the analysis for the quintic to other one Kähler parameter Calabi–Yau threefolds, whose mirror W has a Picard–Fuchs system with three regular singular points. Note that this type of CY has been completely classified [55] starting from the Riemann–Hilbert reconstruction of the Picard–Fuchs equations given the monodromies and imposing the special geometry property as well as integrality conditions on the solutions of the latter. There are 13 cases whose mirrors can be realized as hypersurfaces and complete intersections in weighted projective spaces with trivial fundamental group,[14] and are well-known in the mirror symmetry literature. A fourteenth case is related to a degeneration of a two-parameter model as pointed out in [56]. Five are obtained as a free discrete orbifold of the former CY.

We focus on the 13 former cases. In the notation of [25] these complete intersections of degree $(d_1, d_2, \cdots, d_k)$ in weighted projective spaces $\mathbb{P}^n(w_1, \ldots, w_l)$, Calabi–Yau manifolds are abbreviated as $X_{d_1, d_2, \cdots, d_k}$ $(w_1, \ldots, w_l)$. For example, the familiar quintic manifold is denoted as $X_5(1^5)$. The list of 13 such examples is the following:

$$X_5(1^5) : \mathbf{a} = (\frac{1}{5}, \frac{2}{5}, \frac{3}{5}, \frac{4}{5}), \quad X_6(1^4, 2) : \mathbf{a} = (\frac{1}{6}, \frac{2}{6}, \frac{4}{6}, \frac{5}{6}),$$

$$X_8(1^4, 4) : \mathbf{a} = (\frac{1}{8}, \frac{3}{8}, \frac{5}{8}, \frac{7}{8}),$$

$$X_{10}(1^3, 2, 5) : \mathbf{a} = (\frac{1}{10}, \frac{3}{10}, \frac{7}{10}, \frac{9}{10}), \quad X_{3,3}(1^6) : \mathbf{a} = (\frac{1}{3}, \frac{1}{3}, \frac{2}{3}, \frac{2}{3}),$$

---

[14] We also have some results on the cases with non-trivial fundamental group. They are available on request.

$$X_{4,2}(1^6) : \mathbf{a} = (\frac{1}{4}, \frac{1}{2}, \frac{1}{2}, \frac{3}{4}), \quad X_{3,2,2}(1^7) : \mathbf{a} = (\frac{1}{3}, \frac{1}{2}, \frac{1}{2}, \frac{2}{3}),$$

$$X_{2,2,2,2}(1^8) : \mathbf{a} = (\frac{1}{2}, \frac{1}{2}, \frac{1}{2}, \frac{1}{2}), \quad X_{4,3}(1^5, 2) : \mathbf{a} = (\frac{1}{4}, \frac{1}{3}, \frac{2}{3}, \frac{3}{4}),$$

$$X_{4,4}(1^4, 2^2) : \mathbf{a} = (\frac{1}{4}, \frac{1}{4}, \frac{3}{4}, \frac{3}{4}), \quad X_{6,2}(1^5, 3) : \mathbf{a} = (\frac{1}{6}, \frac{1}{2}, \frac{1}{2}, \frac{5}{6}),$$

$$X_{6,4}(1^3, 2^2, 3) : \mathbf{a} = (\frac{1}{6}, \frac{1}{4}, \frac{3}{4}, \frac{5}{6}), \quad X_{6,6}(1^2, 2^2, 3^2) : \mathbf{a} = (\frac{1}{6}, \frac{1}{6}, \frac{5}{6}, \frac{5}{6}).$$

The examples satisfy the Calabi–Yau condition $\sum_i d_i = \sum_i w_i$ required by the vanishing of the first Chern class. The components of the vector $\mathbf{a}$ specify the Picard–Fuchs operators for the mirror manifolds with $h^{2,1} = 1$,

$$\{(\psi\frac{\partial}{\partial\psi})^4 - \psi^{-1}\prod_{i=1}^{4}(\psi\frac{\partial}{\partial\psi} - a_i)\}\Pi = 0. \tag{81}$$

The indices of the Picard–Fuchs equation satisfy $\sum_{i=1}^{4} a_i = 2$ and we have arranged a_i in increasing order for later convenience.

There are three singular points in the moduli space. The maximally unipotent point is the large complex structure modulus limit $\psi \sim \infty$ that has three logarithmic solutions for the Picard–Fuchs equation. The conifold point is $\psi = 1$ with three power series solutions and one logarithmic solution for the Picard–Fuchs equation. If the indices a_i are not degenerate, the Picard–Fuchs equation around the orbifold point $\psi = 0$ has four powers series solutions with the leading terms going like ψ^{a_i}. Each degeneration of the indices generates a logarithmic solution for the Picard–Fuchs equation around the orbifold point $\psi = 0$.

4.1 The Integration of the Anomaly Equation

We can straightforwardly generalize the formalism in [37] to the above class of models. The mirror map is normalized as $t = \log(\prod_i d_i^{d_i} / \prod_i w_i^{w_i}\psi) + \mathcal{O}(1/\psi)$, so that the classical intersection number in the prepotential is $F^{(0)} = \kappa/6t^3 + \cdots$ where $\kappa = \prod_i d_i / \prod_i w_i$. The generators of the topological string amplitudes are defined accordingly,

$$A_p := \frac{(\psi\partial\psi)^p G_{\psi\bar{\psi}}}{G_{\psi\bar{\psi}}}, \quad B_p := \frac{(\psi\partial\psi)^p e^{-K}}{e^{-K}}, \quad (p = 1, 2, 3, \cdots)$$

$$C := C_{\psi\psi\psi}\psi^3, \quad X := \frac{1}{1 - \psi}. \tag{82}$$

Here the familiar three-point Yukawa coupling is $C_{\psi\psi\psi} \sim \psi^{-2}/1 - \psi$ and as in the case of the quintic we denote $A := A_1$ and $B := B_1$. The generators satisfy the derivative relations

$$\psi \partial_\psi A_p = A_{p+1} - AA_p, \quad \psi \partial_\psi B_p = B_{p+1} - BB_p, \quad \psi \partial_\psi X = X(X-1),$$

and the recursion relations

$$B_4 = \left(\sum_i a_i\right)XB_3 - \left(\sum_{i<j} a_i a_j\right)XB_2 + \left(\sum_{i<j<k} a_i a_j a_k\right)XB - \left(\prod_i a_i\right)X,$$

$$A_2 = -4B_2 - 2AB - 2B + 2B^2 - 2A + 2XB + XA - r_0 X - 1, \tag{83}$$

where the first equation can be derived from the Picard–Fuchs equation (81) and the second equation is derived from the special geometry relation up to a holomorphic ambiguity denoted by the constant r_0. One can fix the constant r_0 by expanding the generators around any of the singular points in the moduli space. For example, as we will explain, the asymptotic behaviours of various generators around the orbifold point are $A_p = (a_2 - a_1 - 1)^p + \mathcal{O}(\psi)$, $B_p = a_1^p + \mathcal{O}(\psi)$ and $X = 1 + \mathcal{O}(\psi)$, so we find the constant is

$$r_0 = a_1(1 - a_1) + a_2(1 - a_2) - 1. \tag{84}$$

The polynomial topological amplitudes P_g are defined by $P_g^{(n)} := C^{g-1}\psi^n C_{\psi^n}^{(g)}$ and satisfy the recursion relations with initial data

$$P_{g=0}^{(3)} = 1,$$

$$P_{g=1}^{(1)} = \left(\frac{\chi}{24} - 2\right)B - \frac{A}{2} + \frac{1}{12}(X - 1) + \frac{s_1}{2},$$

$$P_g^{(n+1)} = \psi \partial_\psi P_g^{(n)} - [n(A + 1) + (2 - 2g)(B - \frac{X}{2})]P_g^{(n)}, \tag{85}$$

where χ is the Euler character of the Calabi–Yau space and $s_1 = 2c - 5/6$ is a constant that can be fixed by the second Chern class of the Calabi–Yau; see Appendix A. We provide the list of constants for the 13 cases of one-parameter Calabi–Yau in the following Table 1.

After changing variables to a convenient basis from (A, B, B_2, B_3, X) to (u, v_1, v_2, v_3, X) as the following:

Table 1. Euler numbers and the constant s_1

CY	$X_5(1^5)$	$X_6(1^4, 2)$	$X_8(1^4, 4)$	$X_{10}(1^3, 2, 5)$	$X_{3,3}(1^6)$
χ	-200	-204	-296	-288	-144
s_1	$\frac{10}{3}$	$\frac{8}{3}$	$\frac{17}{6}$	2	$\frac{11}{3}$
CY	$X_{4,2}(1^6)$	$X_{3,2,2}(1^7)$	$X_{2,2,2,2}(1^8)$	$X_{4,3}(1^5, 2)$	$X_{4,4}(1^4, 2^2)$
χ	-176	-144	-128	-156	-144
s_1	$\frac{23}{6}$	$\frac{25}{6}$	$\frac{9}{2}$	$\frac{19}{6}$	$\frac{5}{2}$
CY	$X_{6,2}(1^5, 3)$	$X_{6,4}(1^3, 2^2, 3)$	$X_{6,6}(1^2, 2^2, 3^2)$		
χ	-256	-156	-120		
s_1	$\frac{7}{2}$	$\frac{11}{6}$	1		

$$B = u, \quad A = v_1 - 1 - 2u, \quad B_2 = v_2 + uv_1,$$
$$B_3 = v_3 - uv_2 + uv_1 X - (r_0 + 1)uX.$$

One can follow [37] and use the BCOV formalism to show that P_g is a degree $3g - 3$ inhomogeneous polynomial of v_1, v_2, v_3, X with the assigned degrees $1, 2, 3, 1$ for v_1, v_2, v_3, X respectively. The BCOV holomorphic anomaly equation becomes

$$\left(\frac{\partial}{\partial v_1} - u\frac{\partial}{\partial v_2} - u(u + X)\frac{\partial}{\partial v_3}\right)P_g = -\frac{1}{2}\left(P_{g-1}^{(2)} + \sum_{r=1}^{g-1} P_r^{(1)} P_{g-r}^{(1)}\right). \tag{86}$$

As in the case of the quintic, the holomorphic anomaly equation determines the polynomial P_g up to a holomorphic ambiguity

$$f = \sum_{i=0}^{3g-3} a_i X^i. \tag{87}$$

4.2 The Boundary Behaviour

The constant term a_0 is fixed by the known constant map contribution in the A-model expansion

$$F_{\text{A-model}}^g = \lim_{t \to \infty} \omega_0^{2(g-1)} \left(\frac{1 - \psi}{\kappa \psi}\right)^{g-1} P_g, \tag{88}$$

where ω_0 is the power series solution in the large complex structure modulus limit.

The main message here is that structure of the conifold expansion is universal. For all cases of Calabi–Yau spaces, the Picard–Fuchs equation around $z = \psi - 1$ has four solutions that go like $\omega_0 = 1 + \mathcal{O}(z)$, $\omega_1 = z + \mathcal{O}(z^2)$, $\omega_2 = z^2 + \mathcal{O}(z^3)$ and $\omega_4 = \omega_1 \log(z) + \mathcal{O}(z^4)$. We define a dual mirror map $t_D = \frac{\omega_1}{\omega_0} = z + \mathcal{O}(z^2)$, expand the topological string amplitudes in terms of t_D and impose the following gap structure in the conifold expansion:

$$F_{\text{conifold}}^{(g)} = \lim_{\bar{z} \to 0} \omega_0^{2(g-1)} \left(\frac{1 - \psi}{\psi}\right)^{g-1} P_g$$
$$= \frac{(-1)^{g-1} B_{2g}}{2g(2g - 2)t_D^{2g-2}} + \mathcal{O}(t_D^0). \tag{89}$$

This fixes $2g - 2$ coefficients in the holomorphic ambiguity.

On the other hand, we discover a rich variety of singularity structures around the orbifold point $\psi = 0$. A natural symplectic basis of solutions of the Picard–Fuchs equation (81) is picked out by the fractional powers of the leading terms. As in the case of quintic for our purpose we only need the first two solutions ω_0, ω_1. The leading behaviours are

1. If $a_2 > a_1$, then we have $\omega_0 = \psi^{a_1}(1 + \mathcal{O}(\psi))$, $\omega_1 = \psi^{a_2}(1 + \mathcal{O}(\psi))$.
2. If $a_2 = a_1$, then we have $\omega_0 = \psi^{a_1}(1 + \mathcal{O}(\psi))$, $\omega_1 = \omega_0(\log(\psi) + \mathcal{O}(\psi))$.

In both the cases we can define a mirror map around the orbifold point as $s = \frac{\omega_1}{\omega_0}$. Using the behaviours of Kähler potential $e^{-K} \sim \omega_0$ and metric $G_{\psi\bar\psi} \sim \partial_\psi s$ in the holomorphic limit, we can find the asymptotic expansion of various generators. In both the cases, $a_2 > a_1$ and $a_2 = a_1$, the leading behaviours are non-singular:

$$A_p = (a_2 - a_1 - 1)^p + \mathcal{O}(\psi)$$
$$B_p = a_1^p + \mathcal{O}(\psi). \tag{90}$$

This can be used to fix a holomorphic ambiguity (84) relating A_2 to other generators. On the other hand, since the constant (84) can be also derived at other singular points of the moduli space, this also serves as a consistency check that we have chosen the correct basis of solutions ω_0, ω_1 at the orbifold point.

The orbifold expansion of the topological string amplitudes are

$$F^{(g)}_{\text{orbifold}} = \lim_{\bar\psi \to 0} \omega_0^{2(g-1)} \left(\frac{1-\psi}{\psi}\right)^{g-1} P_g \sim \frac{P_g}{\psi^{(1-2a_1)(g-1)}} \tag{91}$$

We can expand the polynomial P_g around the orbifold point $\psi = 0$. Generically, P_g is power series of ψ starting from a constant term, so $F^{(g)}_{\text{orbifold}} \sim 1/\psi^{(1-2a_1)(g-1)}$. Since $\sum_i a_i = 2$ and a_1 is the smallest, we know $a_1 \le 1/2$ and the topological string amplitude around orbifold point is generically singular. Interestingly, we find the singular behaviour of the topological strings around the orbifold point is not universal and falls into four classes:

1. This class includes all four cases of hypersurfaces and four other cases of complete intersections. They are the Calabi–Yau spaces $X_5(1^5)$, $X_6(1^4, 2)$, $X_8(1^4, 4)$, $X_{10}(1^3, 2, 5)$, $X_{3,3}(1^6)$, $X_{2,2,2,2}(1^8)$, $X_{4,4}(1^4, 2^2)$, $X_{6,6}(1^2, 2^2, 3^2)$. For these cases the singularity at the orbifold point is cancelled by the series expansion of the polynomial P_g. The requirement of cancellation of singularity in turn imposes

$$\lceil (1 - 2a_1)(g - 1) \rceil \tag{92}$$

conditions on the holomorphic ambiguity in P_g. Taking into account the A-model constant map condition and the boundary condition from conifold expansion, we find the number of unknown coefficients at genus g is

$$[2a_1(g - 1)]. \tag{93}$$

Notice for the Calabi–Yau $X_{2,2,2,2}(1^8)$ the cancellation is trivial since in this case $a_1 = 1/2$ and topological strings around the orbifold point are generically non-singular, so this does not impose any boundary conditions.

2. This class of Calabi–Yau spaces include $X_{4,2}(1^6)$ and $X_{6,2}(1^5, 3)$. We find the singularity at the orbifold point is not cancelled by the polynomial P_g, but it has a gap structure that closely resembles the conifold expansion. Namely, when we expand the topological strings in terms of the orbifold mirror map $s = \omega_1/\omega_0$, we find

$$F^{(g)}_{\text{orbifold}} = \frac{C_g}{s^{2(g-1)}} + \mathcal{O}(s^0) \tag{94}$$

where in our normalization convention, the constant is $C_g = B_{2g}/2^{5g-4}$ $(g-1)g$ for the $X_{4,2}(1^6)$ model and $C_g = B_{2g}/3^{3(g-1)}4(g-1)g$ for the $X_{6,2}(1^5, 3)$ model. Since the mirror map goes like $s \sim \psi^{a_2-a_1}$ and the P_g is a power series of ψ, this imposes

$$\lceil 2(a_2 - a_1)(g - 1) \rceil \tag{95}$$

conditions on the holomorphic ambiguity in P_g. In both cases of $X_{4,2}(1^6)$ and $X_{6,2}(1^5, 3)$, we have $a_2 = 1/2$, so the number of un-fixed coefficients is the same as the models in the first class, namely

$$\lceil 2a_1(g - 1) \rceil. \tag{96}$$

3. This class of Calabi–Yau spaces include only the Calabi–Yau $X_{3,2,2}(1^7)$. For this model, the singularity around the orbifold point does not cancel, so there is no boundary condition imposed on the holomorphic ambiguity in P_g. At genus g we are simply left with $g - 1$ unknown coefficients.

4. This class of Calabi–Yau spaces include $X_{4,3}(1^5, 2)$ and $X_{6,4}(1^3, 2^2, 3)$. For this class of models, the singularity at the orbifold point is partly cancelled. Specifically, we find $P_g/\psi^{(1-2a_1)(g-1)}$ is not generically regular at $\psi \sim 0$, but $P_g/\psi^{(1-2a_2)(g-1)}$ is always regular. This then imposes

$$\lceil (1 - 2a_2)(g - 1) \rceil \tag{97}$$

conditions on holomorphic ambiguity in P_g and the number of unknown coefficients at genus g is now

$$\lceil 2a_2(g - 1) \rceil. \tag{98}$$

It looks like we have the worst scenarios in the two models $X_{2,2,2,2}(1^8)$ and $X_{3,2,2}(1^7)$, where we essentially get no obvious boundary conditions at the orbifold point $\psi = 0$. However after a closer examination, we find some patterns in the leading coefficients of the orbifold expansion as the followings. In our normalization convention, we find the leading constant coefficients of $X_{2,2,2,2}(1^8)$ model is

$$F^{(g)}_{\text{orbifold}} = \frac{(21 \cdot 2^{2g-2} - 5)(-1)^g B_{2g} B_{2g-2}}{2^{2g-3} g(2g - 2)(2g - 2)!} + \mathcal{O}(\psi) \tag{99}$$

whereas for $X_{3,2,2}(1^7)$ model the leading coefficients are

$$F_{\text{orbifold}}^{(g)} = \frac{(7 \cdot 2^{2g-2} - 1)B_{2g}}{2^{4g-3}3^{2g-2}(g-1)g} \frac{1}{s^{2g-2}} + \mathcal{O}(\frac{1}{s^{2g-8}}) \tag{100}$$

These leading coefficients provide one more boundary condition for the models $X_{2,2,2,2}(1^8)$ and $X_{3,2,2}(1^7)$, although this is not much significant at large genus. On the other hand, we observe the leading coefficients of $X_{2,2,2,2}(1^8)$ model are similar to the constant map contribution of Gromov–Witten invariants except the factor of $(21 \cdot 2^{2g-2} - 5)$, whereas the leading coefficients of $X_{3,2,2}(1^7)$ model are similar to the conifold expansion except the factor of $(7 \cdot 2^{2g-2} - 1)$. These non-trivial factors cannot be simply removed by a different normalization of variables and therefore contain useful information. In fact in the latter case of $X_{3,2,2}(1^7)$ model, the factor of $(7 \cdot 2^{2g-2} - 1)$ will motivate our physical explanations of the singularity in a moment.

As shown in [25], see also Sect. 5, for a fixed genus g, the Gopakumar–Vafa invariants n_d^g are only non-vanishing when the degree d is bigger than a_g, where a_g is a model dependent number with weak genus dependence, in particular for large g one has $d_{\min} - 1 = a_g \sim \sqrt{g}$. So long as the number of zeros in the low degree Gopakumar–Vafa invariants are bigger than the number of unknown coefficients that we determine above using all available boundary conditions, we have a redundancy of data to compute the topological strings recursively genus by genus and are able to make non-trivial checks of our computations. For all of the 13 cases of one-parameter Calabi–Yau spaces, we are able to push the computation to very high genus. So far our calculations are limited only by the power of our computational facilities.

We now propose a "phenomenological" theory of the singularity structures at the point with rational branching. Our underlying philosophy is that a singularity of $F^{(g)}$ in the moduli space can only be generated if there are charged massless states near this point of moduli space. This is already familiar from the behaviours at infinity $\psi = \infty$ and conifold point $\psi = 1$. At infinity $\psi = \infty$ the relevant charged states are massive $D2 - D0$ brane-bound states and therefore $F^{(g)}$ are regular, whereas at the conifold point there is a massless charged state from a $D3$ brane wrapping a vanishing three-cycle and this generates the gap like singularity at the conifold point as we have explained. We should now apply this philosophy to the much richer behaviour at the orbifold point $\psi = 0$. We discuss the four classes of models in the same order as mentioned above.

1. We argue for this class of models the $F^{(g)}$ are regular at the orbifold point because there is no massless charged state. A necessary condition would be that the mirror map parameter s is non-zero at the orbifold point, since as we have learned there are D-branes wrapping cycles whose charge and mass are measured by s. This is clear for the complete intersection cases, namely $X_{3,3}(1^6)$, $X_{2,2,2,2}(1^8)$, $X_{4,4}(1^4, 2^2)$, $X_{6,6}(1^2, 2^2, 3^2)$, because for these models the first two indices of the Picard–Fuchs equation is degenerate $a_2 = a_1$, therefore the mirror map goes like

$$s \sim \log \psi \to \infty \qquad (101)$$

and we see that the D-branes are very massive and generate exponentially small corrections just like the situation at infinity $\psi = \infty$. As for the hypersurface cases $X_5(1^5)$, $X_6(1^4, 2)$, $X_8(1^4, 4)$, $X_{10}(1^3, 2, 5)$, we comment that although naively the mirror map goes like $s \sim \psi^{a_2 - a_1} \to 0$, there is a change of basis under which the generators are invariant as explained in Appendix A.3 and, which could make the periods goes like $\omega_0 \sim \omega_1 \sim \psi^{a_1}$, therefore the mirror map becomes finite at the orbifold point. This is also consistent with the basis at orbifold point we obtained by analytic continuation from infinity $\psi = \infty$. In fact we checked that the regularity of $F^{(g)}$ is not affected when we take s to be the ratio of generic arbitrary linear combinations of the periods ω_i, $i = 0, 1, 2, 3$.

2. We argue this class of models have a conifold like structure because of the same mechanism we have seen for the conifold point $\psi = 1$. This is consistent with the fact that the degenerate indices in these cases, e.g. models $X_{4,2}(1^6)$ and $X_{6,2}(1^5, 3)$, are the middle indices, namely we have $a_2 = a_3 = \frac{1}{2}$. Therefore the Picard–Fuchs equation constrains one of periods to be a power series proportional to

$$\omega_1 \sim \psi^{\frac{1}{2}} \qquad (102)$$

Since there is another period that goes like $\omega_0 \sim \psi^{a_1}$, the mirror map goes like $s \sim \psi^{1/2 - a_1} \to 0$ and it is not possible to change the basis in a way such that the mirror map is finite. Integrating out a charged nearly massless particle generates the gap like conifold singularity as we have explained.

3. For the model $X_{3,2,2}(1^7)$ we argue that there are two massless states near the orbifold point. Since the middle indices also degenerate $a_2 = a_3 = 1/2$, we can apply the same reasoning from the previous case and infer that the mirror map goes like $s \sim \psi^{1/6}$ and there must be at least one charged massless state from a D3 brane wrapping vanishing three-cycle. However the situation is now more complicated. We do not find a gap structure in the expansion of $F^{(g)}_{\text{orbifold}}$ and the leading coefficients differ from the usual conifold expansion by a factor of $(7 \cdot 2^{2g-2} - 1)$ as observed in (100). These can be explained by postulating that there are two massless particles in this case whose masses are m and $2m$. This fits nicely with the 2^{2g-2} power in the intriguing factor of $(7 \cdot 2^{2g-2} - 1)$ and also explains the absence of gap structure by the possible interactions between the two massless particles.

4. Finally, we discuss the cases of models $X_{4,3}(1^5, 2)$ and $X_{6,4}(1^3, 2^2, 3)$. The indices a_i are not degenerate in these cases. What makes these models different from the hypersurface cases is the fact that the ratio a_2/a_1 is now not an integer in these cases. This makes it difficult to change the basis such that the mirror map is finite. We conjecture that the mirror map indeed goes like $s \sim \psi^{a_2 - a_1}$ and therefore there exist massless particle(s)

whose masses are proportional to s and who are responsible for generating the singularity of $F^{(g)}$ at the orbifold point. This is consistent with our analytic continuation analysis in Appendix A.4. We note that a necessary non-trivial consequence of this scenario would be that the $F^{(g)}_{\text{orbifold}}$ is no more singular than $1/s^{2g-2}$, i.e. the product

$$s^{2g-2} F^{(g)}_{\text{orbifold}} \sim \frac{P_g}{\psi^{(1-2a_2)(g-1)}} \tag{103}$$

should be regular. This is precisely what we observe experimentally.

5 Symplectic Invariants at Large Radius

The coefficients of the large radius expansion of the $\mathcal{F}^{(g)} = \lim_{\bar{t}\to\infty} F^{(g)}(t,\bar{t})$ have an intriguing conjectural interpretation as symplectic invariants of M. First of all we have

$$\mathcal{F}^{(g)}(q) = \sum_\beta r_\beta^{(g)} q^\beta \,, \tag{104}$$

where $r_\beta^{(g)} \in \mathbb{Q}$ are the *Gromow–Witten* invariants of holomorphic maps. Secondly the *Gopakumar–Vafa invariants* [36] count the cohomology of the $D_0 - D_2$ bound state moduli space, see also [25], and are related to the

$$\mathcal{F}(\lambda, t) = \sum_{g=0}^{\infty} \lambda^{2g-2} \mathcal{F}^{(g)}(t)$$

$$= \frac{c(t)}{\lambda^2} + l(t) + \sum_{g=0}^{\infty} \sum_{\beta \in H_2(M,\mathbb{Z})} \sum_{m=1}^{\infty} n_\beta^{(g)} \frac{1}{m} \left(2\sin\frac{m\lambda}{2} \right)^{2g-2} q^{\beta m} \,. \tag{105}$$

Here $c(t)$ and $l(t)$ are some cubic and linear polynomials in t, which follow from the leading behaviour of $\mathcal{F}^{(0)}$ and $\mathcal{F}^{(1)}$ as explained in (37) and (20). With $q_\lambda = e^{i\lambda}$ we can write a product form[15] for the partition function $Z^{\text{hol}} = \exp(\mathcal{F}^{\text{hol}})$

$$Z^{\text{hol}}_{\text{GV}}(M, \lambda, q) = \prod_\beta \left[\left(\prod_{r=1}^{\infty} (1 - q_\lambda^r q^\beta)^{r n_\beta^{(0)}} \right) \right.$$

$$\left. \prod_{g=1}^{\infty} \prod_{l=0}^{2g-2} (1 - q_\lambda^{g-l-1} q^\beta)^{(-1)^{g+r} \binom{2g-2}{l} n_\beta^{(g)}} \right] \tag{106}$$

in terms of the *Gopakumar–Vafa invariants* $n_\beta^{(g)}$. Based on the partition functions there is a conjectural relation of the latter to the *Donaldson–Thomas*

---

[15] Here we dropped the $\exp(c(t)/\lambda^2 + l(t))$ factor of the classical terms at genus $0, 1$.

invariants $\tilde{n}_{\beta}^{(g)}$, which are invariants of the moduli space $I_k(M, \beta)$ of ideal sheaves $\mathcal{I}$ on M. Defining $Z_{\mathrm{DT}}^{\mathrm{hol}}(M, q_\lambda, q) = \sum_{\beta, k \in \mathbb{Z}} \tilde{n}_{\beta}^{(k)} q_\lambda^k q^\beta$ one expects [57]

$$Z_{\mathrm{GV}}^{\mathrm{hol}}(M, q_\lambda, q) M(q_\lambda)^{\frac{\chi(M)}{2}} = Z_{\mathrm{DT}}^{\mathrm{hol}}(M, -q_\lambda, q) , \qquad (107)$$

where the McMahon function is defined as $M(q_\lambda) := \prod_{n \geq 0} 1/(1 - q_\lambda^n)^n$. We will give below the information about the $\mathcal{F}^{(g)}$ in terms of the *Gopakumar–Vafa invariants* and give a more detailed account of the data of the symplectic invariants on the webpage [45].

5.1 Castelnuovo's Theory and the Cohomology of the BPS State Moduli Space

Let us give checks of the numbers using techniques of algebraic geometry and the description of the BPS moduli space and its cohomology developed in [25, 36]. The aim is to check the gap condition in various geometric settings, namely hypersurfaces and complete intersections in (weighted) projective spaces discussed before. According to [25, 36] the BPS number of a given charge, i.e. degree d, can be calculated from cohomology of the moduli space $\hat{\mathcal{M}}$ of a $D_2 - D_0$ brane system. The latter is the fibration of the Jacobian $T^{2\tilde{g}}$ of a genus $\tilde{g}$ curve over its moduli space of deformations $\mathcal{M}$. Curves of arithmetic genus $g < \tilde{g}$ are degenerate curves, in the simplest case with $\delta = \tilde{g} - g$ nodes. Their BPS numbers are calculated using the Euler numbers of relative Hilbert schemes $\mathcal{C}^{(i)}$ of the universal curve ($\mathcal{C}^{(0)} = \mathcal{M}$, $\mathcal{C}^{(1)}$ is the universal curve, etc.) in simple situations as follows:

$$n_d^g = n_d^{\tilde{g}-\delta} = (-1)^{\dim(\mathcal{M})+\delta} \sum_{p=0}^{\delta} b(\tilde{g} - p, \delta - p) e(\mathcal{C}^{(p)}),$$

$$b(g, k) = \tfrac{2}{k!}(g - 1) \prod_{i=1}^{k-1} (2g - (k + 2) + i), \qquad b(g, 0) = 0. \qquad (108)$$

As explained in [25] curves in projective spaces meeting the quintic are either plane curves in $\mathbb{P}^2$, curves in $\mathbb{P}^3$, or $\mathbb{P}^4$. In all case one gets from Castelnuovo theory a bound on g, which grows for large d like $g(d) \sim d^2$. For a detailed exposition of curves in projective space see [58]. Using this information one can determine which curves above is realized and contributes to the BPS numbers. These statements generalize to the hypersurfaces and complete intersections with one Kähler modulus in weighted projective spaces. In particular the qualitative feature $g(d) \sim d^2$ of the bound for large d carries over. We note for later convenience that to go from a smooth curve of genus $\tilde{g}$ to a curve with arithmetic genus $g = \tilde{g} - \delta$ by enforcing δ nodes we get from (108)

$$n_d^{\tilde{g}-1} = (-1)^{\dim(\mathcal{M})+1} \left(e(\mathcal{C}) + (2\tilde{g} - 2) e(\mathcal{M}) \right)$$

$$n_d^{\tilde{g}-2} = (-1)^{\dim(\mathcal{M})+1} \left(e(\mathcal{C}^{(2)}) + (2\tilde{g} - 4) e(\mathcal{C}) + \tfrac{1}{2}(2\tilde{g} - 2)(2\tilde{g} - 5) e(\mathcal{M}) \right) .$$

$$(109)$$

5.2 D-Branes on the Quintic

One consequence of our global understanding of the $F^{(g)}$ is that we can make detailed statements about the 'number' of D-brane states for the quintic at large radius. We focus on $d = 5$, because there is a small numerical flaw in the analysis of [25], while the right numerics confirms the gap structure quite significantly. In this case the complete intersection with multidegree $(1, 1, 5)$ is a plane curve with genus $\tilde{g} = (d-1)(d-2)/2 = 6$, while the other possibilities have at most genus $g = 2$. Curves of $(g = \tilde{g}, d) = (6, 5)$ are therefore smooth plane curves with $\delta = 0$ and according to (108) their BPS number is simply $n_d^g = (-1)^{\dim \mathcal{M}} e(\mathcal{M})$. Since $d = 5$ their moduli space is simply the moduli space of $\mathbb{P}^2$'s in $\mathbb{P}^4$, $\mathcal{M}$ is the Grassmannian[16] $\mathbb{G}(2, 4)$. Grassmannians $\mathbb{G}(k, n)$ have dimensions $(k + 1)(n - k)$ and their Euler number can be calculated most easily by counting toric fixed points to be $\chi(\mathbb{G}(k, n)) = \binom{n+1}{k+1}$. We get $n_5^6 = (-1)^6 10 = 10$.

For the $(g, d) = (5, 5)$ curves we have to determine the Euler number of the universal curve $\mathcal{C}$, which is a fibration $\pi : \mathcal{C} \to \mathcal{M}$ over $\mathcal{M}$. To get an geometric model for $\mathcal{C}$ we consider the projection $\tilde{\pi} : \mathcal{C} \to X$. The fibre over a point $p \in X$ is the set of $\mathbb{P}^2$'s in $\mathbb{P}^4$ which contain the point p. This is described as the space of $\mathbb{P}^1$'s in $\mathbb{P}^3$, i.e. $\mathbb{G}(1, 3)$ with [17] $\chi(\mathbb{G}(1, 3)) = 6$. As the fibration $\tilde{\pi}$ is smooth we obtain $e(\mathcal{C}) = \chi(X)\chi(\mathbb{G}(1, 3)) = -200 \cdot 6 = -1200$. Applying now (109) we get $n_5^5 = (-1)^5(-1200 + (2 \cdot 6 - 2)10) = 1100$.

The calculation of n_4^4 requires the calculation of $e(\mathcal{C}^{(2)})$. The model for $\mathcal{C}^{(2)}$ is constructed from the fibration $\tilde{\pi} : \mathcal{C}^{(2)} \to \mathrm{Hilb}^2(X)$ as follows. A point in $\mathrm{Hilb}^2(X)$ are either two distinct points or one point of multiplicity 2 with distinct tangent direction. In both cases the fiber over $P \in \mathrm{Hilb}^2(X)$ is an $\mathbb{P}^2$ passing though 2 points in $\mathbb{P}^4$, which is a $\mathbb{P}^2$. The fibration is smooth and it remains to calculate the Euler number of the basis. There are nice product formulas for the Euler number of symmetric products of surfaces modded out by S_n. For surfaces it is more cumbersome. We calculate the Euler number $e(\mathrm{Sym}^2(X)) = \binom{-199}{2}$. $\mathrm{Hilb}^2(X)$ is the resolution of the orbifold $\mathrm{Sym}^2(X)$, which has the diagonal X as fix point set. The resolution replaces each point in the fixed point set by $\mathbb{P}^2$. Simple surgery and the smooth fibration structure of $\mathcal{C}^{(2)}$ gives hence $e(\mathcal{C}^{(2)}) = 3(e(\mathrm{Sym}^2(X) + (3 - 1)e(X)) = 58500$, which by (109) yields $n_5^4 = 58500 + (2 \cdot 6 - 4)(-1200) + 35 \cdot 10 = 49250$.

The approach becomes more difficult with the number of free points δ and at $\delta = 4$ it is currently not known how to treat the singularities of the Hilbert scheme.

---

[16] The space of $\mathbb{P}^k$'s in $\mathbb{P}^n$, which we call $\mathbb{G}(k, n)$, is also the space of $k + 1$ complex dimensional subspaces in an $n + 1$ dimensional complex vector space, which is often alternatively denote as $G(k + 1, n + 1)$.

[17] $\mathbb{G}(1, 3)$ is Plücker embedded in $\mathbb{P}^5$ as a quadric (degree 2). From the adjunction formula we also get $\chi(\mathbb{G}(1, 3)) = 6$.

On the other hand smooth curves at the "edge" of the Castelnuovo bound are of no principal problem. Example, using adjunction for a smooth complete intersection of degree $(d_1, \ldots, d_r)$ in a (weighted) projective space $\mathbb{WCP}^n(w_1, \ldots, w_{n+1})$ in Appendix A, we calculate $\chi = (2 - 2g)$ and see that the degree 10 genus 16 curve is the complete intersection $(1, 2, 5)$. The moduli space is calculated by counting the independent deformations of that complete intersection. The degree five constraint lies on the quintic, the linear constraint has five parameters. The identification by the $\mathbb{C}^*$ action of the ambient $\mathbb{P}^4$ shows that these parameters lie in a $\mathbb{P}^4$. This constraint allows to eliminate one variable from the generic quadratic constraint which has hence 10 parameters and a $\mathbb{P}^9$ as moduli space. So we check in Table 2 the entry $n_{10}^{16} = (-1)^{13} 5 \cdot 10 = -50$.

Let us discuss the upper bound on the genus at which we can completely fix the F_g given simply the bound (80). We claim that this bound is $g \leq 51$. At degree 20 there is a smooth complete intersection curve $(1, 4, 5)$ of that genus. We first check that this is the curve of maximal genus in degree 20. The Castelnuovos bound for curves in $\mathbb{P}^4$ shows that they have smaller genus [58]. We further see from the discussion in [58] that for curves in $\mathbb{P}^3$ not on quadric and a cubic, which would have the wrong degree, the Castelnuovos's bound is saturated for the complete intersection $(1, 4, 5)$. For $g = 51$ (80) indicates that the gap, constant map contribution and regularity at the orbifold fixes 131 of the 151 unknown coefficients in (55). The vanishing of $n_d^{51} = 0$, $1 \leq d \leq 19$ and the value of $n_{20}^{51} = (-1)^{4+34} \chi(\mathbb{P}^4) \chi(\mathbb{P}^{34}) = 165$ for the Euler number of the moduli space of the smooth curve give us the rest of the data.

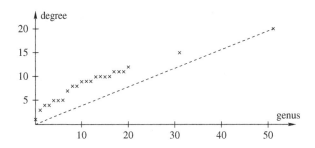

Castelnuovo's bound for higher genus curves on the quintic. The dashed line correspond roughly (up to taking the floor) to the number of coefficients in f_g (55) which are not fixed by constant map contribution, conifold and orbifold boundary conditions.

It is of course no problem to calculate form the B-model the higher genus amplitudes to arbitrary degree. For completeness we report the first non-trivial numbers for $g = 18 - 20$ in Table 3.

Table 2. BPS invariants n_d^g on the Quintic hypersurface in $\mathbb{P}^4$. See also Table 3

g	d=1	d=2	d=3	d=4	d=5	d=6
0	2875	609250	317206375	242467530000	229305888887625	248249742118022000
1	0	0	609250	3721431625	12129909700200	31147299733286500
2	0	0	0	534750	75478987900	8717081396383250
3	0	0	0	8625	-15663750	315644162875
4	0	0	0	0	49250	-7529331750
5	0	0	0	0	1100	-3079125
6	0	0	0	0	10	-34500
7	0	0	0	0	0	0

g	d=7	d=8	d=9
0	295091050570845639250	375632160937476603550000	503840510416985243645106250
1	71578406022880761750	154990541752961568418125	324064464310279585657008750
2	5185462556617269625	22516841063105917766750	81464921786339566502560125
3	111468926053022750	13034659840858345000	9523213659169217568991500
4	245477430615250	25517502254834226750	50772349651443561498250
5	-1917984531500	4656989619570625	10280743594493108319750
6	1300955250	-471852100909500	30884164195870217250
7	4874000	2876330661125	-135197508177440750
8	0	-1670397000	1937652990971125
9	0	-6092500	-12735865055000
10	0	0	18763368375
11	0	0	5502750
12	0	0	60375
13	0	0	0

(continued)

Table 2. (continued)

g	d=10	d=11
0	704288164978454686113488249750	1017913203569692432490203659468875
1	662863774391411409674240657630 0	1336442091735463067608016312923750
2	261910639528673259095451374 50	775720627148503750199049691449750
3	529399661897916624420404068 25	245749672908222069999611527634750
4	564669022311863868292985660 0	448475557200658307168403004753 75
5	30265304636080268273129787 5	46950866098449138653717762000 0
6	6948750094748611384962730	26778976421684176016869138162 5
7	4017951999615823390768 00	735709924295207023870887000 0
8	−25301032766083303150	7274265159936800289770125 0
9	1155593062739271425	140965985795732693440000
10	−179762095292424700	7228507120311700920 00
11	1504440957417 80	−18998955257482171250
12	−4540926631 50	3536502289027385 00
13	50530375	−40417087803245 00
14	−286650	225623064943 75
15	−5700	−299380132 50
16	−50	−7357125
17	0	−86250
18	0	0

Table 3. Some higher degree genus 18–20 BPS numbers for the quintic. Note that we can calculate all Donaldson–Thomas invariants for $d = 1, \ldots, 12$ exactly

d	g=18	g=19
$\vdots$	$\vdots$	$\vdots$
11	0	0
12	-3937166500	-13403500
13	285683687197594125	-2578098061480250
14	-9507695749687326805250	273001250682019210000
15	6438165666769014564325336250	$-34230433710262920027769700$
16	15209062594213864261318125134875	15209062594213864261318125134875

d	g=20
$\vdots$	$\vdots$
11	0
12	0
13	10690009494250
14	-5920586255923315250
15	15368486208424999875838025
16	$-10368247303939805037092472905000$

5.3 D-Brane States on Hypersurfaces in Weighted Projective Space

Similarly, for the sextic in $\mathbb{P}^4(1^4, 2)$, the degree $(1, 2, 6)$ complete intersection curve has genus $g = 10$ and degree $d = \prod_i d_i / \prod_k w_k = 6$ in the weighted projective space. Its moduli space is $\mathbb{P}^3$ for the degree one constraint, i.e. we can eliminate x_4 form the quadric and the seven coefficeints of the monomials $x_1^2, x_1 x_2, x_1 x_3, x_2^2, x_2 x_3, x_3^2, x_5$ form a $\mathbb{P}^6$. This yields $n_6^{10} = (-1)^9 4 \cdot 7 = -28$ in Table 4.

There are further checks for the $\mathbb{P}^4(1^4, 4)$ invariants listed in Table 5. The complete intersection $(1^2, 8)$ has total degree 2 and genus $g = 3$. The two linear constraints describe a $\mathbb{P}^1$ in $\mathbb{P}^3$, i.e. an $\mathbb{G}(1, 3)$ with Euler number 6 and dimension 4, which yields $n_2^3 = 6$. Similarly we have a $g = 7$ complete intersection $(1, 2, 8)$ of degree 4, whose moduli space is $\mathbb{P}^3$ times $\mathbb{P}^5$ hence $n_4^7 = 24$.

For the degree 10 hypersurface in $\mathbb{P}^4(1^3, 2, 5)$ we check the BPS invariants listed in Table 6: From the degree $(1, 1, 10)$ hypersuface of degree 1 complete intersection with $g = 2$. The moduli space of the linear constraints are just the one of point in $\mathbb{P}^2$, i.e. $\mathbb{P}^2$, hence $n_1^2 = 3$. The degree $(1, 2, 10)$ complete intersection with total degree 2 and genus 4 has the moduli space of the linear constraint, which is $\mathbb{P}^2$ and of the quadratic constraint is $\mathbb{P}^3$ (from the coefficients of the monomials $x_1^2, x_1 x_2, x_2^2, x_4$), yielding $n_2^4 = -12$. Finally the $(1, 3, 10)$ complete intersection with genus 7 and degree 3, has a moduli space $\mathbb{P}^2$ times $\mathbb{P}^5$ (from the coefficients of the monomials $x_1^3, x_1^2 x_2, x_1 x_2^2, x_1 x_4, x_2^3, x_2 x_4$) and $n_3^7 = -18$.

Table 4. BPS invariants n_d^g on the Sextic hypersurface in $\mathbb{P}^4(1^4, 2)$

g	d=1	d=2	d=3	d=4	d=5
0	7884	6028452	11900417220	34600752005688	24595034333130080
1	0	7884	145114704	177304432885	17149900584158168
2	0	0	17496	10801446444	571861298748384
3	0	0	576	-14966100	1985113680408
4	0	0	6	-47304	-21559102992
5	0	0	0	0	22232340
6	0	0	0	0	63072
7	0	0	0	0	0

g	d=6	d=7	d=8
0	513797193321737210316	2326721904320912944749252	11284058913384803271372834984
1	147664736456952923604	1197243574587406496495592	9381487423491392389034886369
2	137531000198040005556	233127389355701229349884	3246006977306701566424657380
3	411536108778637626	19655355035815833642912	561664896373523768591774196
4	1094535956564124	628760082868148062854	48641994707889298118864544
5	-18316495265688	3229011644338336680	1863506489926528403683191
6	207237771936	-18261998133124302	20758968356323626025164
7	-583398600	51363614205788	10040615124628834206
8	-146988	-8041642037676	1129699628821681740
9	-3168	54521267292	-38940584273866593
10	-28	-43329384	904511824896888
11	0	-110376	-12434437188576
12	0	0	76595605884

Table 5. BPS invariants n_d^g on the Octic hypersurface in $\mathbb{P}^4(1^4, 4)$

g	d=1	d=2	d=3	d=4	d=5
0	29504	128834912	1423720546880	23193056024793312	467876474625249316800
1	0	41312	21464350592	1805292092705856	101424054914016355712
2	0	864	−16551744	12499667277744	5401493537244872896
3	0	6	−177024	−174859503824	20584473699930496
4	0	0	0	396215800	−674562224718848
5	0	0	0	301450	12063928269056
6	0	0	0	4152	−86307810432
7	0	0	0	24	37529088
8	0	0	0	0	354048
⋮	⋮	⋮	⋮	⋮	

Table 6. BPS invariants n_d^g on the degree 10 hypersurface in $\mathbb{P}^4(1^3, 2, 5)$

g	d=1	d=2	d=3	d=4
0	231200	12215785600	1700894366474400	3501546588513246560000
1	280	207680960	161279120326840	103038403740897786400
2	3	−537976	1264588024791	8495973047204168640
3	0	−1656	−46669244594	61218893443516800
4	0	−12	630052679	−2460869494476896
5	0	0	−1057570	145198012290472
6	0	0	−2646	−5611087226688
7	0	0	−18	125509540304
8	0	0	0	−1268283512
⋮	⋮	⋮	⋮	⋮

These checks in different geometrical situations establish quite impressively the universality of the gap structure at the conifold expansion.

5.4 D-Branes on Complete Intersections

Here we summarize our results on one modulus complete intersections in (weighted) projective space. More complete results are available in [45]. Again we can check many BPS invariants associated with the smooth curves.

Let us check, e.g. in Table 7 the $n_9^{10} = 15$. According to (114) we see that at degree 9 there is a smooth genus 10 curve, given by a complete intersection of multi degree $(1^2, 3^2)$ in $\mathbb{P}^5$. Their moduli space is the Grassmannian $\mathbb{G}(3, 5)$, which has Euler number 15 and dimension 8, hence $n_9^{10} = 15$. In a very similar way it can be seen that the n_8^9 comes form a complete intersection curve of degree $(4, 2, 1^2)$ with the same moduli space, so $n_8^9 = 15$ in Table 8. Grassmannians related to complete intersection are also identified with the moduli spaces of the following smooth curves: The total degree six curve of genus 7 in Table 11 is a CI of multi degree $(1^2, 3, 4)$. Its moduli space is a

Table 7. n_d^g for the degree (3,3) complete intersection in $\mathbb{P}^5$

g	d=1	d=2	d=3	d=4	d=5	d=6
0	1053	52812	6424326	1139448384	249787892583	62660964509532
1	0	0	3402	5520393	4820744484	3163476682080
2	0	0	0	0	5520393	23395810338
3	0	0	0	0	0	6852978
4	0	0	0	0	0	10206
5	0	0	0	0	0	0

g	d=7	d=8	d=9
0	17256453900822009	5088842568426162960	1581250717976557887945
1	1798399482469092	944929890853230501	473725044069553679454
2	42200615912499	50349477671013600	47431893998882182563
3	174007524240	785786604262830	1789615720312984368
4	−484542	2028116431098	21692992151427138
5	158436	−784819773	36760497856020
6	0	372762	−61753761036
7	0	6318	−5412348
8	0	0	39033
9	0	0	1170
10	0	0	15
11	0	0	0

$\mathbb{G}(2,4)$ explaining $n_6^7 = 10$. The degree 4 curve of genus 5 in Table 12 is of multi degree $(1^2, 4^2)$ and has moduli space $\mathbb{G}(1,3)$ yielding $n_4^5 = 6$. The degree 4 curve of genus 5 in Table 13 is of multi degree $(1^2, 2, 6)$ and has moduli space $\mathbb{G}(2,4)$ yielding $n_4^5 = 10$. The degree 2 curve of genus 3 in Table 14 is

Table 8. n_d^g for the degree (4,2) complete intersection in $\mathbb{P}^5$

g	d=1	d=2	d=3	d=4	d=5	d=6
0	1280	92288	15655168	3883902528	1190923282176	417874605342336
1	0	0	0	−672	16069888	174937485184
2	0	0	0	−8	7680	12679552
3	0	0	0	0	0	276864
4	0	0	0	0	0	0

g	d=7	d=8	d=9
0	160964588281789696	66392895625625639488	28855060316616488359936
1	19078577926517760	14088192680381290336	9895851364631438617600
2	494602061689344	853657285175383648	11377794220513866498304
3	2016330670592	14859083841009280	4928601231129292368
4	−285585152	37334304102560	679351051885623552
5	591360	−46434384200	1103462757073920
6	7680	−8285120	−4031209095680
7	0	67208	370290688
8	0	1520	−2270720
9	0	15	−25600
10	0	0	0

of multi degree $(1^2, 4, 6)$ and has moduli space $\mathbb{G}(0,2) = \mathbb{P}^2$ yielding $n_2^3 = 3$. The moduli space of the degree 4 genus 7 curve $(1, 2, 4, 6)$ is an $\mathbb{P}^2$ times the moduli space $\mathbb{P}^4$ of quadrics in $\mathbb{WCP}^3(1^2, 2^2)$, so that $n_4^7 = (-1)^6 3 \cdot 5 = 15$. For the degree $(6, 6)$ complete intersection in $\mathbb{WCP}^3(1^2, 2^2, 3^2)$, see Table 15, we have a degree 1 genus 2 intersection $(1^2, 6^2)$, whose moduli space is a point hence $n_1^2 = 1$, a degree 2 genus 4 intersection $(1, 2, 6^2)$, whose moduli space is $\mathbb{P}^1$ times the moduli space $\mathbb{P}^2$ of quadrics in $\mathbb{WCP}^2(1, 2^2)$ hence $n_2^4 = -6$, a degree 3 genus 7 intersection $(1, 3, 6^2)$, whose moduli space is $\mathbb{P}^1$ times the moduli space $\mathbb{P}^4$ of cubics in $\mathbb{WCP}^4(1, 2^2, 3^2)$ hence $n_3^7 = -10$ and finally a degree 4 genus 11 intersection $(1, 4, 6^2)$, whose moduli space is $\mathbb{P}^1$ times the moduli space $\mathbb{P}^5$ of quadrics in $\mathbb{WCP}^4(1, 2^2, 3^2)$ hence $n_4^{11} = 12$.

There are many further checks that are somewhat harder to perform. Example, we note that there is a genus 1 degree 3 curve in the $(3, 2, 2)$ CI in $\mathbb{P}^6$, which comes from a complete intersection $(1^4, 3)$. Now the moduli space of this complete intersection in $\mathbb{P}^6$ is $\mathbb{G}(2, 6)$. However not all $\mathbb{P}^2$ parametrized by $\mathbb{G}(2, 6)$, which contain the cubic, are actually in the two quadrics of the $(3, 2, 2)$ CICY. We can restrict to those $\mathbb{P}^2$, which fulfil these constraints, by considering the simultaneous zeros of sections of two rank six bundles of quadratic forms on the moving $\mathbb{P}^2$. These are a number of points, which is calculated by the integral of the product of the Chern classes of these rank 6 bundles over $\mathbb{G}(2, 6)$. Indeed we obtain, for example with "Schubert" [59]

$$n_3^1 = (-1)^0 \int_{\mathbb{G}(2,6)} c_6^2(\mathrm{Sym}(2, Q)) = 64, \tag{110}$$

which confirms the corresponding entry in Table 9.

Table 9. n_d^g for the $(3,2,2)$ complete intersection in $\mathbb{P}^6$

g	d=1	d=2	d=3	d=4	d=5	d=6
0	720	22428	1611504	168199200	21676931712	3195557904564
1	0	0	64	265113	198087264	89191835056
2	0	0	0	0	10080	180870120
3	0	0	0	0	0	-3696
4	0	0	0	0	0	-56
5	0	0	0	0	0	0

g	d=7	d=8	d=9
0	517064870788848	89580965599606752	16352303769375910848
1	32343228035424	10503104916431241	3201634967657293024
2	315217101456	280315384261560	178223080602086784
3	199357344	1430336342574	2915033921871456
4	30240	194067288	8888143990672
5	0	795339	-233104896
6	0	0	4857552
7	0	0	384
8	0	0	0

6 Conclusions

In this chapter we solve the topological string B-model on compact Calabi–Yau M using the modularity of the F_g, the wave function transformation property of Z and the boundary information imposed by effective action considerations. The method pushes the calculation to unprecedented high genus amplitudes, e.g. for the quintic the boundary condition count (80) together with the simplest vanishing arguments at large volume fixes the amplitudes up to genus $g = 51$. Beyond that the prime mathematical problem to overcome in this region of the moduli space is to understand the degeneration of more then four points in the relative Hilbert scheme of the universal curves in a threefold.[18] Similar problems have been encountered in [60], where it was suggested to fix a very similar ambiguity to a anholomorphic SL(2, $\mathbb{Z}$)-modular elliptic index of a $D4 - D2 - D0$ brane system [60, 61, 62]. There one uses SL(2, $\mathbb{Z}$) invariance of the index and a dual dilute gas approximation in $AdS_3 \times S^2 \times M$ to fix the coeffcients of the ring of modular forms. The construction of the moduli-space of the $D4 - D2 - D0$ brane system uses rational GW invariants and implies non-trivial relations among them [60]. Such considerations could in principal provide further boundary conditions at large radius.

Our sharpest tool is the global control of Z over $\mathcal{M}(M)$ and we expect that by a closer analysis of the RR-spectrum at the orbifold of compact Calabi–Yau, we will be able to recover at least the $\lceil 2/d(g - 1) \rceil$ conditions that one loses relative to the local cases [13] and solve the model completely. We obtained not only the Gromov–Witten, the Donaldson–Thomas and Gopakumar–Vafa invariants at infinity, but also the local expansion at the conifold, the Gepner point and other more exotic singularities with one or more massless states. The leading singular terms in the effective action reflect the massless states. The branch locus of the 13 parameter models has an intriguing variety of such light spectra and we can learn from the effective action about the singularity and vice versa. Stability properties of theses states have been analysed in Appendix A.4.

Most importantly our exact expansions do contain further detailed information of the towers massive RR-states at these points. We described them in natural local variables. The information from different genera should be of great value for the study of stable even D-brane bound states on compact Calabi–Yau as it is the content of the supersymmetric index of [36], which is protected under deformations of the complex structure. Non-compact Calabi–Yau such as the resolution of $\mathbb{C}^n/G$, with $G \in$ SL($n, \mathbb{C}$) have no complex moduli. The issue does not arise and the situation is better understood, e.g. see [63, 64] for reviews.

---

[18] As a motivation and check for the task to develop the theory of Hilbert schemes for 3folds we calcultated the invariants explicitly to high genus. For the quintic to genus 20 and for all other the results up genus 12 are available at [45].

Table 10. n_d^g for the degree (2,2,2,2) complete intersection in $\mathbb{P}^7$

g	d=1	d=2	d=3	d=4	d=5	d=6	d=7
0	512	9728	416256	25703936	1957983744	170535923200	16300354777600
1	0	0	0	14752	8782848	2672004608	615920502784
2	0	0	0	0	0	1427968	2440504320
3	0	0	0	0	0	0	86016
4	0	0	0	0	0	0	0

g	d=8	d=9	d=10	d=11
0	1668063096387072	179845756064329728	2020649798389154816	2347339059011866069504
1	123699143093152	22984995833484288	4071465816864581632	698986176207439627264
2	1628589698304	702453851520512	236803123487243776	68301097513852719616
3	2403984384	4702943495168	4206537025629952	24824154746807982 08
4	-37632	2449622016	16316531089408	29624281509824512
5	-672	258048	2777384448	73818807399424
6	0	0	4283904	1153891840
7	0	0	0	26348544
8	0	0	0	0

Table 11. n_d^g for the degree (4,3) complete intersection in $\mathbb{WCP}^5(1^5, 2)$

g	d=1	d=2	d=3	d=4	d=5	d=6
0	1944	223560	64754568	27482893704	14431471821504	8675274727197720
1	0	27	161248	381704265	638555324400	891094220317561
2	0	0	0	227448	3896917776	20929151321496
3	0	0	0	81	155520	75047188236
4	0	0	0	0	5832	−40006768
5	0	0	0	0	0	26757
6	0	0	0	0	0	816
7	0	0	0	0	0	10
8	0	0	0	0	0	0

Table 12. n_d^g for the degree (4,4) complete intersection in $\mathbb{WCP}^5(1^4, 2^2)$

g	d=1	d=2	d=3	d=4	d=5	d=6
0	3712	982464	683478144	699999511744	887939257620352	1289954523115535040
1	0	1408	6953728	26841854688	88647278203648	266969312909257728
2	0	0	3712	148208928	2161190443904	17551821510538560
3	0	0	0	−12432	7282971392	362668189458048
4	0	0	0	384	−14802048	773557598272
5	0	0	0	6	−22272	−7046285440
6	0	0	0	0	0	6367872
7	0	0	0	0	0	11264
8	0	0	0	0	0	0

One can also use the explicit expansions to study the integrable theories that have been associated with the local expansion of the topological string on Calabi–Yau manifold, such as the $c = 1$ string at the conifold or the quiver gauge theories at the orbifold, matrix models, e.g. at the ADE singularities and new ones for more exotic singularities such as the branch points of the complete intersections Calabi–Yau manifolds that we discussed here.

The ability to obtain the imprint of the BPS spectrum on the effective action everywhere on the moduli space is of phenomenological interest as flux compactifications drive the theory to attractor points inside the moduli space.

Our expressions are governed by the representation of the modular group of the Calabi–Yau on almost holomorphic forms, which we explicitly constructed from the periods, without having much of an independent theory about them. The simpler case of the torus suggests that such forms and their extensions should play a role in the study of virtually any physical amplitude–open or closed–in compactifications on the Calabi–Yau space, even as conjectured in the hypermultiplet sector [65].

One may finally wonder whether the topological string B-model is an integrable theory that is genuinely associated with this new and barely explored class of modular forms on Calabi–Yau spaces moduli spaces, whereas most known integrable models are associated with abelian varieties. As it

Table 13. n_d^g for the degree (6,2) complete intersection in $WCP^5(1^5, 3)$

g	d=1	d=2	d=3	d=4	d=5	d=6
0	4992	2388768	2732060032	4599616564224	957971384706240	22839268002374163616
1	0	-504	1228032	7927566480	633074010435840	3666182351842338408
2	0	-4	14976	-13098688	3921835430016	128614837503143532
3	0	0	0	87376	-5731751168	482407033529880
4	0	0	0	1456	-7098624	-3978452463012
5	0	0	0	10	-59904	1776341072
6	0	0	0	0	0	18680344
7	0	0	0	0	0	-7176
8	0	0	0	0	0	-36
9	0	0	0	0	0	0

Table 14. n_d^g for the degree (6,4) complete intersection in $\mathbb{WCP}^5(1^3, 2^2, 3)$

g	d=1	d=2	d=3	d=4	d=5
0	15552	27904176	133884554688	950676829466832	8369111295497240640
1	8	258344	5966034472	126729436388624	2512147219945401752
2	0	128	36976576	4502079839576	264945385369932352
3	0	3	−64432	15929894952	9786781718701824
4	0	0	−48	−272993052	42148996229312
5	0	0	0	800065	−592538522344
6	0	0	0	1036	14847229472
7	0	0	0	15	−148759496
8	0	0	0	0	160128
9	0	0	0	0	96
10	0	0	0	0	0

Table 15. n_d^g for the degree (6,6) complete intersection in $\mathbb{WCP}^5(1^2, 2^2, 3^2)$

g	d=1	d=2	d=3	d=4
0	67104	847288224	28583248229280	1431885139218997920
1	360	40692096	4956204918600	616199133098321280
2	1	291328	254022248925	102984983365762128
3	0	−928	1253312442	6925290146728800
4	0	−6	−39992931	104226246583368
5	0	0	867414	−442845743788
6	0	0	−1807	53221926192
7	0	0	−10	−3192574724
8	0	0	0	111434794
9	0	0	0	−1752454
10	0	0	0	3054
11	0	0	0	12
12	0	0	0	0

was noted in [14, 28, 29] in the complex moduli space extended by the dilaton, called extended phase space, one has rigid special Kähler geometry and many aspects of the sympletic transformations and its metaplectic realization are easily understood in the extended phase space. There are two maps $\Phi^{(i)} : \mathcal{M} \rightarrow T_{IJ}^{(k)}$, $I, J = 1, \ldots, h^3/2$ from the complex moduli space to tensors in the extended phase space on which $\begin{pmatrix} A & B \\ C & D \end{pmatrix} \in \mathrm{Sp}(h^3, \mathbb{Z})$ acts projectively like $T^{(k)} \mapsto (AT^{(k)} + B)(CT^k + D)^{-1}$. For the holomorphic object $\tau_{IJ} = \partial_I \partial_J F^{(0)} =: T_{IJ}^{(1)}(t)$, which is mostly discussed in this context of the metaplectic transfomations [14, 28, 29], Im(τ) is indefinite, while for the non-holomorphic object $\mathcal{N}_{IJ} = \bar{\tau}_{IJ} + 2i\mathrm{Im}\tau_{IK}X^K\mathrm{Im}\tau_{IL}X^L/X^L\mathrm{Im}\tau_{KL}X^L =: T_{IJ}^{(2)}(t, \bar{t})$ comes from the kinetic term in the 10d action whose reduction involves the Hodge-star on M. Its imaginary part Im$(\mathcal{N}) > 0$ is the kinetic term for the vector multiplets and is hence positive definite. In other words $\text{Im}\Phi^{(2)}$ defines a map to the Siegel upper space. $\Phi^{(2)}$ should relate Siegel modular forms for admittedly very exotic subgroups [24] of $\text{Sp}(4, \mathbb{Z})$ to Calabi–Yau amplitudes. Such Siegel modular forms for abelian varieties are also associated with $N = 2$ Seiberg–Witten (gauge) theories, while the modular forms on Calabi–Yau studied in this underline $N = 2$ exact terms in $N = 2$ supergravity. The map $\Phi^{(2)}$ could be a manifestation of a gravity-gauge theory correspondence for 4d theories with $N = 2$ supersymmetry.

It is no principal problem to generalize this to multi-moduli Calabi–Yau as long as the Picard–Fuchs equations are known. These have different, more general singularities with interesting local effective actions. In K3 fibrations which have at least two moduli, the modular properties are much better understood and in fact the ambiguity in the fibre is completely fixed by heterotic string calculations, see [69] for recent progress. Moreover these cases have $N = 2$ field theory limits, which contain further information, which might be sufficient to solve these models [66].

A Appendices

A.1 Classical Intersection Calculations Using the Adjunction Formula

The adjunction formula[19] for the total Chern class of a for dimension $m = n - r$ smooth complete intersections M of multi degree $d_1, \ldots, d_r$ in a weighted projective space $\mathbb{WCP}^n(w_1, \ldots, w_{n+1})$ is

$$c(T_M) = \sum_i c_i(T_M) = \frac{c(T_{\mathbb{WCP}})}{c(\mathcal{N})} = \frac{\prod_{i=1}^{n+1}(1 + w_i K)}{\prod_{k=1}^{r}(1 + d_k K)} = \sum_i c_i K^i \,, \qquad (111)$$

where $c(T_{\mathbb{WCP}}) = \sum_i c_i(T_{\mathbb{WCP}}) = \prod_{i=1}^{n+1}(1 + w_i K)$ is the total Chern class of the weighted projective space, K is its Kähler class and $c(\mathcal{N}) = \prod_{k=1}^{r}(1 + d_k K)$ is the total Chern class of the normal bundle.

Integration of a top form $\omega = x J^m$ with $J = K|_M$ over M is obtained by integration along the normal direction as

$$\int_M \omega = \int_{\mathbb{WCP}} \omega \wedge c_r(\mathcal{N}) = \frac{x}{\prod_{k=1}^{n+1} w_k} \prod_{k=1}^{r} d_k \,. \qquad (112)$$

Here we used the normalization $\int_{\mathbb{WCP}} K^n = \frac{1}{\prod_{k=1}^{n+1} w_k}$. This yields the first line below:

---

[19] See [67] for a pedagogical account of these matters.

$$\kappa = \int_M J^m = \frac{\prod_{k=1}^r d_k}{\prod_{i=1}^{r+1} w_i} \tag{113}$$

$$\chi = \int_M c_3(T_M) = \frac{c_3}{\prod_{k=1}^{n+1} w_k} \prod_{k=1}^r d_k \tag{114}$$

$$c = \frac{1}{24} \int_M c_2 \wedge J = \frac{1}{24} \frac{c_2}{\prod_{k=1}^{n+1} w_k} \prod_{k=1}^r d_k \tag{115}$$

$$a = \frac{1}{2} \int_M i_* c_1(D) \wedge J = \frac{1}{2} \int_{\mathrm{WCP}} \frac{c(T_M)}{(1+J)} \wedge J^{r+1}$$

$$= \frac{\mathrm{coeff}\left(\frac{c(T_M)}{(1+J)}, J^{m-1}\right)}{2 \cdot \prod_{k=1}^{n+1} w_k} \tag{116}$$

Combining (111, 112) one gets the line 2 and 3. The leading t terms in F_0 can be obtained by calculating $Z(M)$ using (38), while the last line follows from the calculation of $Z(D)$ assuming that the D_4-brane is supported on D the restriction of the hyperplane class[20] of WCP to M and the Gysins formula for smooth embeddings [67].

A.2 Tables of Gopakumar–Vafa Invariants

We list the tables of BPS invariants for all the Calabi–Yau models computed in this chapter.

A.3 Invariance of the Generators Under a Change of the Basis

We have seen that the topological strings can be written as polynomials of the generators v_1, v_2, v_3 and X. In the holomorphic limit, these generators can be computed from the first two solutions ω_0, ω_1 of the Picard–Fuchs equation. In the following we prove that under an arbitrary linear change of basis in the space spanned by ω_0 and ω_1, these generators and therefore the topological strings are actually invariant. This is true anywhere in the moduli space. In particular, this partly explains why the gap structure in the conifold expansion is not affected by a change of basis of ω_0 as we observed in all cases.

Since $X = \frac{1}{1-\psi}$ is independent of the basis ω_0 and ω_1, it is trivially invariant. In the holomorphic limit, The Kahler potential and metric go like $e^{-K} \sim \omega_0$ and $G_{\psi\bar\psi} \sim \partial_\psi t$, where $t = \omega_1/\omega_0$ is the mirror map. The generators u and v_i are related to A_i and B_i, which we recall were defined as

$$A_p := \frac{(\psi\partial_\psi)^p G_{\psi\bar\psi}}{G_{\psi\bar\psi}}, \quad B_p := \frac{(\psi\partial_\psi)^p e^{-K}}{e^{-K}}, \quad (p = 1, 2, 3, \cdots) \tag{117}$$

---

[20] a is physically less relevant, as it does not affect the effective action. Its value $a = \frac{11}{2}$ obtained for the quintic from (116) checks with [30].

So a different normalization of basis ω_0 of ω_1, as well as a change of basis in $\omega_1 \to \omega_1 + b_1\omega_0$ obviously do not change the generators A_i and B_i, and therefore the generators u and v_i are also invariant.

We now tackle the remaining less trivial situation, namely a change of basis in ω_0 as the following

$$\omega_0 \to \tilde{\omega}_0 = \omega_0 + b_1\omega_1 \tag{118}$$

where b_1 is an arbitrary constant. We denote the Kähler potential, metric, mirror map and various generators in the new basis by a tilde symbol. It is straightforward to relate them to variables in the original basis. We find the following relations for the mirror map

$$s = \frac{\omega_1}{\tilde{\omega}_0} = \frac{t}{1 + b_1 t}$$

$$\partial_\psi s = \frac{\partial_\psi t}{(1 + b_1 t)^2} \tag{119}$$

and the generators A and B

$$\tilde{A} = \frac{\psi \partial_\psi \tilde{G}_{\psi\bar{\psi}}}{\tilde{G}_{\psi\bar{\psi}}} = A - \frac{2 b_1 \psi \partial_\psi t}{1 + b_1 \psi}$$

$$\tilde{B} = \frac{\psi \partial_\psi \tilde{\omega}_0}{\tilde{\omega}_0} = B + \frac{b_1 \psi \partial_\psi t}{1 + b_1 \psi} \tag{120}$$

So we find the generators A and B, as well as the generator $u = B$ are *not* invariant under the change of basis (118). However, we recall the generator v_1 is defined as

$$v_1 = 1 + A + 2B \tag{121}$$

Using the equations in (120) we find the generator v_1 is invariant, namely $\tilde{v}_1 = v_1$. To see v_2 and v_3 are invariant, we use the derivative relations

$$\psi \partial_\psi v_1 = -v_1^2 - 2v_2 - (1 + r_0)X + v_1 X \tag{122}$$

$$\psi \partial_\psi v_2 = -v_1 v_2 + v_3 \tag{123}$$

where r_0 is a constant that appears in the relation of generator A_2 to lower generators. These derivative relations are exact and independent of the choice of the basis in asymptotic expansion. We have shown that v_1 and X in the first equations (122) are invariant under a change of the basis (118), therefore the generator v_2 appearing on the right-hand side must be also invariant. Applying the same logic to the second equation (123) we find that the generator v_3 is also invariant.

Our proof explains why a change of basis like (118) does not change the gap structure around the conifold point and seems to be related to the SL_2 orbit theorem of [53, 54]. Under a change of basis, the mirror map at the conifold point is $\tilde{t}_D = \frac{\omega_0 t_D}{\tilde{\omega}_0}$ and has the asymptotic leading behaviour $\tilde{t}_D \sim t_D \sim \mathcal{O}(\psi)$. Recall in the holomorphic limit, the conifold expansion is

$$F^{(g)}_{\text{conifold}} = \omega_0^{2(g-1)} \left(\frac{1-\psi}{\psi}\right)^{g-1} P_g(v_1, v_2, v_3, X). \tag{124}$$

As we have shown the generators v_i and therefore P_g are invariant, so in the new basis

$$\tilde{F}^{(g)}_{\text{conifold}} = \left(\frac{\tilde{\omega}_0}{\omega_0}\right)^{2(g-1)} F^{(g)}_{\text{conifold}} = \left(\frac{t_D}{\tilde{t}_D}\right)^{2(g-1)} F^{(g)}_{\text{conifold}}. \tag{125}$$

It is clear if there is a gap structure in one basis $F^{(g)}_{\text{conifold}} = \frac{(-1)^{g-1} B_{2g}}{2g(2g-2)t_D^{2g-2}} + \mathcal{O}(t_D^0)$, the same gap structure will be also present in the other basis,

$$\tilde{F}^{(g)}_{\text{conifold}} = \frac{(-1)^{g-1} B_{2g}}{2g(2g-2)\tilde{t}_D^{2g-2}} + \mathcal{O}(\tilde{t}_D^0). \tag{126}$$

The asymptotic expansion in sub-leading terms $\mathcal{O}(t_D^0)$ and $\mathcal{O}(\tilde{t}_D^0)$ will be different and can be computed by the relation between t_D and $\tilde{t}_D$.

Around the conifold point there is another power series solution to the Picard–Fuchs equation that goes like $\omega_2 \sim \mathcal{O}(\psi^2)$. We also observe that the gap structure is not affected by a change of the basis

$$\omega_0 \to \omega_0 + b_2 \omega_2 \tag{127}$$

It appears to be much more difficult to prove this observation, since now the generator v_i is not invariant under this change of basis. A proof of our observation would depend on the specific details of the polynomial P_g and probably requires a deeper conceptual understanding of the conifold expansion. We shall leave this for future investigation.

A.4 Symplectic Basis, Vanishing Cycles and Massless Particles

We can study in more details the analytic continuation of the symplectic basis of the periods to the orbifold point $\psi = 0$. For the four hypersurface cases and two other complete intersection models $X_{4,3}(1^5, 2)$ and $X_{6,2}(1^3, 2^2, 3)$, the indices a_i ($i = 1, 2, 3, 4$) of the Picard–Fuchs equation are not degenerate at the orbifold point, so there are four power series with the leading behaviour of ψ^{a_i} and the analytic continuation procedure is similar to the quintic case. In our physical explanation of the singularity structure of higher genus topological string amplitudes in these models, we claim that for the hypersurface cases there is no stable massless charged state around the orbifold point, whereas for models $X_{4,3}(1^5, 2)$ and $X_{6,2}(1^3, 2^2, 3)$ there are nearly massless charged states of mass $m \sim \psi^{a_2 - a_1}$. Since the charge and the mass of a D-brane wrapping cycle is determined by the mirror map parameter, which is the ratio of two symplectic periods, it is only possible to have massless particles if there is a rational linear combination of the periods with the leading behaviour of ψ^{a_k}, $k > 1$.

For a complete intersection of degree $(d_1, \cdots, d_r)$ in weighted projective space $\mathbb{WCP}^n(w_1, \cdots, w_{n+1})$ with non-degenerate indices a_k, the natural basis of solutions at the orbifold point is

$$\omega_k^{\mathrm{orb}} = \psi^{a_k} \frac{\prod_{i=1}^r \Gamma(d_i a_k)}{\prod_{i=1}^{n+1} \Gamma(w_i a_k)}$$

$$\sum_{n=0}^{\infty} (c_0 \psi)^n \frac{\prod_{i=1}^{n+1} \Gamma(w_i(n+a_k))}{\prod_{i=1}^r \Gamma(d_i(n+a_k))}, \quad k = 1, 2, 3, 4 \qquad (128)$$

where $c_0 = \prod_{i=1}^r d_i^{d_i} / \prod_{i=1}^{n+1} w_i^{w_i}$. We can write the sum of the series as a contour integral[21] enclosing the positive real axis and analytically continue to the negative real axis to relate the above basis of solutions to the known symplectic basis at infinity $\psi = \infty$. We find the following relation, generalizing the result for quintic case,

$$\omega_k^{\mathrm{orb}} = (2\pi i)^4 \frac{\prod_{i=1}^r \Gamma(d_i a_k)}{\prod_{i=1}^{n+1} \Gamma(w_i a_k)} \Big\{ \frac{\alpha_k F_0}{1 - \alpha_k} - \frac{\alpha_k F_1}{(1 - \alpha_k)^2}$$

$$+ \frac{\alpha_k[\kappa(1 + \alpha_k) - 2a(1 - \alpha_k)]}{2(1 - \alpha_k)^3} X_1$$

$$+ \frac{\alpha_k[12c(1 - \alpha_k)^2 + \kappa(1 + 4\alpha_k + \alpha_k^2)]}{6(1 - \alpha_k)^4} X_0 \Big\} \qquad (129)$$

here $\alpha_k = \exp(2\pi i a_k)$ and κ, c, a are from the classical intersection calculations in Appendix A.1. We find for all cases the symplectic form and Kahler potential have the same diagonal behaviour as the case of the quintic

$$\omega = dF_k \wedge dX_k = s_1 d\omega_1^{\mathrm{orb}} \wedge d\omega_4^{\mathrm{orb}} + s_2 d\omega_2^{\mathrm{orb}} \wedge d\omega_3^{\mathrm{orb}} \qquad (130)$$

and $e^{-K} = \sum_{k=1}^4 r_k \omega_k^{\mathrm{orb}} \overline{\omega_k^{\mathrm{orb}}}$, for some constants s_1, s_2 and r_k.

It is straightforward to invert the transformation (129) and study the asymptotic behaviour of the geometric symplectic basis (F_0, F_1, X_0, X_1) around the orbifold point. Generically a linear combination of the symplectic periods (F_0, F_1, X_0, X_1) is proportional to the power of ψ with the lowest index, namely, for generic coefficients c_1, c_2, c_3, c_4 we have

$$c_1 X_0 + c_2 X_1 + c_3 F_1 + c_4 F_0 \sim \omega_1^{\mathrm{orb}} \sim \psi^{a_1}. \qquad (131)$$

---

[21] Further useful properties of the periods of the one parameter models have been established in [68].

In order for massless particles to appear at the orbifold point, there must be a $Sp(4, \mathbb{Z})$ transformation of the symplectic basis such that one of the periods goes to zero faster than in the generic situation, namely we should have integer coefficients $n_k \in \mathbb{Z}$ satisfying

$$n_1 X_0 + n_2 X_1 + n_3 F_1 + n_4 F_0 \sim \omega_2^{\text{orb}} \sim \psi^{a_k}, \quad k > 1. \tag{132}$$

We find (132) is impossible for three of the hypersurface cases $X_5(1^5)$, $X_8(1^4, 4)$ and $X_{10}(1^3, 2, 5)$, but possible for the Sextic hypersurface $X_6(1^4, 2)$ and the complete intersection cases $X_{4,3}(1^5, 2)$ and $X_{6,2}(1^3, 2^2, 3)$. Specifically, for the Sextic hypersurface $X_6(1^4, 2)$, the condition for (132) with $k = 2$

$$n_1 + 3n_3 = 0, \quad 3n_1 + 4(n_2 + n_4) = 0 \tag{133}$$

whereas for the complete intersections $X_{4,3}(1^5, 2)$ and $X_{6,2}(1^3, 2^2, 3)$, the conditions for (132) are

$$n_1 + 3n_3 = 0, \quad n_1 = 4n_2 + 8n_4, \tag{134}$$

and

$$n_1 + 2n_3 = 0, \quad n_2 + \frac{7}{24}n_3 + n_4 = 0. \tag{135}$$

respectively.

This fact that there are massless integer charged states possible in the models $X_{4,3}(1^5, 2)$ and $X_{4,3}(1^5, 2)$ is entirely consistent with our physical picture, which explained the singular behaviour of the $F^{(g)}$ form the effective action point of view. What is very interesting is the fact that the period degeneration at the branch point of the hypersurface models, $X_5(1^5)$, $X_6(1^4, 2)$, $X_8(1^4, 4)$ and $X_{10}(1^3, 2, 5)$ is at genus zero very similar to the cases $X_{4,3}(1^5, 2)$ and $X_{6,2}(1^3, 2^2, 3)$. In particular the periods of the models have no logarithmic singularities. The leading behaviour of the higher genus expansion, which we obtain from the global properties, indicates that the BPS states, which are possibly massless by (133, 134, 135), are stable in $X_{4,3}(1^5, 2)$ and $X_{6,2}(1^3, 2^2, 3)$ models, but not stable in the $X_6(1^4, 2)$ model.

Acknowledgments

We thank M. Aganagic, V. Bouchard, T. Grimm, S. Katz, M. Kontsevich, M. Marino, C. Vafa, S. T. Yau and D. Zagier for discussions. Sheldon Katz helped us with the verifications of the BPS numbers and Cumrun Vafa with remarks on the draft. Don Zagier's comments on [12] triggered many ideas here. We thank the MSRI in Berkeley and AK thanks in particular the Simons Professorship Program. MH/AK thank also the Simons Workshops in Mathematics and Physics 05/06 for its hospitality.

References

1. A. Klemm and E. Zaslow, "Local mirror symmetry at higher genus," (1999) [arXiv:hep-th/9906046].
2. M. Aganagic, A. Klemm, M. Marino and C. Vafa, "The topological vertex," Commun. Math. Phys. **254**, 425 (2005) [arXiv:hep-th/0305132].
3. E. Witten, "Chern-simons gauge theory as a string theory," Prog. Math. **133**, 637 (1995) [arXiv:hep-th/9207094].
4. R. Gopakumar and C. Vafa, "On the gauge theory/geometry correspondence," Adv. Theor. Math. Phys. **3**, 1415 (1999) [arXiv:hep-th/9811131].
5. R. Dijkgraaf and C. Vafa, "Matrix models, topological strings, and supersymmetric gauge theories," Nucl. Phys. B **644**, 3 (2002) [arXiv:hep-th/0206255].
6. M. Aganagic, R. Dijkgraaf, A. Klemm, M. Marino and C. Vafa, "Topological strings and integrable hierarchies," Commun. Math. Phys. **261**, 451 (2006) [arXiv:hep-th/0312085].
7. C. Vafa, "Two dimensional Yang-Mills, black holes and topological strings," (2004) arXiv:hep-th/0406058.
8. H. Ooguri, A. Strominger and C. Vafa, "Black hole attractors and the topological string," Phys. Rev. D **70**, 106007 (2004) [arXiv:hep-th/0405146].
9. H. Ooguri, C. Vafa and E. P. Verlinde, "Hartle-Hawking wave-function for flux compactifications," Lett. Math. Phys. **74**, 311 (2005) [arXiv:hep-th/0502211].
10. K. Costello, "Topological conformal field theories and Calabi-Yau categories," Adv. Math. 210 (2007), no. 1, 165–214. [math.QA/0412149].
11. M. Kontsevich, "TQFTs and geometry of pre-Frobenius manifolds" and "A_∞ categories and their (co)homology theories," Lecures at the "Erwin Schroedinger Institute," Vienna 2006.
12. M.-x. Huang and A. Klemm, "Holomorphic anomaly in gauge theories and matrix models," JHEP 0709:054 (2007) [arXiv:hep-th/0605195].
13. M.-x. Huang and A. Klemm, Modularity versus Holomarphicity in gauge theories and matrix models, to appear.
14. M. Aganagic, V. Bouchard and A. Klemm, "Topological strings and (almost) modular forms," Commun. Math. Phys. 277:771–819 (2008) [arXiv:hep-th/0607100].
15. R. Dijkgraaf, "Mirror Symmetry and elliptic curves," in *The moduli Space of Curves*, Progr. Math. **129**, 149 (1995); K. Kaneko and D. B. Zagier, "A generalized Jacobi theta functions and quasimodular forms," Progr. Math. **129**, 165 (1995).
16. N. Seiberg and E. Witten, "Electric - magnetic duality, monopole condensation, and confinement in N=2 supersymmetric Yang-Mills theory," Nucl. Phys. B **426**, 19 (1994) [Erratum-ibid. B **430**, 485 (1994)] [arXiv:hep-th/9407087].
17. M. Bershadsky, S. Cecotti, H. Ooguri and C. Vafa, "Kodaira-Spencer theory of gravity and exact results for quantum string amplitudes," Commun. Math. Phys. **165**, 311 (1994) [arXiv:hep-th/9309140].
18. D. Ghoshal and C. Vafa, "C = 1 string as the topological theory of the conifold," Nucl. Phys. B **453**, 121 (1995) [arXiv:hep-th/9506122].
19. M. Marino and G. W. Moore, "Counting higher genus curves in a Calabi-Yau manifold," Nucl. Phys. B **543**, 592 (1999) [arXiv:hep-th/9808131].
20. S. Hosono, M. H. Saito and A. Takahashi, Adv. Theor. Math. Phys. **3** (1999) 177 [arXiv:hep-th/9901151].

21. S. Hosono, M. H. Saito and A. Takahashi, Adv. Theor. Math. Phys. 3:177–208 (1999). [arXiv:math.ag/0105148].
22. M. Bershadsky, S. Cecotti, H. Ooguri and C. Vafa, "Holomorphic anomalies in topological field theories," Nucl. Phys. B **405**, 279 (1993) [arXiv:hep-th/9302103].
23. K. Hori, S. Katz, A. Klemm, R. Pandharipande, R. Thomas, C. Vafa, R. Vavil and E. Zaslow, "*Mirror Symmetry*", Am. Math. Soc. (2003).
24. Y.-H. Chen, Y. Yang, N. Yui, "Monodromy of Picard-Fuchs differential equations for Calabi-Yau threefolds," J. Reine Angew. Math. 616 (2008) 167–203. [arXiv:math.AG/0605675].
25. S. Katz, A. Klemm and C. Vafa, "M-theory, topological strings and spinning black holes," Adv. Theor. Math. Phys. **3**, 1445 (1999) [arXiv:hep-th/9910181].
26. E. Witten, "Two-Dimensional Gravity and intersection theory on moduli spaces," Surv. in Diff. Geom. **1**, 243 (1991).
27. E. Witten, "Quantum background independence in string theory," Nucl. Phys. B **372**, 187 (1992), [arXiv:hep-th/9306122].
28. E. P. Verlinde, "Attractors and the holomorphic anomaly," (2004) arXiv:hep-th/0412139.
29. M. Gunaydin, A. Neitzke and B. Pioline, "Topological wave functions and heat equations," JHEP 0612:070, (2006) arXiv:hep-th/0607200.
30. S. Hosono, A. Klemm, S. Theisen and S. T. Yau, "Mirror symmetry, mirror map and applications to complete intersection Calabi-Yau spaces," Nucl. Phys. B **433**, 501 (1995) [arXiv:hep-th/9406055].
31. A. Strominger, "Massless black holes and conifolds in string theory," Nucl. Phys. B **451**, 96 (1995) [arXiv:hep-th/9504090].
32. C. Vafa, "A Stringy test of the fate of the conifold," Nucl. Phys. B **447**, 252 (1995) [arXiv:hep-th/9505023].
33. R. Gopakumar and C. Vafa, "Branes and fundamental Groups," Adv. Theor. Math. Phys. 2:399–411, (1998) [arXiv:hep-th/9712048].
34. A. Klemm and P. Mayr, "Strong coupling singularities and non-Abelian gauge symmetries in N = 2 string theory," Nucl. Phys. B **469**, 37 (1996) [arXiv:hep-th/9601014].
35. R. Gopakumar and C. Vafa, "M-theory and topological strings. I," (1998) [arXiv:hep-th/9809187].
36. R. Gopakumar and C. Vafa, "M-theory and topological strings. II," (1998) [hep-th/9812127].
37. S. Yamaguchi and S. T. Yau, "Topological string partition functions as polynomials," J. High Energy Phys. **0407**, 047 (2004) [arXiv:hep-th/0406078].
38. P. Candelas, X. C. De La Ossa, P. S. Green and L. Parkes, "A pair of Calabi-Yau manifolds as an exactly soluble superconformal theory," Nucl. Phys. B **359**, 21 (1991).
39. A. Ceresole, R. D'Auria, S. Ferrara and A. Van Proeyen, "Duality transformations in supersymmetric Yang-Mills theories coupled to supergravity," Nucl. Phys. B **444**, 92 (1995) [arXiv:hep-th/9502072].
40. R. Minasian and G. W. Moore, "K-theory and Ramond-Ramond charge," J. High Energy Phys. **9711**, 002 (1997) [arXiv:hep-th/9710230].
41. Y. K. Cheung and Z. Yin, "Anomalies, branes, and currents," Nucl. Phys. B **517**, 69 (1998) [arXiv:hep-th/9710206].
42. I. Brunner, M. R. Douglas, A. E. Lawrence and C. Romelsberger, "D-branes on the quintic," J. High Energy Phys. **0008**, 015 (2000) [arXiv:hep-th/9906200].

43. D. E. Diaconescu and C. Romelsberger, "D-branes and bundles on elliptic fibrations", Nucl. Phys. B **574**, 245 (2000) [arXiv:hep-th/9910172].
44. P. Mayr, "Phases of supersymmetric D-branes on Kaehler manifolds and the McKay correspondence," J. High Energy Phys. **0101**, 018 (2001) [arXiv:hep-th/0010223].
45. http://uw.physics.wisc.edu/~strings/aklemm/highergenusdata/
46. C. Faber and R. Pandharipande, "Hodge Integrals and Gromov-Witten theory," Invent. Math. 139 (2000), no. 1, 173–199. [arXiv:math.ag/9810173].
47. W. Chen and Y. Ruan, "Orbifold Gromow-Witten theory", Orbifolds in mathmatics and physics Madison WI 2001, Contemp. Math., **310**, Amer. Math. Soc., Providence, RI, 2002 25–85.
48. J. Bryan, T. Graber, R. Pandharipande, "The orbifold quantum cohomology of C^2/Z_3 and Hurwitz-Hodge integrals," J. Algebraic Geom. 17 (2008), no. 1, 1–28. [arXiv:math.AG/0510335].
49. T. Coates, A. Corti, H. Iritani and H.-H. Tseng, "Wall-Crossing in Toric Gromow-Witten Theory I: Crepant Examples", (2006) [arXiv:math.AG/0611550].
50. A. B. Giventhal, "Symplectic geometry of Frobenius structures," Frobenius manifolds, Aspects. Math., E36, Vieweg, Wiesbaden, (91–112) 2004, T. Coates, "Giventhals Lagrangian Cone and S^1-Equivariant Gromow-Witten Theory," [arXiv:math.AG.0607808].
51. A. Klemm and S. Theisen, "Considerations of one modulus Calabi-Yau compactifications: Picard-Fuchs equations, Kahler potentials and mirror maps," Nucl. Phys. B **389**, 153 (1993) [arXiv:hep-th/9205041].
52. V. I. Arnold, S. M. Gusein-Zade and A. N. Varchencko, "Singularities of diefferential maps I & II," Birkhäuser, Basel (1985).
53. W. Schmidt, "Variaton of Hodge structure: The singularities of the period mapping," Inventiones Math. **22** (1973) 211–319.
54. E. Cattani, A. Kaplan and W. Schmidt, "Degenerarions of Hodge structures," Ann. of Math. **123** (1986) 457–535.
55. C. Doran and J Morgan, "Mirror Symmetry. V, 517–537, AMS/IP Stud. Adv. Math., 38, Amer. Math. Soc., Providence, RI, (2006).
56. A. Klemm, M. Kreuzer, E. Riegler and E. Scheidegger, "Topological string amplitudes, complete intersection Calabi-Yau spaces and threshold corrections," J. High Energy Phys. **0505**, 023 (2005) [arXiv:hep-th/0410018].
57. D. Maulik, N. Nekrasov, A. Okounkov and R. Pandharipande, "Gromov-Witten theory and Donaldson-Thomas theory I," Compos. Math. 142 (2006), no. 5, 1263–1285 [arXiv:math.AG/0312059].
58. J. Harris, "Curves in Projective Space," Sem. de Math. Sup. Les Presses de l'University de Monteral (1982).
59. S. Katz, S. Strømme and J-M. Økland: "Schubert: A Maple package for intersection theory and enumerative geometry," (2001) http://www.uib.no/People/nmasr/schubert/.
60. D. Gaiotto, A. Strominger and X. Yin, "The M5-brane elliptic genus: Modularity and BPS states," JHEP 0708:070 (2007) [arXiv:hep-th/0607010].
61. J. de Boer, M. C. N. Cheng, R. Dijkgraaf, J. Manschot and E. Verlinde, "A farey tail for attractor black holes," J. High Energy Phys. **0611**, 024 (2006) [arXiv:hep-th/0608059].
62. F. Denef and G. Moore, "Split states, entropy enigmas, holes and halos" hep-th/0702146

63. T. Bridgeland, "Derived categories of coherent sheaves," Survey article for ICM 2006, [arXiv:math.AG/0602129].
64. P. S. Aspinwall, "D-branes on Calabi-Yau manifolds," Boulder (2003), Progress in string theory, 1–152 [arXiv:hep-th/0403166].
65. M. Rocek, C. Vafa and S. Vandoren, "Hypermultiplets and topological strings," J. High Energy Phys. **0602**, 062 (2006) [arXiv:hep-th/0512206].
66. T. W. Grimm, A. Klemm, M. Marino and M. Weiss, "Direct integration of the topological string", JHEP **0708**, 058 (2007).
67. W. Fulton, "Intersection Theory," Springer (1998).
68. C. I. Lazaroiu, "Collapsing D-branes in one-parameter models and small/large radius duality," Nucl. Phys. B **605**, 159 (2001) [arXiv:hep-th/0002004].
69. A. Klemm and M. Marino, "Counting BPS states on the Enriques Calabi-Yau," Commun. Math. Phys. 280:27–76, (2008) [arXiv:hep-th/0512227].

Gauge Theory, Mirror Symmetry, and the Geometric Langlands Program

A. Kapustin

California Institute of Technology, Pasadena, CA 91125, USA

Abstract. I provide an introduction to the recent work on the Montonen–Olive duality of $\mathcal{N} = 4$ super-Yang–Mills theory and the Geometric Langlands Program.

1 Introduction

The Langlands Program is a far-reaching collection of theorems and conjectures about representations of the absolute Galois group in certain fields. For a recent accessible review see [1]. V. Drinfeld and G. Laumon [2] introduced a geometric analogue which deals with representations of the fundamental group of a Riemann surface C, or, more generally, with equivalence classes of homomorphisms from $\pi_1(C)$ to a reductive algebraic Lie group $G_{\mathbb{C}}$ (which we think of as a complexification of a compact reductive Lie group G). From the geometric viewpoint, such a homomorphism corresponds to a flat connection on a principal $G_{\mathbb{C}}$ bundle over C. The Geometric Langlands Duality associates to an irreducible flat $G_{\mathbb{C}}$ connection a certain D-module on the moduli stack of holomorphic $^L G$-bundles on C. Here $^L G$ is, in general, a different compact reductive Lie group called the Langlands dual of G. The group $^L G$ is defined by the condition that the lattice of homomorphisms from $U(1)$ to a maximal torus of G be isomorphic to the weight lattice of $^L G$. For example, the dual of $SU(N)$ is $SU(N)/\mathbb{Z}_N$, the dual of $Sp(N)$ is $SO(2N + 1)$, while the groups $U(N), E_8, F_4$ and G_2 are self-dual.

The same notion of duality for Lie groups appeared in the work of Goddard et al. on the classification of magnetic sources in gauge theory [3]. These authors found that magnetic sources in a gauge theory with gauge group G are classified by irreducible representations of the group $^L G$. On the basis of this, Montonen and Olive [4] conjectured that Yang–Mills theories with gauge groups G and $^L G$ might be isomorphic on the quantum level. Later Osborn [5] noticed that the Montonen–Olive conjecture is more likely to hold for $\mathcal{N} = 4$ supersymmetric version of the Yang–Mills theory. There is currently much circumstantial evidence for the MO conjecture, but no proof.

Kapustin, A.: *Gauge Theory, Mirror Symmetry, and the Geometric Langlands Program.* Lect. Notes Phys. **757**, 103–124 (2009)
DOI 10.1007/978-3-540-68030-7_4

It has been suggested by Atiyah soon after the work of Goddard et al. that Langlands duality might be related to the MO duality, but only recently the precise relation has been found [6]. In these lectures I will try to explain the main ideas of [6]. For a detailed derivations and a more extensive list of references the reader is referred to the original study. I will not discuss the ramified version of Geometric Langlands Duality; for that the reader is referred to [7].

2 Montonen–Olive Duality

I will begin by reviewing the Montonen–Olive conjecture. Consider first non-supersymmetric Yang–Mills theory with an action

$$I(A) = \int_X \left(-\frac{1}{e^2} \operatorname{Tr} F \wedge \star F + \frac{i\theta}{8\pi^2} \operatorname{Tr} F \wedge F \right)$$

Here X is a Riemannian four-manifold, $F = dA + A^2 \in \Omega^2(\mathrm{ad}(E))$ is the curvature 2-form of a connection $A = A_\mu dx^\mu$ on a principal G-bundle E over X and G is a compact reductive Lie group. The theory has two real parameters, e and θ. Since θ is the coefficient of a topological invariant (the instanton number), it does not enter the classical equations of motion obtained by varying the above action. In fact, neither does e, since the equations of motion read

$$D \star F = 0,$$

where D stands for covariant derivative. The Bianchi identity

$$DF = 0$$

is also independent of e and θ. But on the quantum level one should consider a path integral

$$Z = \sum_E \int \mathcal{D}A \, e^{-I(A)},$$

and it does depend on both e and θ. Note that in quantum theory one sums over all isomorphism classes of E.

The gauge field A can be coupled naturally to sources which transform in a (unitary) representation R of G. These are so-called electric sources, as they create a Coulomb-like field of the form

$$A_0^a \sim T^a \frac{e^2}{r}$$

where $T^a, a = 1, \dim G$, are generators of G in the representation R. Here we took $X = \mathbb{R} \times \mathbb{R}^3$, r is the distance from the origin in $\mathbb{R}^3$, and we assumed that the worldline of the source is given by $r = 0$.[1] It was noticed by Goddard et al. [3] that magnetic sources are labelled by irreducible representations of a different group which they called the magnetic gauge group. As a matter of fact, the magnetic gauge group coincides with the Langlands dual of G, so we will denote it $^L G$. A static magnetic source in Yang–Mills theory should create a field of the form

$$F = \star_3 \, d\left(\frac{\mu}{2r}\right),$$

where μ is an element of the Lie algebra $\mathfrak{g}$ of G defined up to adjoint action of G and $\star_3$ is the Hodge star operator on $\mathbb{R}^3$. Goddard et al. showed that μ is "quantized". More precisely, using gauge freedom one can assume that μ lies in a particular Cartan subalgebra $\mathfrak{t}$ of $\mathfrak{g}$, and then it turns out that μ must lie in the coweight lattice of G, which, by definition, is the same as the weight lattice of $^L G$.[2] Furthermore, μ is defined up to an action of the Weyl group, so possible values of μ are in one-to-one correspondence with highest weights of $^L G$.

On the basis of this observation, Montonen and Olive [4] conjectured that Yang–Mills theories with gauge groups G and $^L G$ are isomorphic on the quantum level and that this isomorphism exchanges electric and magnetic sources. Thus the Montonen–Olive duality is a non-abelian version of electric-magnetic duality in Maxwell theory.

In order for the energy of electric and magnetic sources to transform properly under MO duality, one has to assume that for $\theta = 0$ the dual gauge coupling is

$$\tilde{e}^2 = \frac{16\pi^2 n_{\mathfrak{g}}}{e^2}.$$

Here the integer $n_{\mathfrak{g}}$ is 1, 2, or 3 depending on the maximal multiplicity of edges in the Dynkin diagram of $\mathfrak{g}$ [8, 9]; for simply laced groups $n = 1$. This means that MO duality exchanges weak coupling ($e \to 0$) and strong coupling ($e \to \infty$). For this reason, it is extremely hard to prove the MO duality conjecture. For general θ, it is useful to define

$$\tau = \frac{\theta}{2\pi} + \frac{4\pi i}{e^2}$$

The parameter τ takes values in the upper half-plane and under MO duality transforms as

$$\tau \to -\frac{1}{n_{\mathfrak{g}}\tau} \tag{1}$$

---

[1] One should think of A_0 as an operator acting in the enlarged Hilbert space which is the tensor product of the Hilbert space of the gauge fields and the representation space on which T acts.

[2] The coweight lattice of G is defined as the lattice of homomorphisms from $U(1)$ to a maximal torus T of G. The weight lattice of G is the lattice of homomorphisms from T to $U(1)$.

The Yang–Mills theory has another, much more elementary symmetry, which does not change the gauge group:

$$\tau \rightarrow \tau + k.$$

Here k is an integer which depends on the geometry of X and G. For example, if $X = \mathbb{R}^4$, then $k = 1$ for all G. Together with the MO duality, these transformations generate some subgroup of $PSL(2, \mathbb{R})$. In what follows we will mostly set $\theta = 0$ and will discuss only the $\mathbb{Z}_2$ subgroup generated by the MO duality.

To summarize, if the MO duality were correct, then the partition function would satisfy

$$Z(X, G, \tau) = Z(X, {}^L G, -\frac{1}{n_{\mathfrak{g}} \tau}).$$

Of course, the partition function is not a very interesting observable. Isomorphism of two quantum field theories means that we should be able to match all observables in the two theories. That is, for any observable O in the gauge theory with gauge group G we should be able to construct an observable $\tilde{O}$ in the gauge theory with gauge group ${}^L G$ so that all correlators agree:

$$\langle O_1 \ldots O_n \rangle_{X,G,\tau} = \langle \tilde{O}_1 \ldots \tilde{O}_n \rangle_{X,{}^L G,-1/(n_{\mathfrak{g}} \tau)}.$$

At this point I should come clean and admit that the MO duality as stated above is not correct. The most obvious objection is that the parameters e and $\tilde{e}$ are renormalized and the relation like (1) is not compatible with renormalization. However, it was pointed out later by Osborn [5] (who was building on the work of Witten and Olive [10]) that the duality makes much more sense in $\mathcal{N} = 4$ super-Yang–Mills theory. This is a maximally supersymmetric extension of Yang–Mills theory in four dimensions and it has a remarkable property that the gauge coupling is not renormalized at any order in perturbation theory. Furthermore, Osborn was able to show that certain magnetically charged solitons in $\mathcal{N} = 4$ SYM theory have exactly the same quantum numbers as gauge bosons. (The argument assumes that the vacuum breaks spontaneously the gauge group G down to its maximal abelian subgroup, so that both magnetically charged solitons and the corresponding gauge bosons are massive.) Later strong evidence in favour of the MO duality for $\mathcal{N} = 4$ SYM was discovered by Sen [11] and Vafa and Witten [12]. Nowadays MO duality is often regarded as a consequence of string dualities. One particular derivation which works for all G is explained in [13]. Nevertheless, the MO duality is still a conjecture, not a theorem. In what follows we will assume its validity and deduce from it the main statements of the Geometric Langlands Program.

Apart from the connection 1-form A, $\mathcal{N} = 4$ SYM theory contains six scalar fields $\phi_i, i = 1, \ldots, 6$, which are sections of ad(E), four spinor fields $\bar{\lambda}_a, a = 1, \ldots, 4$, which are sections of ad$(E) \otimes \mathcal{S}_-$ and four spinor fields $\lambda^a, a = 1, \ldots, 4$ which are sections of ad$(E) \otimes \mathcal{S}_+$. Here $\mathcal{S}_\pm$ are the two spinor bundles over X. The fields A and ϕ_i are bosonic (even), while the spinor fields are fermionic (odd). In Minkowski signature the fields $\bar{\lambda}_a$ and λ^a are complex-conjugate, but in Euclidean signature they are independent.

The action of $N = 4$ SYM theory has the form

$$I_{\mathcal{N}=4} = I_{YM} + \frac{1}{e^2} \int_X \left(\sum_i \mathrm{Tr}\, D\phi_i \wedge \star D\phi_i + \mathrm{vol}_X \sum_{i<j} \mathrm{Tr}[\phi^i, \phi^j]^2 \right) + \ldots$$

where dots denote terms depending on the fermions. The action has $Spin(6) \simeq SU(4)$ symmetry under which the scalars ϕ_i transform as a vector, the fields λ_a transform as a spinor and $\bar{\lambda}^a$ transform as the dual spinor. This symmetry is present for any Riemannian X and is known as the R-symmetry. If X is $\mathbb{R}^4$ with a flat metric, the action also has translational and rotational symmetries, as well as 16 supersymmetries $\bar{Q}_{a\alpha}$ and $Q^a_{\dot{\alpha}}$, where $a = 1, \ldots, 4$ and the $Spin(4)$ spinor indices α and $\dot{\alpha}$ run from 1 to 2. As is clear from the notation, $\bar{Q}_a$ and Q^a transform as spinors and dual spinors of the R-symmetry group $Spin(6)$; they also transform as spinors and dual spinors of the rotational group $Spin(4)$.

One can show that under the MO duality all bosonic symmetry generators are mapped trivially, while supersymmetry generators are multiplied by a τ-dependent phase:

$$\bar{Q}_a \to e^{i\phi/2} \bar{Q}_a, \quad Q^a \to e^{-i\phi/2} Q^a, \quad e^{i\phi} = \frac{|\tau|}{\tau}$$

This phase will play an important role in the next section.

3 Twisting $\mathcal{N} = 4$ Super-Yang–Mills Theory

In order to extract mathematical consequences of MO duality, we are going to turn $N = 4$ SYM theory into a topological field theory. The procedure for doing this is called topological twisting [14].

Topological twisting is a two-step procedure. On the first step, one chooses a homomorphism ρ from $Spin(4)$, the universal cover of the structure group of TX, to the R-symmetry group $Spin(6)$. This enables one to redefine how fields transform under $Spin(4)$. The choice of ρ is constrained by the requirement that after this redefinition some of supersymmetries become scalars, i.e. transform trivially under $Spin(4)$. Such supersymmetries survive when X is taken to be an arbitrary Riemannian manifold. In contrast, if we consider ordinary $\mathcal{N} = 4$ SYM on a curved X, it will have supersymmetry only if X admits a covariantly constant spinor.

It is easy to show that there are three inequivalent choices of ρ satisfying these constraints [12]. The one relevant for the Geometric Langlands Program is identifies $Spin(4)$ with the obvious $Spin(4)$ subgroup of $Spin(6)$. After redefining the spins of the fields accordingly, we find that one of the left-handed supersymmetries and one of the right-handed supersymmetries become scalars. We will denote them Q_l and Q_r respectively.

On the second step, one notices that Q_l and Q_r both square to zero and anticommute (up to a gauge transformation). Therefore one may pick any linear combination of Q_l and Q_r

$$Q = uQ_l + vQ_r,$$

and declare it to be a BRST operator. That is, one considers only observables which are annihilated by Q (and are gauge-invariant) modulo those which are Q-exact. This is consistent because any correlator involving Q-closed observables, one of which is Q-exact, vanishes. From now on, all observables are assumed to be Q-closed. Correlators of such observables will be called topological correlators.

Clearly, the theory depends on the complex numbers u, v only up to an overall scaling. Thus we get a family of twisted theories parametrized by the projective line $\mathbb{P}^1$. Instead of the homogenous coordinates u, v, we will mostly use the affine coordinate $t = v/u$ which takes values in $\mathbb{C} \cup \{\infty\}$. All these theories are diffeomorphism-invariant, i.e. do not depend on the Riemannian metric. To see this, one writes an action (which is independent of t) in the form

$$I = \{Q, V\} + \frac{i\Psi}{4\pi} \int_X \operatorname{Tr} F \wedge F$$

where V is a gauge-invariant function of the fields and Ψ is given by

$$\Psi = \frac{\theta}{2\pi} + \frac{t^2 - 1}{t^2 + 1} \frac{4\pi i}{e^2}$$

All the metric dependence is in V and since changing V changes the action by Q-exact terms, we conclude that topological correlators are independent of the metric.

It is also apparent that for fixed t topological correlators are holomorphic functions of Ψ and this dependence is the only way e^2 and θ may enter. In particular, for $t = i$ we have $\Psi = \infty$, independently of e^2 and θ. This means that for $t = i$ topological correlators are independent of e^2 and θ.

To proceed further, we need to describe the field content of the twisted theory. Since the gauge field A is invariant under $Spin(6)$ transformations, it is not affected by the twist. As for the scalars, four of them become components of a 1-form ϕ with values in $\operatorname{ad}(E)$ and the other two remain sections of $\operatorname{ad}(E)$; we may combine the latter into a complex scalar field σ which is a section of the complexification of $\operatorname{ad}(E)$. The fermionic fields in the twisted theory are a pair of 1-forms ψ and $\tilde{\psi}$, a pair of 0-forms η and $\tilde{\eta}$, and a 2-form χ, all taking values in the complexification of $\operatorname{ad}(E)$.

What makes the twisted theory manageable is that the path integral localizes on Q-invariant field configurations. One way to deduce this property is to note that as a consequence of metric-independence, semiclassical (WKB) approximation is exact in the twisted theory. Thus the path-integral localizes on absolute minima of the Euclidean action. On the other hand, such configurations are exactly Q-invariant configurations.

The condition of Q-invariance is a set of partial differential equations on the bosonic fields A, ϕ and σ. We will only state the equations for A and ϕ, since the equations for σ generically imply that $\sigma = 0$:

$$(F - \phi \wedge \phi + tD\phi)^+ = 0, \quad (F - \phi \wedge \phi - t^{-1}D\phi)^- = 0, \quad D \star \phi = 0.$$

Here subscripts $+$ and $-$ denote self-dual and anti-self-dual parts of a 2-form.

If t is real, these equations are elliptic. A case which will be of special interest is $t = 1$; in this case the equations can be rewritten as

$$F - \phi \wedge \phi + \star D\phi = 0, \quad D \star \phi = 0.$$

They resemble both the Hitchin equations in 2d [15] and the Bogomolny equations in 3d [16] (and reduce to them in special cases). The virtual dimension of the moduli space of these equations is zero, so the partition function is the only non-trivial observable if X is compact without boundary. However, for applications to the Geometric Langlands Program it is important to consider X which are non-compact and/or have boundaries.

Another interesting case is $t = i$. To understand this case, it is convenient to introduce a complex connection $\mathcal{A} = A + i\phi$ and the corresponding curvature $\mathcal{F} = d\mathcal{A} + \mathcal{A}^2$. Then the equations are equivalent to

$$\mathcal{F} = 0, \quad D \star \phi = 0.$$

The first of these equations is invariant under the complexified gauge transformations, while the second one is not. It turns out that the moduli space is unchanged if one drops the second equation and considers the space of solutions of the equation $\mathcal{F} = 0$ modulo $G_\mathbb{C}$ gauge transformations. More precisely, according to a theorem of Corlette [17], the quotient by $G_\mathbb{C}$ gauge transformations should be understood in the sense of Geometric Invariant Theory, i.e. one should distinguish stable and semistable solutions of $\mathcal{F} = 0$ and impose a certain equivalence relation on semistable solutions. The resulting moduli space is called the moduli space of stable $G_\mathbb{C}$ connections on X and will be denoted $\mathcal{M}_{flat}(G, X)$. Thus for $t = i$ the path integral of the twisted theory reduces to an integral over $\mathcal{M}_{flat}(G, X)$. This is an indication that twisted $\mathcal{N} = 4$ SYM with gauge group G has something to do with the study of homomorphisms from $\pi_1(X)$ to $G_\mathbb{C}$.

Finally, let us discuss how MO duality acts on the twisted theory. The key observation is that MO duality multiplies Q_l and Q_r by $e^{\pm i\phi/2}$, and therefore multiplies t by a phase:

$$t \mapsto \frac{|\tau|}{\tau} t.$$

Since $\mathrm{Im}\tau \neq 0$, the only points of the $\mathbb{P}^1$ invariant under the MO duality are the "poles" $t = 0$ and $t = \infty$. On the other hand, if we take $t = i$ and $\theta = 0$, then the MO duality maps it to a theory with $t = 1$ and $\theta = 0$ (it also replaces G with ^LG). As we will explain below, it is this special case of the MO duality that gives rise to the Geometric Langlands Duality.

4 Reduction to Two Dimensions

From now on we specialize to the case $X = \Sigma \times C$ where C and Σ are Riemann surfaces. We will assume that C has no boundary and has genus $g > 1$, while Σ may have a boundary. In our discussion we will mostly work locally on Σ and its global structure will be unimportant.

Topological correlators are independent of the volumes of C and Σ. However, to exploit localization, it is convenient to consider the limit in which the volume of C goes to zero. In the spirit of the Kaluza–Klein reduction, we expect that in this limit the 4d theory becomes equivalent to a 2d theory on Σ. In the untwisted theory, this equivalence holds only in the limit $\mathrm{vol}(C) \to 0$, but in the twisted theory the equivalence holds for any volume.

It is easy to guess the effective field theory on Σ. One begins by considering the case $\Sigma = \mathbb{R}^2$ and requiring the field configuration to be independent of the coordinates on Σ and to have zero energy. One can show that such a generic configuration has $\sigma = 0$, while ϕ and A are pulled-back from C and satisfy

$$F - \phi \wedge \phi = 0, \quad D\phi = 0, \quad D \star \phi = 0.$$

Here all quantities as well as the Hodge star operator refer to objects living on C. These equations are known as Hitchin equations [15] and their space of solutions modulo gauge transformations is called the Hitchin moduli space $\mathcal{M}_H(G, C)$. The space $\mathcal{M}_H(G, C)$ is a non-compact manifold of dimension $4(g - 1) \dim G$ with singularities.[3] From the physical viewpoint, $\mathcal{M}_H(G, C)$ is the space of classical vacua of the twisted $\mathcal{N} = 4$ SYM on $C \times \mathbb{R}^2$.

In the twisted theory, only configurations with vanishingly small energies contribute. In the limit $\mathrm{vol}(C) \to 0$, such configurations will be represented by slowly varying maps from Σ to $\mathcal{M}_H(G, C)$. Therefore we expect the effective field theory on Σ to be a topological sigma-model with target $\mathcal{M}_H(G, C)$.

Before we proceed to identify more precisely this topological sigma-model, let us note that $\mathcal{M}_H(G, C)$ has singularities coming from solutions of Hitchin equations which are invariant under a subgroup of gauge transformations. In the neighbourhood of such a classical vacuum, the effective field theory is not equivalent to a sigma-model, because of unbroken gauge symmetry. In fact, it is difficult to describe the physics around such vacua in purely 2d terms. We will avoid this difficulty by imposing suitable conditions on the boundary of Σ ensuring that we stay away from such dangerous points.

---

[3] We assumed $g > 1$ precisely to ensure that virtual dimension of $\mathcal{M}_H(G, C)$ is positive.

The most familiar examples of topological sigma-models are A-and B-models associated with a Calabi–Yau manifold M [18]. Both the models are obtained by twisting a supersymmetric sigma-model with target M. Both the models are topological field theories (TFTs), in the sense that correlators do not depend on the metric on Σ. In addition, the A-model depends on the symplectic structure of M, but not on its complex structure, while the B-model depends on the complex structure of M, but not on its symplectic structure.

As explained above, we expect that our family of 4d TFTs, when considered on a 4-manifold of the form $\Sigma \times C$, becomes equivalent to a family of topological sigma-models with target $\mathcal{M}_H(G, C)$. To connect this family to ordinary A-and B-models, we note that $\mathcal{M}_H(G, C)$ is a (non-compact) hyper-Kähler manifold. That is, it has a $\mathbb{P}^1$ worth of complex structures compatible with a certain metric. This metric has the form

$$ds^2 = -\int_C \text{Tr} \left(\delta A \wedge \star \delta A + \delta \phi \wedge \star \delta \phi \right)$$

where $(\delta A, \delta \phi)$ is a solution of the linearized Hitchin equations representing a tangent vector to $\mathcal{M}_H(G, C)$. If we parameterize the sphere of complex structures by a parameter $w \in \mathbb{C} \cup \{\infty\}$, the basis of holomorphic differentials is

$$\delta A_{\bar{z}} - w\,\delta\phi_{\bar{z}}, \quad \delta A_z + w^{-1}\delta\phi_z.$$

By varying w, we get a family of B-models with target $\mathcal{M}_H(G, C)$. Similarly, since for each w we have the corresponding Kähler form on $\mathcal{M}_H(G, C)$, by varying w we get a family of A-models with target $\mathcal{M}_H(G, C)$. However, the family of topological sigma-models obtained from the twisted $\mathcal{N} = 4$ SYM does not coincide with either of these families. The reason for this is that a generic A-model or B-model with target $\mathcal{M}_H(G, C)$ depends on the complex structure on C and therefore cannot arise from a TFT on $\Sigma \times C$.

As explained in [6], this puzzle is resolved by recalling that for a hyper-Kähler manifold M there are twists other than ordinary A or B twists. In general, twisting a supersymmetric sigma-model requires picking two complex structures on the target. If we are given a Kähler structure on M, one can choose the two complex structures to be the same (B-twist) or opposite (A-twist). But for a hyper-Kähler manifold there is a whole sphere of complex structures and by independently varying the two complex structures one gets $\mathbb{P}^1 \times \mathbb{P}^1$ worth of 2d TFTs. They are known as generalized topological sigma-models, since their correlators depend on a generalized complex structure on the target M [19, 20]. The notion of a generalized complex structure was introduced by Hitchin [21] and it includes complex and symplectic structures are special cases.

It turns out that for $M = \mathcal{M}_H(G, C)$ there is a one-parameter subfamily of this two-parameter family of topological sigma-models which does not depend on the complex structure or Kähler form of C. It is this subfamily which appears as a reduction of the twisted $\mathcal{N} = 4$ SYM theory. Specifically, the two complex structures on $\mathcal{M}_H(G, C)$ are given by

$$w_+ = -t, \quad w_- = t^{-1}.$$

Note that one gets a B-model if and only if $w_+ = w_-$, i.e. if $t = \pm i$. One gets an A-model if and only if t is real. All other values of t correspond to generalized topological sigma-models.

Luckily to understand Geometric Langlands Duality we mainly need the two special cases $t = i$ and $t = 1$. The value $t = i$ corresponds to a B-model with complex structure J defined by complex coordinates

$$A_z + i\phi_z, \quad A_{\bar{z}} + i\phi_{\bar{z}}$$

on $\mathcal{M}_H(G, C)$. These are simply components of the complex connection $\mathcal{A} = A + i\phi$ along C. In terms of this complex connection two out of three Hitchin equations are equivalent to

$$\mathcal{F} = d\mathcal{A} + \mathcal{A}^2 = 0.$$

This equation is invariant under complexified gauge transformations. The third equation $D \star \phi = 0$ is invariant only under G gauge transformations. By a theorem of Donaldson [22], one can drop this equation at the expense of enlarging the gauge group from G to $G_{\mathbb{C}}$. (More precisely, one also has to identify certain semistable solutions of the equation $\mathcal{F} = 0$.) This is analogous to the situation in four dimensions. Thus in complex structure J the moduli space $\mathcal{M}_H(G, C)$ can be identified with the moduli space $\mathcal{M}_{flat}(G, C)$ of stable flat $G_{\mathbb{C}}$ connections on C. It is apparent that J is independent of the complex structure on C, which implies that the B-model at $t = i$ is also independent of it.

The value $t = 1$ corresponds to an A-model with a symplectic structure

$$\omega_K = \frac{2}{e^2} \int_C \mathrm{Tr}\, \delta\phi \wedge \delta A.$$

It is a Kähler form of a certain complex structure K on $\mathcal{M}_H(G, C)$. Note that ω_K is exact and independent of the complex structure on C.

Yet another complex structure on $\mathcal{M}_H(G, C)$ is $I = JK$. It will make an appearance later, when we discuss Homological Mirror Symmetry for $\mathcal{M}_H(G, C)$. In this complex structure, $\mathcal{M}_H(G, C)$ can be identified with the moduli space of *stable holomorphic Higgs bundles*. Recall that a (holomorphic) Higgs bundle over C (with gauge group G) is a holomorphic G-bundle E over C together with a holomorphic section φ of $\mathrm{ad}(E)$. In complex dimension one, any principal G bundle can be thought of as a holomorphic bundle, and Hitchin equations imply that $\varphi = \phi^{1,0}$ satisfies

$$\bar{\partial}\varphi = 0.$$

This gives a map from $\mathcal{M}_H(G, C)$ to the set of gauge-equivalence classes of Higgs bundles. This map becomes one-to-one if we limit ourselves to stable or semistable Higgs bundles and impose a suitable equivalence relationship on semistable ones. This gives an isomorphism between $\mathcal{M}_H(G, C)$ and $\mathcal{M}_{Higgs}(G, C)$.

It is evident that the complex structure I, unlike J, does depend on the choice of complex structure on C. Therefore the B-model for the moduli space of stable Higgs bundles cannot be obtained as a reduction of a 4d TFT.[4] On the other hand, the A-model for $\mathcal{M}_{Higgs}(G, C)$ is independent of the choice of complex structure on C, because the Kähler form ω_I is given by

$$\omega_I = \frac{1}{e^2} \int_C \mathrm{Tr} \left(\delta A \wedge \delta A - \delta \phi \wedge \delta \phi \right).$$

In fact, the A-model for $\mathcal{M}_{Higgs}(G, C)$ is obtained by letting $t = 0$. This special case of reduction to 2d has been first discussed in [24].

5 Mirror Symmetry for the Hitchin Moduli Space

Now we are ready to infer the consequences of the MO duality for the topological sigma-model with target $\mathcal{M}_H(G, C)$. For $\theta = 0$, the MO duality identifies twisted $\mathcal{N} = 4$ SYM theory with gauge group ^LG and $t = i$ with a similar theory with gauge group G and $t = 1$. Therefore the B-model with target $\mathcal{M}_{flat}(^LG, C)$ and the A-model with target $(\mathcal{M}(G, C), \omega_K)$ are isomorphic.

Whenever we have two Calabi–Yau manifolds M and M' such that the A-model of M is equivalent to the B-model of M', we say that M and M' are a mirror pair. Thus MO duality implies that $\mathcal{M}_{flat}(^LG, C)$ and $\mathcal{M}_H(G, C)$ (with symplectic structure ω_K) are a mirror pair. This mirror symmetry was first proposed in [25].

It has been argued by Strominger et al. (SYZ) [26] that whenever M and M' are mirror to each other, they should admit Lagrangian torus fibrations over the same base $\mathbf{B}$ and these fibrations are dual to each other, in a suitable sense. In the case of Hitchin moduli spaces, the SYZ fibration is easy to identify [6]. One simply maps a solution (A, ϕ) of the Hitchin equations to the space of gauge-invariant polynomials built from $\varphi = \phi^{1,0}$. For example, for $G = SU(N)$ or $G = SU(N)/\mathbb{Z}_N$ the algebra of gauge-invariant polynomials is generated by

$$\mathcal{P}_n = \mathrm{Tr}\, \varphi^n \in H^0(C, K_C^n), \quad n = 2, \ldots, N,$$

so the fibration map maps $\mathcal{M}_H(G, C)$ to the vector space

$$\oplus_{n=2}^N H^0(C, K_C^n).$$

---

[4] It can be obtained as a reduction of a "holomorphic-topological" gauge theory on $\Sigma \times C$ [23].

The map is surjective [27], so this vector space is the base space $\mathbf{B}$. The map to $\mathbf{B}$ is known as the Hitchin fibration of $\mathcal{M}_H(G, C)$. It is holomorphic in complex structure I and its fibres are Lagrangian in complex structures J and K. In fact, one can regard $\mathcal{M}_{Higgs}(G, C)$ as a complex integrable system, in the sense that the generic fibre of the fibration is a complex torus which is Lagrangian with respect to a holomorphic symplectic form on $\mathcal{M}_{Higgs}(G, C)$

$$\Omega_I = \omega_J + i\omega_K.$$

(Ordinarily, an integrable system is associated with a real symplectic manifold with a Lagrangian torus fibration.)

For general G one can define the Hitchin fibration in a similar way and one always finds that the generic fibre is a complex Lagrangian torus. Moreover, the bases $\mathbf{B}$ and $^L\mathbf{B}$ of the Hitchin fibrations for $\mathcal{M}_H(G, C)$ and $\mathcal{M}_H(^L G, C)$ are naturally identified.[5]

While the Hitchin fibration is an obvious candidate for the SYZ fibration, can we prove that it really *is* the SYZ fibration? It turns out this statement can be deduced from some additional facts about MO duality.

While we do not know how the MO duality acts on general observables in the twisted theory, the observables $\mathcal{P}_n$ are an exception, as their expectation values parameterize the moduli space of vacua of the twisted theory on $X = \mathbb{R}^4$. The MO duality must identify the moduli spaces of vacua, in a way consistent with other symmetries of the theory and this leads to a unique identification of the algebras generated by $\mathcal{P}_n$ for G and $^L G$. See [6] for details.

To complete the argument, we need to consider some particular topological D-branes for $\mathcal{M}_H(G, C)$ and $\mathcal{M}_H(^L G, C)$. A topological D-brane is a boundary condition for a 2d TFT. The set of all topological D-branes has a natural structure of a category (actually, an A_∞ category). For a B-model with a Calabi–Yau target M, this category is believed to be equivalent to the derived category of coherent sheaves on M. Sometimes we will also refer to it as the category of B-branes on M. For an A-model with target M', we get a category of A-branes on M'. It contains the Fukaya category of M' as a full subcategory.

The simplest example of a B-brane on $\mathcal{M}_{flat}(^L G, C)$ is the structure sheaf of a (smooth) point. What is its mirror? Since each point p lies in some fibre $^L\mathbf{F}_p$ of the Hitchin fibration, its mirror must be an A-brane on $\mathcal{M}_H(G, C)$ localized on the corresponding fiber $\mathbf{F}_p$ of the Langlands-dual fibration. By definition, this means that the Hitchin fibration is the same as the SYZ fibration.

According to Strominger et al., the fibres of the two mirror fibrations over the same base point are T-dual to each other. Indeed, by the usual SYZ argument the A-brane on $\mathbf{F}_p$ must be a rank-one object of the Fukaya category, i.e. a flat unitary rank-1 connection on a topologically trivial line bundle over $\mathbf{F}_p$. The moduli space of such objects must coincide with the moduli space of

---

[5] In some cases G and $^L G$ coincide, but the relevant identification is not necessarily the identity map [6, 9].

the mirror B-brane, which is simply $^L\mathbf{F}_p$. This is precisely what we mean by saying that that $^L\mathbf{F}_p$ and $\mathbf{F}_p$ are T-dual.

In the above discussion we have tacitly assumed that both $\mathcal{M}_H(G,C)$ and $\mathcal{M}_H(^LG,C)$ are connected. The components of $\mathcal{M}_H(G,C)$ are labelled by the topological isomorphism classes of principal G-bundles over C, i.e. by elements of $H^2(C,\pi_1(G)) = \pi_1(G)$. Thus, strictly speaking, our discussion applies literally only when both G and ^LG are simply connected. This is rarely true; for example, among compact simple Lie groups only E_8, F_4 and G_2 satisfy this condition (all these groups are self-dual).

In general, to maintain mirror symmetry between $\mathcal{M}_H(G,C)$ and $\mathcal{M}_H(^LG, C)$, one has to consider all possible flat B-fields on both manifolds. A flat B-field on M is a class in $H^2(M,U(1))$ whose image in $H^3(M,\mathbb{Z})$ under the Bockstein homomorphism is trivial. In the case of $\mathcal{M}_H(G,C)$, the allowed flat B-fields have finite order; in fact, they take values in $H^2(C,Z(G)) = Z(G)$, where $Z(G)$ is the centre of G. It is well known that $Z(G)$ is naturally isomorphic to $\pi_1(^LG)$. One can show that MO duality maps the class $w \in Z(G)$ defining the B-field on $\mathcal{M}_H(G,C)$ to the corresponding element in $\pi_1(^LG)$ labeling the connected component of $\mathcal{M}_H(^LG,C)$ (and vice versa). For example, if $G = SU(N)$, then $\mathcal{M}_H(G,C)$ is connected and has N possible flat B-fields labelled by $Z(SU(N)) = \mathbb{Z}_N$. On the other hand, $^LG = SU(N)/\mathbb{Z}_N$ and therefore $\mathcal{M}_H(^LG,C)$ has N connected components labelled by $\pi_1(^LG) = \mathbb{Z}_N$. There can be no non-trivial flat B-field on $\mathcal{M}_H(^LG,C)$ in this case. See Sect. 7 of [6] for more details.

Let us summarize what we have learned so far. MO duality implies that $\mathcal{M}_H(G,C)$ and $\mathcal{M}_H(^LG,C)$ are a mirror pair, with the SYZ fibrations being the Hitchin fibrations. The most powerful way to formulate the statement of mirror symmetry between two Calabi–Yau manifolds is in terms of the corresponding categories of topological branes. In the present case, we get that the derived category of coherent sheaves on $\mathcal{M}_{flat}(^LG,C)$ is equivalent to the category of A-branes on $\mathcal{M}_H(G,C)$ (with respect to the exact symplectic form ω_K). Furthermore, this equivalence maps a (smooth) point p belonging to the fibre $^L\mathbf{F}_p$ of the Hitchin fibration of $\mathcal{M}_{flat}(^LG,C)$ to the Lagrangian submanifold of $\mathcal{M}_H(G,C)$ given by the corresponding fiber of the dual fibration of $\mathcal{M}_H(G,C)$. The flat unitary connection on F_p is determined by the position of p on $^L\mathbf{F}_p$.[6]

6 From A-Branes to D-Modules

Geometric Langlands Duality says that the derived category of coherent sheaves on $\mathcal{M}_{flat}(^LG,C)$ is equivalent to the derived category of D-modules on the moduli stack $\mathrm{Bun}_G(C)$ of holomorphic G-bundles on C. This equivalence is supposed to map a point on $\mathcal{M}_{flat}(^LG,C)$ to a Hecke eigensheaf on

---

[6] This makes sense only if $^L\mathbf{F}_p$ is smooth. If p is a smooth point but $^L\mathbf{F}_p$ is singular, it is not clear how to identify the mirror A-brane on $\mathcal{M}_H(G,C)$.

Bun$_G(C)$. We have seen that MO duality implies a similar statement, with A-branes on $\mathcal{M}_H(G,C)$ taking place of objects of the derived category of D-modules on Bun$_G(C)$. Our first goal is to explain the connection between A-branes on $\mathcal{M}_H(G,C)$ and D-modules on Bun$_G(C)$. Later we will see how the Hecke eigensheaf condition can be interpreted in terms of A-branes.

The starting point of our argument is a certain special A-brane on $\mathcal{M}_H(G,C)$ which was called the canonical coisotropic brane in [6]. Recall that a submanifold Y of a symplectic manifold M is called coisotropic if for any $p \in Y$ the skew-complement of TY_p in TM_p is contained in TY_p. A coisotropic submanifold of M has dimension larger or equal than half the dimension of M; a Lagrangian submanifold of M can be defined as a middle-dimensional coisotropic submanifold. While the most familiar examples of A-branes are Lagrangian submanifolds equipped with flat unitary vector bundles, it is known from the work [28] that the category of A-branes may contain more general coisotropic submanifolds with non-flat vector bundles. (Because of this, in general the Fukaya category is only a full subcategory of the category of A-branes.) The conditions on the curvature of a vector bundle on a coisotropic A-brane are not understood except in the rank-one case; even in this case they are fairly complicated [28]. Luckily, for our purposes we only need the special case when $Y = M$ and the vector bundle has rank one. Then the condition on the curvature 2-form $F \in \Omega^2(M)$ is

$$(\omega^{-1}F)^2 = -1. \tag{2}$$

Here we regard both F and the symplectic form ω as bundle morphisms from TM to T^*M, so that $I_F = \omega^{-1}F$ is an endomorphism of TM. We also multiplied F by a factor i to make it real, rather than purely imaginary.

The condition (2) says that I_F is an almost complex structure. Using the fact that ω and F are closed 2-forms, one can show that I_F is automatically integrable [28].

Let us now specialize to the case $M = \mathcal{M}_H(G,C)$ with the symplectic form ω_K. Then if we let

$$F = \omega_J = \frac{2}{e^2} \int_C \operatorname{Tr} \delta\phi \wedge \star\delta A,$$

the equation is solved and

$$I_F = \omega_K^{-1}\omega_J = I.$$

Furthermore, since F is exact:

$$F \sim \delta \int_C \operatorname{Tr} \phi \wedge \star\delta A,$$

we can regard F as the curvature of a unitary connection on a trivial line bundle. This connection is defined uniquely if $\mathcal{M}_H(G,C)$ is simply connected;

otherwise any two such connections differ by a flat connection. One can show that flat $U(1)$ connections on a connected component of $\mathcal{M}_H(G, C)$ are classified by elements of $H^1(C, \pi_1(G))$ [6]. Thus we obtain an almost canonical coisotropic A-brane on $\mathcal{M}_H(G, C)$: it is unique up to a finite ambiguity and its curvature is completely canonical.

Next we need to understand the algebra of endomorphisms of the canonical coisotropic brane. From the physical viewpoint, this is the algebra of vertex operators inserted on the boundary of the worldsheet Σ; such vertex operators are usually referred to as open string vertex operators (as opposed to closed string vertex operators which are inserted at interior points of Σ). Actually, the knowledge of the algebra turns out to be insufficient: we would like to work locally in the target space $\mathcal{M}_H(G, C)$ and work with a sheaf of open string vertex operators on $\mathcal{M}_H(G, C)$. The idea of localizing in target space has been previously used by Malikov et al. to define the chiral de Rham complex [29]; we need an open-string version of this construction.

Localizing the path-integral in target space makes sense only if non-perturbative effects can be neglected [30, 31]. The reason is that perturbation theory amounts to expanding about constant maps from Σ to M and therefore the perturbative correlator is an integral over M of a quantity whose value at a point $p \in M$ depends only of the infinitesimal neighbourhood of p. In such a situation it makes sense to consider open-string vertex operators defined only locally on M, thereby getting a sheaf on M. Because of the topological character of the theory, the OPE of Q-closed vertex operators is non-singular, and Q-cohomology classes of such locally defined vertex operators form a sheaf of algebras on M. One difference compared to the closed-string case is that operators on the boundary of Σ have a well-defined cyclic order and therefore the multiplication of vertex operators need not be commutative. The cohomology of this sheaf of algebras is the endomorphism algebra of the brane [6].

One can show that there are no non-perturbative contributions to any correlators involving the canonical coisotropic A-brane [6] and so one can localize the path-integral in $\mathcal{M}_H(G, C)$. But a further problem arises: perturbative results are formal power series in the Planck constant and there is no guarantee of convergence. In the present case, the role of the Planck constant is played by the parameter e^2 in the gauge theory.[7] In fact, one can show that the series defining the multiplication have zero radius of convergence for some locally defined observables.

In order to understand the resolution of this problem, let us look more closely the structure of the perturbative answer. In the classical approximation (leading order in e^2), the sheaf of open-string states is the sheaf of holomorphic functions on $\mathcal{M}_H(G, C)$ in complex structure I [28, 32]. In this complex structure, $\mathcal{M}_H(G, C)$ can be identified with $\mathcal{M}_{Higgs}(G, C)$ and the

---

[7] At first sight, the appearance of e^2 in the twisted theory may seem surprising, but one should remember that the argument showing that the theory at $t = 1$ is independent of e^2 is valid only when the manifold X has no boundary.

holomorphic coordinates are $A_{\bar{z}}$ and ϕ_z. There is a natural grading on this algebra in which ϕ_z has degree 1 and $A_{\bar{z}}$ has degree 0. Note also that the projection $(A_{\bar{z}}, \phi_z) \mapsto A_{\bar{z}}$ defines a map from $\mathcal{M}_H(G, C)$ to $\mathrm{Bun}_G(C)$. If we restrict the target of this map to the subspace of stable G-bundles $\mathcal{M}(G, C)$, then the preimage of $\mathcal{M}(G, C)$ in $\mathcal{M}_H(G, C)$ can be thought of as the cotangent bundle to $\mathcal{M}(G, C)$.

At higher orders, the multiplication becomes non-commutative and incompatible with the grading. However, it is still compatible the associated filtration. That is, the product of two functions on $\mathcal{M}_{Higgs}(G, C)$ which are polynomials in ϕ_z of degrees k and l is a polynomial of degree $k + l$. Therefore the product of polynomial observables defined by perturbation theory is well-defined (it is a polynomial in the Planck constant).

We see that we can get a well-defined multiplication of vertex operators if we restrict to those which depend polynomially on ϕ_z. That is, we have to "sheafify" our vertex operators only along the base of the projection to $\mathcal{M}(G, C)$, while along the fibres the dependence is polynomial.

Holomorphic functions on the cotangent bundle of $\mathcal{M}(G, C)$ polynomially depending on the fibre coordinates can be thought of as symbols of differential operators acting on holomorphic functions on $\mathcal{M}(G, C)$, or perhaps on holomorphic sections of a line bundle on $\mathcal{M}(G, C)$. One may therefore suspect that the sheaf of open-string vertex operators is isomorphic to the sheaf of holomorphic differential operators on some line bundle $\mathcal{L}$ over $\mathcal{M}(G, C)$. It is shown in [6] that this is indeed the case and the line bundle $\mathcal{L}$ is a square root of the canonical line bundle of $\mathcal{M}(G, C)$. (It does not matter which square root one takes.)

Now we can finally explain the relation between A-branes and (twisted) D-modules on $\mathcal{M}(G, C) \subset \mathrm{Bun}_G(C)$. Given an A-brane β, we can consider the space of morphisms from the canonical coisotropic brane α to the brane β. It is a left module over the endomorphism algebra of α. Better still, we can sheafify the space of morphisms along $\mathcal{M}(G, C)$ and get a sheaf of modules over the sheaf of differential operators on $K^{1/2}$, where K is the canonical line bundle of $\mathcal{M}(G, C)$. This is the twisted D-module corresponding to the brane β.

In general it is rather hard to determine the D-module corresponding to a particular A-brane. A simple case is when β is a Lagrangian submanifold defined by the condition $\phi = 0$, i.e. the zero section of the cotangent bundle $\mathcal{M}(G, C)$. In that case, the D-module is simply the sheaf of sections of $K^{1/2}$. From this example, one could suspect that the A-brane is simply the characteristic variety of the corresponding D-module. However, this is not so in general, since in general A-branes are neither conic nor even holomorphic subvarieties of $\mathcal{M}_{Higgs}(G, C)$. For example, a fiber of the Hitchin fibration $\mathbf{F}_p$ is holomorphic but not conic. It is not clear how to compute the D-module corresponding to $\mathbf{F}_p$, even when $\mathbf{F}_p$ is a smooth fibre.[8]

---

[8] The abelian case $G = U(1)$ is an exception, see [6] for details.

We conclude this section with two remarks. First, the relation between A-branes is most readily understood if we replace the stack $\text{Bun}_G(C)$ by the space of stable G-bundles $\mathcal{M}(G, C)$. Second, from the physical viewpoint it is more natural to work directly with A-branes rather than with corresponding D-modules. In some sense it is also more natural from the mathematical viewpoint, since both the derived category of $\mathcal{M}_{flat}(^LG, C)$ and the category of A-branes on $\mathcal{M}_H(G, C)$ are "topological", in the sense that they do not depend on the complex structure on C. The complex structure on C appears only when we introduce the canonical coisotropic brane (its curvature F manifestly depends on the Hodge star operator on C).

7 Wilson and 't Hooft Operators

In any gauge theory one can define Wilson loop operators:

$$\text{Tr}_R P \exp \int_\gamma A = \text{Tr}\, R \left(\text{Hol}(A, \gamma) \right),$$

where R is a finite-dimensional representation of G, γ is a closed loop in M and $P \exp \int$ is simply a physical notation for holonomy. The Wilson loop is a gauge-invariant function of the connection A and therefore can be regarded as a physical observable. Inserting the Wilson loop into the path-integral is equivalent to inserting an infinitely massive particle travelling along the path γ and having internal "colour-electric" degrees of freedom described by representation R of G. For example, in the theory of strong nuclear interactions we have $G = SU(3)$ and the effect of a massive quark can be modelled by a Wilson loop with R being a three-dimensional irreducible representation. The vacuum expectation value of the Wilson loop can be used to distinguish various massive phases of the gauge theory [33]. Here however we will be interested in the algebra of Wilson loop operators, which is insensitive to the long-distance properties of the theory.

The Wilson loop is not BRST-invariant and therefore is not a valid observable in the twisted theory. But it turns out that for $t = \pm i$ there is a simple modification which *is* BRST-invariant:

$$W_R(\gamma) = \text{Tr}_R P \exp \int_\gamma (A \pm i\phi) = \text{Tr}\, R \left(\text{Hol}(A \pm i\phi), \gamma) \right)$$

The reason is that the complex connection $\mathcal{A} = A \pm i\phi$ is BRST-invariant for these values of t. There is nothing similar for any other value of t.

We may ask how the MO duality acts on Wilson loop operators. The answer is to a large extent fixed by symmetries, but turns out to be rather non-trivial [34]. The difficulty is that the dual operator cannot be written as a function of fields, but instead is a *disorder operator*. Inserting a disorder operator into the path-integral means changing the space of fields over which one integrates. For example, a disorder operator localized on a closed curve γ is defined by specifying a singular behaviour for the fields near γ. The disorder operator dual to a Wilson loop has been first discussed by G. 't Hooft [35] and is defined as follows [34]. Let μ be an element of the Lie algebra $\mathfrak{g}$ defined up to adjoint action of G and let us choose coordinates in the neighbourhood of a point $p \in \gamma$ so that γ is defined by the equations $x^1 = x^2 = x^3 = 0$. Then we require the curvature of the gauge field to have a singularity of the form

$$F \sim \star_3 \, d \left(\frac{\mu}{2r} \right),$$

where r is the distance to the origin in the 123 hyperplane and $\star_3$ is the Hodge star operator in the same hyperplane. For $t = 1$ Q-invariance requires the 1-form Higgs field ϕ to be singular as well:

$$\phi \sim \frac{\mu}{2r} \, dx^4.$$

One can show that such an ansatz for F makes sense (i.e. one can find a gauge field whose curvature is F) if and only if μ is a Lie algebra homomorphism from $\mathbb{R}$ to $\mathfrak{g}$ obtained from a Lie group homomorphism $U(1) \to G$ [3]. To describe this condition in a more suggestive way, let us use the gauge freedom to conjugate μ to a particular Cartan subalgebra $\mathfrak{t}$ of $\mathfrak{g}$. Then μ must lie in the coweight lattice $\Lambda_{cw}(G) \subset \mathfrak{t}$, i.e. the lattice of homomorphisms from $U(1)$ to the maximal torus T corresponding to $\mathfrak{t}$. In addition, one has to identify points of the lattice which are related by an element of the Weyl group $\mathcal{W}$. Thus 't Hooft loop operators are classified by elements of $\Lambda_{cw}(G)/\mathcal{W}$.

By definition, $\Lambda_{cw}(G)$ is identified with the weight lattice $\Lambda_w({}^LG)$ of ${}^LG$. But elements of $\Lambda_w({}^LG)/\mathcal{W}$ are in one-to-one correspondence with irreducible representations of ${}^LG$. This suggests that MO duality maps the 't Hooft operator corresponding to a coweight $B \in \Lambda_{cw}(G)$ to the Wilson operator corresponding to a representation ${}^LR$ with highest weight in the Weyl orbit of $B \in \Lambda_w({}^LG)$. This is a reinterpretation of the the Goddard–Nuyts–Olive argument discussed in Sect. 2 in terms of operators rather than states.

To test this duality, one can compare the algebra of 't Hooft operators for gauge group G and Wilson operators for gauge group ${}^LG$. In the latter case, the operator product is controlled by the algebra of irreducible representations of ${}^LG$. That is, we expect that as the loop γ' approaches γ, we have

$$W_R(\gamma) W_{R'}(\gamma') \sim \bigoplus_{R_i \subset R \otimes R'} W_{R_i}(\gamma),$$

where R and R' are irreducible representations of G and the sum on the right-hand side runs over irreducible summands of $R \otimes R'$.

In the case of 't Hooft operators the computation of the operator product is much more non-trivial [6]. It can be related to computing the L^2 cohomology of Schubert cells of the affine Grassmannian Gr_G. Assuming that the L^2

cohomology coincides with the intersection cohomology of the closure of the cell, the prediction of the MO duality reduces to the statement of the geometric Satake correspondence, which says that the tensor category of equivariant perverse sheaves on Gr_G is equivalent to the category of finite-dimensional representations of $^L G$ [36, 37, 38]. This provides a new and highly non-trivial check of the MO duality.

The geometric Satake correspondence can be thought of as a local version of the Geometric Langlands Duality. The reason is that irreducible objects in the category of equivariant perverse coherent sheaves on Gr_G are naturally associated with Hecke operations on G-bundles on C which modify the G-bundle at a single point.

Similarly, from the gauge-theoretic viewpoint one can think about 't Hooft operators as functors from the category of A-branes on $\mathcal{M}_H(G, C)$ to itself. To understand how this comes about, consider a loop operator (Wilson or 't Hooft) in the twisted $\mathcal{N} = 4$ SYM theory (at $t = i$ or $t = 1$, respectively). As usual, we take the four-manifold X to be $\Sigma \times C$ and let the curve γ be of the form $\gamma_0 \times p$, where $p \in C$ and γ_0 is a curve on Σ. Let $\partial\Sigma_0$ be a connected component of $\partial\Sigma$ on which we specify a boundary condition corresponding to a given brane β. This brane is either a B-brane in complex structure J or an A-brane in complex structure K, depending on whether $t = i$ or $t = 1$. Now suppose γ_0 approaches $\partial\Sigma_0$. The "composite" of $\partial\Sigma_0$ with boundary condition β and the loop operator can be thought of as a new boundary condition for the topological sigma-model with target $\mathcal{M}_H(G, C)$. It depends on $p \in C$ as well as other data defining the loop operator. One can show that this "fusion" operation defines a functor from the category of topological branes to itself [6].

In the case of the Wilson loop, it is very easy to describe this functor. In complex structure J, we can identify $\mathcal{M}_H(G, C)$ with $\mathcal{M}_{flat}(G, C)$. On the product $\mathcal{M}_{flat}(G, C) \times C$ there is a universal G-bundle which we call $\mathcal{E}$. For any $p \in C$ let us denote by $\mathcal{E}_p$ the restriction of $\mathcal{E}$ to $\mathcal{M}_{flat}(G, C) \times p$. For any representation R of G we can consider the operation of tensoring coherent sheaves on $\mathcal{M}_{flat}(G, C)$ with the associated holomorphic vector bundle $R(\mathcal{E}_p)$. One can show that this is the functor corresponding to the Wilson loop in representation R inserted at a point $p \in C$. We will denote this functor $W_R(p)$. The action of 't Hooft loop operators is harder to describe, see Sects. 9 and 10 of [6] for details. In particular, it is shown there that 't Hooft operators act by Hecke transformations.

Consider now the structure sheaf $\mathcal{O}_x$ of a point $x \in \mathcal{M}_{flat}(^L G, C)$. For any representation $^L R$ of $^L G$ the functor corresponding to $W_{^L R}(p)$ maps $\mathcal{O}_x$ to the sheaf $\mathcal{O}_x \otimes {}^L R(\mathcal{E}_p)_x$. That is, $\mathcal{O}_x$ is simply tensored with a vector space $^L R(\mathcal{E}_p)_x$. One says that $\mathcal{O}_x$ is an eigenobject of the functor $W_{^L R}(\mathcal{E}_p)$ with eigenvalue $^L R(\mathcal{E}_p)_x$. The notion of an eigenobject of a functor is a categorification of the notion of an eigenvector of a linear operator: instead of an element of a vector space one has an object of a $\mathbb{C}$-linear category, instead of a linear operator one has a functor from the category to itself and instead of a complex number (eigenvalue) one has a complex vector space $^L R(\mathcal{E}_p)_x$.

We conclude that $\mathcal{O}_x$ is a common eigenobject of all functors $W_{L_R}(p)$ with eigenvalues $^L R(\mathcal{E}_p)_x$. Actually, since we can vary p continuously on C and the vector spaces $^L R(\mathcal{E}_p)_x$ are naturally identified as one varies p along any path on C, it is better to say that the eigenvalue is a flat $^L G$-bundle $^L R(\mathcal{E})_x$. Tautologically, this flat vector bundle is obtained by taking the flat principal $^L G$-bundle on C corresponding to x and associating it with a flat vector bundle via the representation $^L R$.

Applying the MO duality, we may conclude that the A-brane on $\mathcal{M}_H(G, C)$ corresponding to a fibre of the Hitchin fibration is a common eigenobject for all 't Hooft operators, regarded as functors on the category of A-branes. The eigenvalue is the flat $^L G$ bundle on C determined by the mirror of the A-brane. This is the gauge-theoretic version of the statement that the D-module on $\mathrm{Bun}_G(C)$ corresponding to a point on $\mathcal{M}_{flat}(^L G, C)$ is a Hecke eigensheaf.

Acknowledgments

This work was supported in part by the DOE grant DE-FG03-92-ER40701.

References

1. E. Frenkel, "Lectures on the Langlands Program and Conformal Field Theory," (2005) [arXiv:hep-th/0512172].
2. G. Laumon, "Correspondance Langlands Geometrique Pour Les Corps Des Fonctions," Duke Math. J. **54**, 309–359 (1987).
3. P. Goddard, J. Nuyts and D. I. Olive, "Gauge Theories and Magnetic Charge," Nucl. Phys. B **125**, 1 (1977).
4. C. Montonen and D. I. Olive, "Magnetic Monopoles as Gauge Particles?," Phys. Lett. B **72**, 117 (1977).
5. H. Osborn, "Topological Charges for N=4 Supersymmetric Gauge Theories and Monopoles of Spin 1," Phys. Lett. B **83**, 321 (1979).
6. A. Kapustin and E. Witten, "Electric-Magnetic Duality and the Geometric Langlands Program," (2006) [arXiv:hep-th/0604151].
7. S. Gukov and E. Witten, "Gauge Theory, Ramification, and the Geometric Langlands Program," (2006) [arXiv:hep-th/0612073].
8. N. Dorey, C. Fraser, T. J. Hollowood and M. A. C. Kneipp, "S-Duality In N=4 Supersymmetric Gauge Theories," Phys. Lett. B **383**, 422 (1996) [arXiv:hep-th/9605069].
9. P. C. Argyres, A. Kapustin and N. Seiberg, "On S-duality for Non-Simply-Laced Gauge Groups," J. High Energy Phys. **0606**, 043 (2006) [arXiv:hep-th/0603048].
10. E. Witten and D. I. Olive, "Supersymmetry Algebras that Include Topological Charges," Phys. Lett. B **78**, 97 (1978).
11. A. Sen, "Dyon – Monopole Bound States, Self-Dual Harmonic Forms on the Multi-Monopole Moduli Space, and SL(2,Z) Invariance in String Theory," Phys. Lett. B **329**, 217 (1994) [arXiv:hep-th/9402032].

12. C. Vafa and E. Witten, "A Strong Coupling Test of S-Duality," Nucl. Phys. B **431**, 3 (1994) [arXiv:hep-th/9408074].

13. C. Vafa, "Geometric Origin of Montonen-Olive Duality," Adv. Theor. Math. Phys. **1**, 158 (1998) [arXiv:hep-th/9707131].

14. E. Witten, "Topological Quantum Field Theory," Commun. Math. Phys. **117**, 353 (1988).

15. N. Hitchin, "The Self-Duality Equations on the Riemann Surface," Proc. Lond. Math. Soc. (3) **55**, 59–126 (1987).

16. E. B. Bogomolny, "Stability of Classical Solutions," Sov. J. Nucl. Phys. **24**, 449 (1976) [Yad. Fiz. **24**, 861 (1976)].

17. K. Corlette, "Flat G-Bundles with Canonical Metrics," J. Diff. Geom. **28**, 361–382 (1988).

18. E. Witten, "Mirror Manifolds And Topological Field Theory," (1991) [arXiv:hep-th/9112056].

19. A. Kapustin, "Topological Strings on Noncommutative Manifolds," Int. J. Geom. Meth. Mod. Phys. **1**, 49 (2004) [arXiv:hep-th/0310057].

20. A. Kapustin and Y. Li, "Topological Sigma-Models with H-Flux and Twisted Generalized Complex Manifolds," (2004) [arXiv:hep-th/0407249].

21. N. Hitchin, "Generalized Calabi-Yau Manifolds," Q. J. Math. **54**, 281–308 (2003).

22. S. K. Donaldson, "Twisted Harmonic Maps and the Self-Duality Equations," Proc. Lond. Math. Soc. (3), **55**, 127–131 (1987).

23. A. Kapustin, "Holomorphic Reduction of N = 2 Gauge Theories, Wilson-'t Hooft Operators, and S-Duality," (2006) [arXiv:hep-th/0612119].

24. M. Bershadsky, A. Johansen, V. Sadov and C. Vafa, "Topological Reduction of 4-d SYM to 2-d Sigma Models," Nucl. Phys. B **448**, 166 (1995) [arXiv:hep-th/9501096].

25. T. Hausel and M. Thaddeus, "Mirror Symmetry, Langlands Duality, and the Hitchin System," Invent. Math. **153**, 197–229 (2003) [arXiv:math.AG/0205236].

26. A. Strominger, S. T. Yau and E. Zaslow, "Mirror symmetry is T-duality," Nucl. Phys. B **479**, 243 (1996) [arXiv:hep-th/9606040].

27. N. Hitchin, "Stable Bundles and Integrable Systems," Duke Math. J. **54**, 91–114 (1987).

28. A. Kapustin and D. Orlov, "Remarks on A-branes, Mirror Symmetry, and the Fukaya Category," J. Geom. Phys. **48**, 84 (2003) [arXiv:hep-th/0109098].

29. F. Malikov, V. Schechtman, and A. Vaintrob, "Chiral De Rham Complex," Comm. Math. Phys. **204**, 439–473 (1999) [arXiv:math.AG/9803041].

30. A. Kapustin, "Chiral De Rham Complex and the Half-Twisted Sigma-Model," (2005) [arXiv:hep-th/0504074].

31. E. Witten, "Two-Dimensional Models with (0,2) Supersymmetry: Perturbative Aspects," (2005) [arXiv:hep-th/0504078].

32. A. Kapustin and Y. Li, "Open String BRST Cohomology for Generalized Complex Branes," Adv. Theor. Math. Phys. **9**, 559 (2005) [arXiv:hep-th/0501071].

33. K. G. Wilson, "Confinement of Quarks," Phys. Rev. D **10**, 2445 (1974).

34. A. Kapustin, "Wilson-'t Hooft Operators in Four-Dimensional Gauge Theories And S-Duality," Phys. Rev. D **74**, 025005 (2006) [arXiv:hep-th/0501015].

35. G. 't Hooft, "On the Phase Transition Towards Permanent Quark Confinement," Nucl. Phys. B **138**, 1 (1978); "A Property of Electric and Magnetic Flux in Nonabelian Gauge Theories," Nucl. Phys. B **153**, 141 (1979).

36. G. Lusztig, "Singularities, Character Formula, and a q-Analog of Weight Multiplicities," in *Analyse Et Topolgie Sur Les Espaces Singuliers II-III,* Asterisque vol. 101–2, 208–229 (1981).
37. V. Ginzburg, "Perverse Sheaves on a Loop Group and Langlands Duality," (1995) [arXiv:alg-geom/9511007].
38. I. Mirkovic and K. Vilonen, "Perverse Sheaves on Affine Grassmannians and Langlands Duality," Math. Res. Lett. **7**, 13–24 (2000) [arXiv:math.AG/9911050].

Homological Mirror Symmetry
and Algebraic Cycles

L. Katzarkov

University of Miami
l.katzarkov@math.miami.edu

Abstract. In this chapter we outline some applications of Homological Mirror Symmetry to classical problems in Algebraic Geometry, like rationality of algebraic varieties and the study of algebraic cycles. Several examples are studied in detail.

1 Introduction

In this chapter we discuss some classical questions of algebraic geometry from the point of view of Homological Mirror Symmetry.

Mirror symmetry was introduced as a special duality between two $N = 2$ super conformal field theories. Traditionally an $N = 2$ super conformal field theory (SCFT) is constructed as a quantization of a σ-model with target a compact Calabi–Yau manifold equipped with a Ricci flat Kähler metric and a closed 2-form – the so-called B-field. We say that two Calabi–Yau manifolds X and Y form a *mirror pair* $X|Y$ if the associated $N = 2$ SCFTs are mirror dual to each other [10].

This chapter contains an outline of a program for applications of ideas of Homological Mirror Symmetry (HMS) to the study of rationality questions and the Hodge Conjecture. These ideas were announced at the conference on HMS, Vienna, 2006. Complete results will appear elsewhere.

We start with a quick introduction of Homological Mirror Symmetry. After that we introduce our construction of the Mirror Landau–Ginzburg Model which differs from constructions in [19] we refer to other papers for the fundamentals of this construction and give some examples and applications.

1. First we outline an application to Birational Geometry. One of the main questions in classical Algebraic Geometry is if a smooth projective n dimensional variety X has a field of meromorphic functions $\mathbb{C}(X)$ isomorphic to the field of meromorphic functions of $\mathbb{C}\mathbb{P}^n$. The following techniques have been used to study this question:

Katzarkov, L.: *Homological Mirror Symmetry and Algebraic Cycles.* Lect. Notes Phys. **757**, 125–152 (2009)
DOI 10.1007/978-3-540-68030-7_5   © Springer-Verlag Berlin Heidelberg 2009

- The technique of intermediate Jacobians introduced by Clemens and Griffiths [8].
- The technique of birational automorphisms of X introduced by Iskovskikh and Manin [16]. As a consequence of it, namely of the Noether–Fano inequality, Pukhlikov and later Ein, Mustata and de Fernex have introduced the technique of log canonical thresholds [11, 12, 32].
- The degeneration techniques introduced by Kollár [26].
- Artin's and Mumford's approach later developed by Saltman and Bogomolov [5].

Examples, where this new technique applies, will be discussed:

- We outline ideas for proving criteria for non-rationality of conic bundles and compare it with a rationality criteria introduced by Iskovskikh [17] and Sarkisov.
- We outline a proof of non-rationality of generic three and four dimensional cubic.

Birational geometry	Homological mirror symmetry
X	$w : Y \to \mathbb{CP}^1$
Blow-up	Adding singular fibers
Blow-down	Taking out singular fibers

2. Then we move to the case of abelian varieties an we study the following correspondence there

Algebraic cycles	Lagrangian cycles
X	$w : Y \to \mathbb{CP}^1$
Tropical classes	Lagrangian submanifolds
Tropical varieties	Fukaya Seidel Lagrangians

The paper is organized as follows. In Sect. 2 we introduce basics of HMS. Connection with Birational Geometry is discussed in Sect. 3. Questions related to Hodge Conjecture appear in Sect. 4. We mainly outline the flavor of the arguments – full details will appear elsewhere.

2 Definitions

We first recall a definition which belongs to Seidel [33]. Historically the idea was introduced first by M. Kontsevich and later by K. Hori. We begin by briefly reviewing Seidel's construction of a Fukaya-type A_∞-category associated with a symplectic Lefschetz pencil see [33, 34].

Let (Y, ω) be an open symplectic manifold of dimension $\dim_\mathbb{R} Y = 2n$. Let $w : Y \to \mathbb{C}$ be a symplectic Lefschetz fibration, i.e. w is a C^∞ complex-valued function with isolated non-degenerate critical points $p_1, \ldots, p_r$ near which w is given in adapted complex local coordinates by $w(z_1, \ldots, z_n) = f(p_i) + z_1^2 + \cdots + z_n^2$ and whose fibers are symplectic submanifolds of Y. Fix a regular value λ_0 of w and consider arcs $\gamma_i \subset \mathbb{C}$ joining λ_0 to the critical value $\lambda_i = f(p_i)$. Using the horizontal distribution which is symplectic orthogonal to the fibers of w, we can transport the vanishing cycle at p_i along the arc γ_i to obtain a Lagrangian disc $D_i \subset Y$ fibered above γ_i, whose boundary is an embedded Lagrangian sphere L_i in the fiber $\Sigma_{\lambda_0} = w^{-1}(\lambda_0)$. The Lagrangian disc D_i is called the *Lefschetz thimble* over γ_i and its boundary L_i is the vanishing cycle associated to the critical point p_i and to the arc γ_i.

After a small perturbation we may assume that the arcs $\gamma_1, \ldots, \gamma_r$ in $\mathbb{C}$ intersect each other only at λ_0 and are ordered in the clockwise direction around λ_0. Similarly we may always assume that the Langrangian spheres $L_i \subset \Sigma_{\lambda_0}$ intersect each other transversely inside Σ_0.

Definition 2.1 Given a coefficient ring R, the R-linear directed Fukaya category $FS(Y, w; \{\gamma_i\})$ is the following A_∞-category: the objects of $FS(Y, w; \{\gamma_i\})$ are the Lagrangian vanishing cycles $L_1, \ldots, L_r$; the morphisms between the objects are given by

$$\text{Hom}(L_i, L_j) = \begin{cases} CF^*(L_i, L_j; R) = R^{|L_i \cap L_j|} & \text{if } i < j \\ R \cdot id & \text{if } i = j \\ 0 & \text{if } i > j; \end{cases}$$

and the differential m_1, composition m_2 and higher order products m_k are defined in terms of Lagrangian Floer homology inside Σ_{λ_0}.

More precisely,

$$m_k : \text{Hom}(L_{i_0}, L_{i_1}) \otimes \cdots \otimes \text{Hom}(L_{i_{k-1}}, L_{i_k}) \to \text{Hom}(L_{i_0}, L_{i_k})[2 - k]$$

is trivial when the inequality $i_0 < i_1 < \cdots < i_k$ fails to hold (i.e. it is always zero in this case, except for m_2 where composition with an identity morphism is given by the obvious formula).

When $i_0 < \cdots < i_k$, m_k is defined by fixing a generic ω-compatible almost-complex structure on Σ_{λ_0} and counting pseudo-holomorphic maps from a disc with $k+1$ cyclically ordered marked points on its boundary to Σ_{λ_0}, mapping the marked points to the given intersection points between vanishing cycles and the portions of boundary between them to $L_{i_0}, \ldots, L_{i_k}$ respectively (see [34]).

It is shown in [34] that the directed Fukaya category $FS(Y, w; \{\gamma_i\})$ is independent of the choice of paths. We will denote this category by $FS(Y, w)$ and will refer to it as the *Fukaya–Seidel* category of w.

Let Y be a complex algebraic variety (or a complex manifold) and let $w: Y \to \mathbb{C}$ be a holomorphic function. Following [31] we define:

Definition 2.2 (Orlov) *The derived* category $D^b(Y, \mathsf{w})$ of a holomorphic potential
$\mathsf{w} : Y \to \mathbb{C}$ is defined as the disjoint union

$$D^b(Y, \mathsf{w}) := \coprod_t D^b_{\mathrm{sing}}(Y_t)$$

of the derived categories of singularities $D^b_{\mathrm{sing}}(Y_t)$ of all fibers $Y_t := \mathsf{w}^{-1}(t)$ of w.

The category $D^b_{\mathrm{sing}}(Y_t) = D^b(Y_t)/\operatorname{Perf}(Y_t)$ is defined as the quotient category of derived category of coherent sheaves on Y_t modulo the full triangulated subcategory of perfect complexes on Y_t. Note that $D^b_{\mathrm{sing}}(Y_t)$ is non-trivial only for singular fibers Y_t.

In the following paper we will use notions defined in [19] Sect. 7.3. Let X be a manifold of general type, i.e. a sufficiently high power of the canonical linear system on X defines a birational map. We will use the above definitions to formulate and motivate an analog of Kontsevich's Homological Mirror Symmetry (HMS) conjecture for such manifolds as well as for Fano manifolds – manifolds whose anticanonical linear system defines a birational map.

A quantum sigma-model with target X is free in the infrared limit, while in the ultraviolet limit it is strongly coupled. In order to make sense of this theory at arbitrarily high energy scales, one has to embed it into some asymptotically free $N = 2$ field theory, for example into a gauged linear sigma-model (GLSM). Here "embedding" means finding a GLSM such that the low-energy physics of one of its vacua is described by the sigma-model with target X. In mathematical terms, this means that X has to be realized as a complete intersection in a toric variety.

The GLSM usually has additional vacua, whose physics is not related to X. Typically, these extra vacua have a mass gap. To learn about X by studying the GLSM, it is important to be able to recognize the extra vacua. Let Z be a toric variety defined as a symplectic quotient of $\mathbb{C}^N$ by a linear action of the gauge group $G \simeq U(1)^k$. The weights of this action will be denoted Q_{ia}, where $i = 1, \ldots, N$ and $a = 1, \ldots, k$. Let X be a complete intersection in Z given by homogeneous equations $G_\alpha(X) = 0$, $\alpha = 1, \ldots, m$. The weights of G_α under the G-action will be denoted $d_{\alpha a}$. The GLSM corresponding to X involves chiral fields Φ_i, $i = 1, \ldots, N$ and Ψ_α, $\alpha = 1, \ldots, m$. Their charges under the gauge group G are given by matrices Q_{ia} and $d_{\alpha a}$, respectively. The Lagrangian of the GLSM depends also on complex parameters t_a, $a = 1, \ldots, k$. On the classical level, the vector t_a is the level of the symplectic quotient and thus parameterizes the complexified Kähler form on Z. The Kähler form on X is the induced one. On the quantum level the parameters t_a are renormalized and satisfy linear renormalization semigroup equations:

$$\mu \frac{\partial t_a}{\partial \mu} = \beta_a := \sum_i Q_{ia} - \sum_\alpha d_{\alpha a}.$$

In the Calabi–Yau case all β_a vanish and the parameters t_a are not renormalized. Here μ denote the massive and massless vacua.

According to Hori–Vafa [20] the Landau–Ginzburg model (Y, w) that is mirror to X will have (twisted) chiral fields Λ_a, $a = 1, \ldots, k$, Y_i, $i = 1, \ldots, N$ and Υ_α, $\alpha = 1, \ldots, m$ and a superpotential is given by

$$\mathsf{w} = \sum_a \Lambda_a \left(\sum_i Q_{ia} Y_i - \sum_\alpha d_{\alpha a} \Upsilon_\alpha - t_a \right) + \sum_i e^{-Y_i} + \sum_\alpha e^{-\Upsilon_\alpha}.$$

The vacua are in one-to-one correspondence with the critical points of w. By definition, massive vacua are those corresponding to non-degenerate critical points. An additional complication is that before computing the critical points one has to partially compactify the target space of the LG model.

One can determine which vacua are "extra" (i.e. unrelated to X) as follows. The infrared limit is the limit $\mu \to 0$. Since t_a depend on μ, so do the critical points of w. A critical point is relevant for X (i.e. is not an extra vacuum) if and only if the critical values of e^{-Y_i} all go to zero as μ goes to zero. In terms of the original variables Φ_i, this means that vacuum expectation values of $|\Phi_i|^2$ go to $+\infty$ in the infrared limit. This is precisely the condition which justifies the classical treatment of vacua in the GLSM. Recall also that the classical space of vacua in the GLSM is precisely X.

Now let us state the analog of the HMS for complete intersections X of general type. We will write $D^b(X)$ for the standard derived category of bounded complexes of coherent sheaves on X and by $\mathrm{Fuk}(X, \omega)$ the standard Fukaya category of a symplectic manifold X with a symplectic form ω.

Conjecture 2.3 Let X be a variety of general type which is realized as a complete intersection in a toric variety and let (Y, w) be the mirror LG model. The derived Fukaya category $\mathrm{Fuk}(X, \omega)$ of X embeds as a direct summand into the category $D^b(Y, \mathsf{w})$ (the category of B-branes for the mirror LG model). If the extra vacua are all massive, the complement of the Fukaya category of X in $D^b(Y, \mathsf{w})$ is very simple: each extra vacuum contributes a direct summand which is equivalent to the derived category of graded modules over a Clifford algebra.

There is also a mirror version of this conjecture:

Conjecture 2.4 The derived category $D^b(X)$ of coherent sheaves on X embeds as a full sub-category into the derived Fukaya-Seidel category $FS(Y, \mathsf{w})$ of the potential w.

With an appropriate generalization of Fukaya–Seidel category to the case of non-isolated singularities we arrive at Table 1 summarizing our previous considerations. The categories $D^b_{\lambda_i, r_i}(Y, \mathsf{w})$ and $FS_{(\lambda_i, r_i)}(Y, \mathsf{w}, \omega)$ appearing in the last row of Table 1 are modifications which contain information only about singular fibers with base points contained in a disc with a radius – a

Table 1. Kontsevich's HMS conjecture

A side	B side
X — compact manifold, ω — symplectic form on X.	X — smooth projective variety over $\mathbb{C}$
$\mathrm{Fuk}(X,\omega) = \left\{ \begin{array}{l} \text{Obj: } (L_i, \mathcal{E}) \\ \text{Mor: } HF(L_i, L_j) \end{array} \right.$ L_i — Lagrangian submanifold of X, $\mathcal{E}$ — flat $U(1)$-bundle on L_i	$D^b(X) = \left\{ \begin{array}{l} \text{Obj: } C_i^\bullet \\ \text{Mor: } Ext(C_j^\bullet, C_j^\bullet) \end{array} \right.$ $C_i^\bullet$ — complex of coherent sheaves on X

(Y, ω) — open symplectic manifold, $\mathsf{w} : Y \to \mathbb{C}$ — a proper C^∞ map with symplectic fibers.	Y — smooth quasi-projective variety over $\mathbb{C}$, $\mathsf{w} : Y \to \mathbb{C}$ — proper algebraic map.
$FS(Y, \mathsf{w}, \omega) = \left\{ \begin{array}{l} \text{Obj: } (L_i, \mathcal{E}) \\ \text{Mor: } HF(L_i, L_j) \end{array} \right.$ L_i — Lagrangian submanifold of Y_{λ_0}, $\mathcal{E}$ — flat $U(1)$-bundle on L_i.	$D^b(Y, \mathsf{w}) = \bigsqcup_t D^b(Y_t) \Big/ \mathrm{Perf}(Y_t)$
$FS_{\lambda_i, r_i}(Y, \mathsf{w}, \omega)$ – the Fukaya-Seidel category of $(Y_{|t-\lambda_i|<r_i}, \mathsf{w}, \omega)$.	$D^b_{\lambda_i, r_i}(Y, \mathsf{w}) = \bigsqcup_{|t-\lambda_i| \leq r_i} D^b(Y_t) \Big/ \mathrm{Perf}(Y_t)$

real number λ, are introduced in order to deal with problems arising from massless vacua.

In this formalism we have to take a Karoubi closure on both sides.

3 The Construction

The standard construction of Landau–Ginzburg models was done for smooth complete intersections in toric varieties in [19]. We describe a new procedure which is a natural continuation of our previous results [2] and of the works of Moishezon and Teicher. Our procedure underlines the geometric nature of HMS and several categorical correspondences come naturally. We will restrict

ourselves to the case of three dimensional Fano manifolds X with the anti-
anticanonical linear system on it. For simplicity we will assume that $|-K_X|$
contains a pencil.

Step 1. Choose a Moishezon degeneration [2] $\mathcal{X} \to \Delta$ of X corresponding
to a projective embedding of X given by a multiple of the ample line bundle
$-K_X$. This means that we choose a generic Noether normalization projection
$X \to \mathbb{P}^3$ in the given projective embedding and then degenerate the covering
map to a totally split cover. The central fiber $\mathcal{X}_0$ of this degeneration is a
configuration of spaces intersecting over rational surfaces and curves. Next
we choose a pencil : $\mathcal{X} \dashrightarrow \mathbb{P}^1$ in the linear system $|-K_{\mathcal{X}/\Delta}|$ so that on
each fiber X_t, $t \in \Delta$, the base of the pencil $_t$: $\mathbf{X}_t \dashrightarrow \mathbb{P}^1$ intersects the
discriminant locus Δ_{X_t} in minimal number of points. As we will see below,
such lines $\mathbb{P}^1 \subset |-K_{X_t}|$ have very strong rigidity properties.

Step 2. The degeneration $\mathcal{X} \to \Delta$ can be seen as a logarithmic structure
(see [13]) $X_0^\dagger$ on X_0 – roughly an union of toric varieties, intersecting over
toric varieties plus monodromy data. The pencil induces a pencil † on the
logarithmic space $X_0^\dagger$.

Step 3. We apply a Legendre transform [13] to $(X_0^\dagger, \boldsymbol{f}^\dagger)$ and obtain a new
logarithmic structure $(Y_0^\dagger, \mathsf{w}^\dagger)$ with a pencil on it. Roughly we take the dual
intersection complex and then for vertices we specify the normal fan structure
and corresponding monodromy.

Step 4. We use the logarithmic structure $Y_0^\dagger$ to smooth Y_0 and get a mirror
degenerating family $\mathcal{Y} \to \Delta$ and a mirror Landau–Ginzburg potential $\mathsf{w} \colon \mathcal{Y} \to$
$\mathbb{P}^1$. Basically, on the generic fiber Y of $\mathcal{Y} \to \Delta$, the potential $\mathsf{w} \colon Y \dashrightarrow \mathbb{P}^1$ is an
anticanonical pencil of mirror manifolds to the anticanonical $K3$ surfaces in X.
As before we move the $\mathbb{P}^1 \subset |-K_Y|$ corresponding to the pencil $\mathsf{w} \colon Y \dashrightarrow \mathbb{P}^1$
so that it intersects the discriminant locus Δ_Y in a minimal number of points
(0 and ∞ among them – in Physics language this is the point of maximal
degeneration and the Gepner point).

The construction above will allow us to see HMS in purely geometric terms.
In many cases it could lead to a proof of HMS and in many cases it could
lead to establishing an isomorphism between the K-theory of the categories
involved. The construction suggests other categorical correspondences. Similar
procedure works in dimensions other than three and applies not only to Fano
manifolds.

Remark 3.1 The construction above works not only on anticanonical pencils but for other pencils as well. We will demonstrate this on examples of
three and four dimensional cubics. In all these examples the lines we chose
to intersect Δ_Y are rather rigidified – the number of intersection points is

given by the eigenvalues of the operator of quantum multiplication by $c_1(X)$ on $H(X)$. The correspondences coming from the associated noncommutative Hodge structures put additional restrictions (see [25]).

Step 5. The construction above allow us to work not only with the X and its Landau–Ginzburg model but with the algebraic cycles in X and the mirror Lagrangians in Y. We briefly describe how this works – for more see [23]. Let us restrict ourselves to the case of a curve C in three dimensional Fano manifold X. The construction from Sect. 2 allows us to follow what happens with the mirror image of any algebraic cycles. In the example of an algebraic curve C and its tropical realization T [29] we get that the image of T under the Legendrian transform is the conormal tori fibration – see the picture bellow. The new lagrangian cycle L is the mirror the curve C.

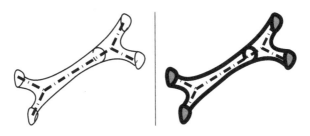

4 Birational Transformations and HMS

4.1 Some Examples

Let X be a smooth projective variety and Z be a smooth subvariety. As a consequence of the weak factorization theorem [35] it is enough to study the Landau–Ginzburg mirrors of blow-ups and blow-downs with smooth centers. Recall the following result.

Theorem 4.1 *(Orlov [30]) Let X be a smooth projective variety and X_Z be a blow up of X in a smooth subvariety Z of codimension k. Then $D^b(X_Z)$ has a semiorthogonal decomposition $(D^b(X), D^b(Z)_{k-1}, \ldots, D^b(Z)_0)$. Here $D^b(Z)_i$ are corresponding twists by $\mathcal{O}(i)$ (see [30]).*

This B-side statement has an A-model counterpart. In a joint work in progress with D. Auroux we consider the following:

Conjecture 4.2 The Landau–Ginzburg mirror (T, g, ω_T) of X_Z is the connected symplectic sum of the Landau–Ginzburg mirror (Y, w, ω_Y) of X and the Landau–Ginsburg mirror (S, f, ω_S) of $Y \times (\mathbb{C}^*)^k$. On the level of categories this means that $FS(T, g, \omega_T)$ has a semiorthogonal decomposition $(FS(T, g, \omega_T), FS(S, f, \omega_S)_{k-1}, \ldots, FS(S, f, \omega_S)_0)$.

Here $FS(S, \boldsymbol{f}, \omega_S)_i$ are categories of vanishing cycles located at $k-1$ roots of unity around infinity.

We discuss some simple examples in order to illustrate the above statement.

Example 4.3 We discuss the LG model mirror to $\mathbb{CP}^3$ blown up in a point. In this case $k = 3$. The LG mirror of the blown-up manifold is a family of K3 surfaces with 6 singular fibers. Four of these fibers correspond to the LG mirror of $\mathbb{CP}^3$ and they are situated near zero. The two remaining fibers are sitting over second roots of unity in the local chart around ∞ – see Fig. 1.

Example 4.4 We discuss LG model mirror to $\mathbb{CP}^3$ blown up in a line. In this case $k = 2$. We get as LG model of the blown-up manifold a family of K3 surfaces with six singular fibers. Four of these fibers correspond to the LG model of $\mathbb{CP}^3$ and they are situated near zero. The two other fibers are very close to each other. We briefly describe the procedure in this case. We start with the LG model for $\mathbb{CP}^3$

$$\mathsf{w} = x + y + z + \frac{1}{xyz}.$$

We add the additional term μxy, with μ corresponding to the volume of the blown up $\mathbb{CP}^1$. The critical points of

$$\mathsf{w} = x + y + z + \frac{1}{xyz} + \mu xy$$

split in two groups described in Fig. 2.

Remark 4.5 The example above is simple but instructive. Indeed the line in $\mathbb{CP}^3$ defined by $x = y = 1$ has its Landau–Ginzburg mirror defined by

$$w = 2 + z + \frac{1}{z} + 1.$$

This suggests that the Landau Ginzburg mirror of $\mathbb{CP}^1$ is contained in Landau Ginzburg mirror of $\mathbb{CP}^3$.

Fig. 1. LG model of $\mathbb{CP}^3$ blown-up at a point

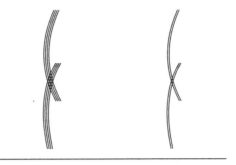

Fig. 2. LG model of $\mathbb{CP}^3$ blown-up at a line

Example 4.6 We discuss the LG model mirror to $\mathbb{CP}^n$ blown-up in $\mathbb{CP}^k$. As in the previous examples it can be seen as the gluing of several Landau–Ginzburg models mirroring to $\mathbb{CP}^n$ and $\mathbb{CP}^k \times (\mathbb{C}^*)^{n-k}$. The result is a LG model with $n + 1$ singular fibers located around zero (corresponding to $\mathbb{CP}^n$) glued to $n - k - 1$ other LG models of $\mathbb{CP}^k$ with singular fibers located around $k - 1$ roots of unity far away from zero – see also [28].

Example 4.7 Let us look at the Landau–Ginzburg model mirror to the manifold X which is the blow-up of $\mathbb{CP}^3$ in a genus two curve embedded as a $(2, 3)$ curve on a quadric surface. We describe the mirror of X. It is the deformation of the image of the compactification D of the map

$$(t : u_1 : u_2 : u_3) \mapsto (t : u_1 : u_2 : u_3 : u_1 \cdot u_2 : u_2 \cdot u_3 : u_1 \cdot u_3)$$

of $\mathbb{CP}^3$ in $\mathbb{CP}^7$.

The function w gives a pencil of degree 8 on the compactified D. We can interpret this pencil as obtained by first taking the Landau–Ginzburg mirror of $\mathbb{CP}^3$ and then adding to it a new singular fiber consisting of three rational surfaces intersecting over degenerated genus two curve Fig. 3. This process is the A-model counterpart of blowing up of the genus two curve in $\mathbb{CP}^3$ and according to Orlov's theorem, stated above, can be used to define HMS for manifolds of general type.

The potential in the equation above is

$$\mathsf{w} = t + u_1 + u_2 + u_3 + \frac{1}{u_1 \cdot u_2} + \frac{1}{u_2 \cdot u_3} + \frac{1}{u_1 \cdot u_3}.$$

Clearing denominators and substituting $t = u_3 = 0$ and $u_1^2 = u_2^3$ we get a singularity. Its resolution produces $D^b(Y, \mathsf{w})$ a quiver category with relations equivalent to the Fukaya category of genus two curve. But observe that this singularity is very different from the one of the $\mathbb{CP}^1$. This is precisely the point we would like to explore. The rational varieties produce simple singularities.

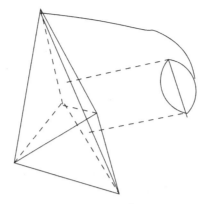

Fig. 3. LG of blow-up of $\mathbb{CP}^3$ in a genus 2 curve

Let us work out the example of LG model of $\mathbb{CP}^4$ blown up in the intersection of a smooth cubic and quadric surfaces. In the case of the cubic surface we have the compactification D of the map

$$(t : u_1 : u_2 : u_3 : u_4) \mapsto (t : u_1 : u_2 : u_3 : u_4 : u_1 \cdot u_2 \cdot u_4 : u_2 \cdot u_3 \cdot u_4 : u_1 \cdot u_2 \cdot u_3).$$

This procedure when restricted to dimension 2 produces LG model from [3]. Indeed after change of coordinates and restricting to dimension 2 we get

$$\mathsf{w} XYZ = c(X + Y + Z)^3$$

where c is a constant.

We notice here that the procedure described above allows us to look at the Landau–Ginzburg models in a different way. Namely we can see that a blow-up of subvariety X in V is nothing else but moving the line in the dual space and creating a new pencil K_Y with one more singular fiber.

We summarize our findings:

Birational transformations	Homological mirror symmetry
X	$\mathsf{w} \colon Y \to \mathbb{CP}^1$
Blowing up of X	Moving to a less singular point in Δ_Y
Blowing up of X	Creating a new singular fiber in K_Y

In the next sections we will consider the following correspondences.

A side	B side
$\mathsf{w} \colon Y \to \mathbb{CP}^1$	X
Singularities – Log canonical thresholds	Rationality of X
Complexity of the sheaf of vanishing cycles	Nonrationality of X

We will also study the relations between:

A side	B side
$w: Y \to \mathbb{CP}^1$	X
Lagrandian cycles	Hodge cycles
Lagrandian thimbles	algebraic cycles

4.2 Degenerations

As in the classical Moishezon–Teicher procedure the degeneration and the smoothing data are remembered by braid factorization and it is interesting to study how braid factorization data recover HMS. We leave this as an open question and move to a rather basic example.

Example 4.8 We describe our procedure in the case of $X = \mathbb{CP}^2$ in Fig. 4.

Example 4.9 We also describe the above procedure in the case $X = \mathbb{CP}^1 \times \mathbb{CP}^1$ and $L = (2, 2)$. This is an instructive example for our purposes. Moishezon degeneration can be summarized by Fig. 5.

Now we apply the Legendre transform to the above degeneration replacing the two affine structures at the ends by $\mathbb{C}^2$, the middle six by $\mathbb{CP}^1 \times \mathbb{C}^1$ and the central point by Del Pezzo surface of degree three in order to get the following degeneration of the Landau–Ginzburg model (cf. Fig. 6).

Figure 7 illustrates the singular fiber over 0 in the Landau–Ginsburg model.

$$\{z_1 z_2 z_3 = 0\} \subset \mathbb{C}^3, \quad W = z_1 + z_2 + z_3$$

$$\downarrow \text{ Smoothing-}$$

$$z_1 z_2 z_3 = t, \quad W = z_1 + z_2 + z_3.$$

Fig. 4. The mirror of $\mathbb{CP}^2$

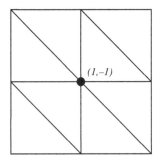

Fig. 5. Moisheson's degeneration

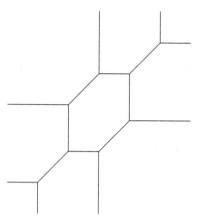

Fig. 6. The intersection complex

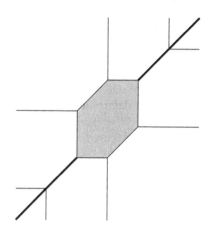

Fig. 7. The fiber over 0 of the LG potential

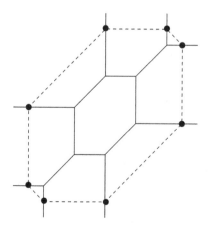

Fig. 8. The fiber over ∞

The fiber over infinity (compare with [3]) is a degenerate elliptic curve consisting of eight rational curves (cf. Fig. 8).

Observe that the procedure described in step 5. can be traced in the above examples – see Figs. 5, 6, 7. The above degeneration procedure allows us to deal with the case of all hypersurfaces applying Moishezon–Teicher degeneration.

5 Studying Non-rationality

In this section we first introduce our main tool.

5.1 The Perverse Sheaf of Vanishing Cycles and a New Technique for Studying Non-Rationality

We suggest a procedure to geometrize the use of categories in the non-rationality questions – perverse sheaf of vanishing cycle. We start by recalling some results from [14]. Suppose M is a complex manifold equipped with a proper holomorphic map $f : M \to \Delta$ onto the unit disc, which is submersive outside of $0 \in \Delta$. For simplicity we will assume that f has connected fibers. In this situation there is a natural deformation retraction $r : M \to M_0$ of M onto the singular fiber $M_0 := f^{-1}(0)$ of f. The restriction $r_t : M_t \to M_0$ of the retraction r to a smooth fiber $M_t := f^{-1}(t)$ is the "specialization to 0" map in topology. The complex of *nearby cocycles* associated with $f : M \to \Delta$ is by definition the complex of sheaves $Rr_{t*}\mathbb{Z}_{M_t} \in D^-(M_0, \mathbb{Z})$. Since for the constant sheaf we have $\mathbb{Z}_{M_t} = r_t^* \mathbb{Z}_{M_0}$, we get (by adjunction) a natural map of complexes of sheaves

$$\mathrm{sp} : \mathbb{Z}_{M_0} \to Rr_{t*}r_t^*\mathbb{Z}_{M_0} = Rr_{t*}\mathbb{Z}_{M_t}.$$

The complex of *vanishing cocycles* for f is by definition the complex $\mathbf{cone}(\mathrm{sp})$ and thus fits in an exact triangle

$$\mathbb{Z}_{M_0} \xrightarrow{\mathrm{sp}} Rr_{t*}\mathbb{Z}_{M_t} \to \mathbf{cone}(\mathrm{sp}) \to \mathbb{Z}_{M_0}[1],$$

of complexes in in $D^-(M_0, \mathbb{Z})$. Note that $\mathbb{H}^i(M_0, Rr_{t*}\mathbb{Z}_{M_t}) \cong H^i(M_t, \mathbb{Z})$ and so if we pass to hypercohomology, the exact triangle above induces a long exact sequence

$$\ldots H^i(M_0, \mathbb{Z}) \xrightarrow{r_t^*} H^i(M_t, \mathbb{Z}) \to \mathbb{H}^i(M_0, \mathbf{cone}(\mathrm{sp})) \to H^{i+1}(M_0, \mathbb{Z}) \to \ldots \quad (1)$$

Since M_0 is a projective variety, the cohomology spaces $H^i(M_0, \mathbb{C})$ carry the canonical mixed Hodge structure defined by Deligne. Also, the cohomology spaces $H^i(M_t, \mathbb{C})$ of the smooth fiber of f can be equipped with the Schmid–Steenbrink limiting mixed Hodge structure which captures essential geometric information about the degeneration $M_t \rightsquigarrow M_0$. With these choices of Hodge structures it is known from the works of Scherk and Steenbrink, that the map r_t^* in (1) is a morphism of mixed Hodge structures and that $\mathbb{H}^i(M_0, \mathbf{cone}(\mathrm{sp}))$ can be equipped with a mixed Hodge structure so that (1) is a long exact sequence of Hodge structures.

Now given a proper holomorphic function $\mathsf{w}: Y \to \mathbb{C}$ we can perform above construction near each singular fiber of w in order to obtain a complex of vanishing cocycles supported on the union of singular fibers of w. We will write $\Sigma \subset \mathbb{C}$ for the discriminant of w $Y_\Sigma := Y \times_{\mathbb{C}} \Sigma$ for the union of all singular fibers of w and $\mathscr{F}^\bullet \in D^-(Y_\Sigma, \mathbb{Z})$ for the complex of vanishing cocycles. Slightly more generally, for any subset $\Phi \subset \Sigma$ we can look at the union Y_Φ of singular fibers of w sitting over points of Φ and at the corresponding complex

$$\mathscr{F}_\Phi^\bullet = \bigoplus_{\sigma \in \Phi} \mathscr{F}_{|Y_\sigma}^\bullet \in D^-(Y_\Phi, \mathbb{Z})$$

of cocycles vanishing at those fibers. In the following we take the hypercohomology $\mathbb{H}^i(Y_\Sigma, \mathscr{F}^\bullet)$ and $\mathbb{H}^i(Y_\Phi, \mathscr{F}_\Phi^\bullet)$ with their natural Scherk–Steenbrink mixed Hodge structure. For varieties with anti-ample canonical class we have:

Theorem 5.1 *Let X be a d-dimensional Fano manifold realized as a complete intersection in some toric variety. Consider the mirror Landau–Ginzburg model $\mathsf{w}: Y \to \mathbb{C}$. Suppose that Y is smooth and that all singular fibers of w are either normal crossing divisors or have isolated singularities.*

Then the Deligne $i^{p,q}$ numbers for the mixed Hodge structure on $\mathbb{H}^\bullet(Y_\Sigma, \mathscr{F}^\bullet)$ satisfy the identity

$$i^{p,q}(\mathbb{H}^\bullet(Y_\Sigma, \mathscr{F}^\bullet)) = h^{d-p,q-1}(X).$$

Similarly we have:

Theorem 5.2 *Suppose that X is a variety with an ample canonical class realized as a complete intersection in a toric variety. Let (Y, w) be the mirror Landau–Ginzburg model. Suppose that all singular fibers of w are either normal crossing divisors or have isolated singularities.*

Then there exists a Zariski open set $U \subset \mathbb{C}$ so that Deligne's $i^{p,q}$ numbers of the mixed Hodge structure on $\mathbb{H}^\bullet(Y_{\Sigma \cap U}, \mathscr{F}_U^\bullet)$ satisfy the identity

$$i^{p,q}(\mathbb{H}^i(Y_{\Sigma \cap U}, \mathscr{F}_{\Sigma \cap U}^\bullet)) = h^{d-p,q-i+1}(X).$$

5.2 Three Dimensional Cubic from the Point of View of HMS

We move one of our main examples – the example of three dimensional cubic. First using the procedure discussed in previous sections we build Landau–Ginzburg model of three dimensional cubic. Theorem 3.2 allows us to recover the Intermediate Jacobian of the cubic. But instead of using intermediate

Fig. 9. The singular set for the LG of a 3d cubic

Jacobian in order to study rationality we will use the perverse sheaf of vanishing cycles for the Landau–Ginzburg model for three dimensional cubic.

Applying the procedure described in the previous sections we get, after smoothing, the following Landau–Ginzburg mirror:

$$xyuvw = (u + v + w)^3.t$$

with potential $x + y$. Here u, v, w are in $\mathbb{CP}^2$ and x, y are in $\mathbb{C}^2$. The singular set W of this Landau–Ginzburg model looks as in Fig. 9.

These singularities are produced as intersection of six surfaces – see Fig. 9.

Topologically the sheaf of vanishing cycles is a fibration of tori over the singular set described above. We desingularize the singular set of $\mathscr{F}$ of rational curves with three points taken out. Then $\mathscr{F}$ restricted on such a rational curve produces an S^1 local system with non-trivial monodromy. Recall that [8] the three-dimensional cubic is a conic bundle over $\mathbb{CP}^2$ with a ramification curve,

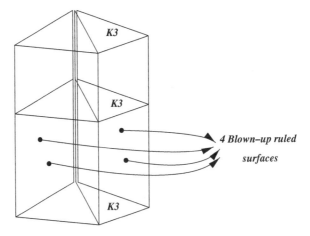

Fig. 10. K3 and arrows

a curve of genus five. The two-sheeted covering corresponds to an odd theta characteristic and this is exactly the reason why $\mathscr{F}$ restricted on such rational curve produces a S^1 local system with non-trivial monodromy. We will briefly work out the case of a conic bundle over $\mathbb{CP}^2$ with a degeneration curve – a curve of degree five and two-sheeted covering corresponding to an even theta characteristic. The first part of the Fig. 10 describes the structure of a conic bundle – two $\mathbb{CP}^1$ fibrations glued over the base. The second part of Fig. 10 describes Moishezon's degeneration of the base.

Following the procedure from the previous sections we get the Landau–Ginzburg mirror. In this case $\mathscr{F}$ restricted on rational curves with 3 points taken out produces a S^1 local system with trivial monodromy only – see Fig. 10.

We arrive at:

5.3 Non-rationality Theorem for Conic Bundles

We start with the conjecture which is a theorem modulo HMS.

Theorem 5.3 *Let X be a three-dimensional conic bundle and $Y \to \mathbb{C}$ is its Landau–Ginzburg mirror. If there exists a rational curve in the singular set such that $\mathscr{F}$ restricted on it produces a local system with non-trivial monodromy, then X is non-rational.*

The idea of the proof of this theorem is as follows. We consider the LG mirror Y of given conic bundle X obtained following the procedure of the previous section. We analyze the singularities of Y. The idea is that if $\mathscr{F}$ restricted on the singular set S produces a local system with non-trivial monodromy then S cannot be obtained as singular fiber obtained via Landau–Ginzburg "mirror partner of the blow up procedure." Using step 5 of the construction we have the following:

Proposition 5.4 The monodromy of $\mathscr{F}$ restricted to the singular set is trivial iff S corresponds to a tropical image of a smooth curve in $\mathbb{CP}^3$. In other words the singular fiber containing S comes as we move singularities from the fiber of Y at infinity to the fiber over zero – see step 5.

The tropical schemes correspond to algebraic curves iff the following conditions are satisfied:

(1) Balance – the sum all vectors coming from an intersection point is zero.

(2) Superabundance – the dimension of the moduli spaces of the tropical version as the same as the one for the existing algebraic cycle realization.

We refer for these conditions [29]. The proof of the above proposition is based on the fact that monodromy of $\mathscr{F}$ restricted to the singular set being trivial implies (1),(2) – for more details see [23].

Remark 5.5 The analysis of the singular set of the three-dimensional cubic demonstrates that balance condition fails.

Remark 5.6 The proof of the above theorem is in a way "mirror image" of Iskovskikh criteria for rationality of conic bundles [17]. Monodromy of $\mathscr{F}$ restricted to the singular set being trivial implies degeneration curve being of special type – at most trigonal or a plane quintic. Via construction described in previous section we see that on LG mirror side these are the cases where we make the monodromy of $\mathscr{F}$ restricted to the singular set being trivial.

The monodromy of $\mathscr{F}$ restricted to the singular set being trivial can be reformulated in a different way. Indeed this condition implies that logcanonical threshold of the pair Y, Y_0, where Y_0 is the zero fiber in Y is equal to one. As a consequence of the theory of Prym varieties [8] we always need to check only one fiber identified by the analysis of $\mathscr{F}$. So we have:

Theorem 5.7 *Let X be a three-dimensional conic bundle and $Y \to \mathbb{C}$ is its Landau–Ginzburg mirror. If the $LC(Y, Y_0) \neq 1$ then X is not rational.*

Remark 5.8 The analysis of the singular set of the LG model for generic three-dimensional cubic X produces $LC(Y, Y_0) = 1/2$ a different proof of nonrationaity of X.

We will briefly discuss one more example – a complete intersections of hypersurfaces of degree $(3, 0)$ and $(2, 2)$ in $\mathbb{CP}^3 \times \mathbb{CP}^2$. This produces a three-dimensional conic bundle X over a smooth cubic surface in $\mathbb{CP}^3$ with a curve of degeneration a curve of genus four. The intermediate Jacobian $J(X)$ cannot be used to detect non-rationality in this case due to the fact that $J(X)$ is isomorphic as a principally polarized abelian variety to a Jacobian of a curve (see [8]). From another point the above non-rationality criterion shows that X is not rational (for more details see [23]).

6 Non-rationality of Generic four Dimensional Cubic

In this section we will explain how the ideas developed in the previous sections apply to the case of the four-dimensional cubic.

We start by degenerating four-dimensional cubic X to three $\mathbb{CP}^4$ and then applying our standard procedure. After smoothing we get the following Landau–Ginzburg Mirror:

$$xyzuvw = (u + v + w)^3 \cdot t$$

with a potential $x + y + z$. Here u, v, w are in $\mathbb{CP}^2$ and x, y, z are in $\mathbb{C}^3$. The singular set W of this Landau–Ginzburg model can be seen on Fig. 11 – it consists of 12 rational surfaces intersecting as shown.

The following conjecture is a theorem modulo HMS.

Theorem 6.1 *The generic four-dimensional cubic X is not rational.*

The proof of this theorem is again based on the analysis of perverse sheaf of vanishing cycles $\mathscr{F}$. For generic four-dimensional cubic there exists an open rational surface in the singular set such that $\mathscr{F}$ restricted on it produces a non-trivial local system with non-trivial monodromy. By the argument above such singular fiber cannot correspond to blowing up an algebraic surface. As it follows from the work of Kulikov there is only one fiber whose vanishing

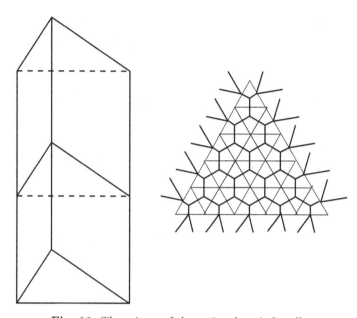

Fig. 11. The mirror of the rational conic bundle

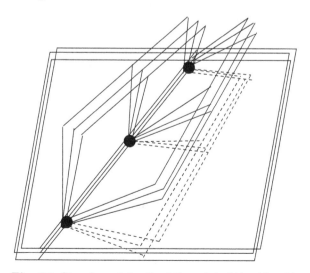

Fig. 12. Singular set for the LG model of the 4d cubic

cycles contribute to the primitive cohomologies of X and our analysis of $\mathscr{F}$ identifies this fiber. (For more details see [23].)

In a similar way as in the case of conic bundles the non-trivilaity of the monodromy translates to the language of $LC(Y, Y_0)$.

Theorem 6.2 *Let X be a four-dimensional cubic and let $Y \to \mathbb{C}$ be its Landau–Ginzburg mirror. If the $LC(Y, Y_0) \neq 1$ then X is not rational.*

Again the above statement depends on the validity of HMS.

Remark 6.3 The analysis of the singular set of the LG model for generic four-dimensional cubic X produces $LC(Y, Y_0) = 1/3$.

Many examples of four-dimensional cubics have been studied by Beauville–Donagi [6], Voisin [36], Hasset [15], Iliev [18], Kuznetsov [27]. We will analyze some of these examples now.

Example 6.4 Example of cubics containing a plane P. In this case the singular set looks as follows – see the figure bellow. As step 5 of the construction suggests that the appearance of only one plane implies that $\mathscr{F}$ has non-trivial monodromy. In this situation $LC(Y, Y_0) = 1/2$. Cubics with one plane contain another Del Pezzo surface of degree 4, whose mirrors can be seen in the figure. They project to a curve of degree 4 in P. The Landau–Ginzburg mirrors of cubics containing a plane P is different from the one of a generic four-dimensional cubic and it is obtained after algebraic degeneration and then smoothing – an earthquake – see [24]. Simillar procedure applies to other Noether–Lefschetz loci in the moduli space of four-dimensional cubics we study in what follows.

Example 6.5 Example of cubic containing two planes P_1 and P_2. In this case the fiber Y_0 of the LG model Y looks as follows:

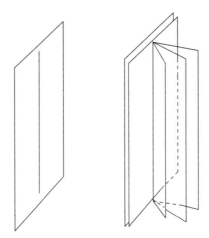

Fig. 13. Singular set for cubics containing a plane

As step 5 of the construction suggests appearance of two planes makes the monodromy of $\mathscr{F}$ trivial and in this situation $LC(Y, Y_0) = 1$.

Theorem 6.6 *Let X be a four-dimensional cubic containing a plane P. Then $LC(Y, Y_0) = 1$ for its LG mirror Y iff the projection of X from P has a section. In this case the singular set of Y is a degenerated K3 surface and X is rational.*

Example 6.7 Example of cubic containing a plane P_1 and being pfaffian. In this particular case we get a section in the projection from P which is a singular del Pezzo surface consisting of elliptic quintic curves – see the figure bellow. Recall that pfaffian cubics are obtained from $\mathbb{CP}^4$ by blowing up a K3 surface and blowing down a scroll over a Del Pezzo surface of degree 5.

Example 6.8 Hassett's examples containing a K3 surface – they are obtained from a singular four-dimensional cubic by changing vanishing cycles. On the LG mirror this results to modifying the Lefschetz thimbles. Let us restrict ourselves to the case of C_{26} – cubics with Fano varieties of lines isomorphic to symmetric power of K3 surface of degree 26.

In this case we start with a LG model of small resolution of a singular cubic. We modify it by creating (see the table below) additional thimbles with intersection form in H^4

	h^2	T
h^2	3	7
T	7	$19 + 2e$

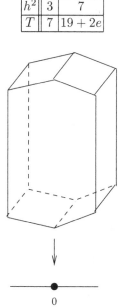

Fig. 14. Singular set for cubics containing two planes

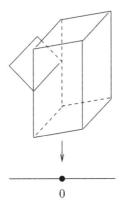

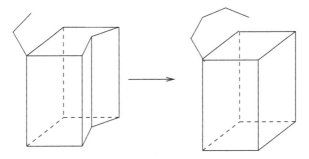

Fig. 15. Singular set for a Pfaffian cubic containing a plane

Fig. 16. Singular set for cubics from C_{26}

This produce a family of three-dimensional thimbles corresponding under mirror Symmetry to a ruled surface T of degree 7 with e the number of double points on T. In case of C_{26}, $e = 3$. Using Torelli theorem for four-dimensional cubics we get:

Modulo HMS we get:

Theorem 6.9 *All cubics in the Hassett's divisor C_{26} are rational.*

The proof follows the construction described above which allows us to recover the scroll we need to blow up.

Example 6.10 Modifications of the singular set of the LG model allows us to construct rational cubics with Fano varieties of lines non-isomorphic to symmetric power of K3 surfaces. In a similar way as explained in a previous example we modify the thimbles so that Fukaya–Seidel category corresponds to a non-commuative Poisson nad complex deformation of a K3 surface. Under this deformation hypercohomology of $\mathscr{F}$ as $LC(Y, Y_0)$ stay unchanged.

In the discussion above we have used HMS in full – we will briefly discuss how one can try proving HMS for cubics – we will extend Seidel's ideas from the case of the quintic. We first have:

Conjecture 6.11 HMS holds for the total space of $\mathcal{O}_{\mathbb{CP}^4}(3)$.

The next step is:

Conjecture 6.12 The Hochschild cohomology $HH^{\bullet}(D_0^b(\text{tot}(\mathcal{O}_{\mathbb{CP}^4}(3)))$ are isomorphic to the space all homogeneous polynomials of degree three in five variables. Here $(D_0^b$ stands for category with support at the zero section.

The Hochschild cohomology $HH^{\bullet}(D_0^b(\text{tot}(\mathcal{O}_{\mathbb{CP}^4}(3))))$ produce all B side deformations of categories of the three dimensional cubic and we have similar phenomenon on the A side as well. For the opposite statement of HMS and more general case see [23].

Remark 6.13 The above argument should work for any degree and for any dimension toric hypersurfaces.

7 Homological Mirror Symmetry and Algebraic Cycles

Algebraic cycles define objects of the category $D^b(X)$. According to Homological Mirror Symmetry they define objects in $FS(Y, \text{w}, \omega)$.

We described our procedure for obtaining this mirror object in step 5 of the construction. Full details will appear in [22], [24]. To summarize briefly, this section suggest a way for Symplectic Geometry applications to the theory of algebraic cycles. These applications are based on the following points:

(1) Most of the Hodge and algebraic cycles can seen in the singular set of the Landau–Ginzburg model. This is a HMS point of view of an idea of Green and Griffiths of an inductive approach to the Hodge conjecture [9].

(2) We enchance this approach by adding HMS and theory of tropical shemes. We briefly demonstrate that images through HMS of algebraic and Hodge cycles do form algebraic sets.

(3) We compare images through HMS of algebraic and Hodge cycles.

7.1 Vanishing Cycles and Tropical Geometry

In this subsection we will give examples of Lagrangian cycles which do not correspond to algebraic cycles. These Lagrangians are not objects of the Fukaya category and some of them still correspond to Hodge cycles in some toric degenration of a higher degree embeding of $\mathbb{CP}^3$.

We start with:

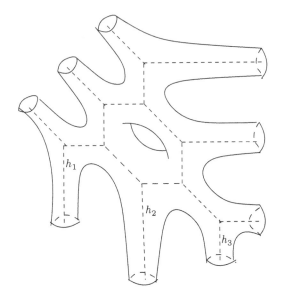

Fig. 17. Non-realizable tropical curve

Example 7.1 Let us consider a smooth plane cubic C in $\mathbb{CP}^3$. We have $dimExt^1(O_C, O_C) = 13$. Now we can associate with C a tropical curve T and let us modify it extending the horns h_1, h_2, h_3 (see Fig. 17). Using step 5 of our construction we get a Lagrangian L and $dimH^1(L) = 14$. This leads to $m_0(L) \neq 0$ and therefore L it does not represent an object in Fukaya Seidel category.

7.2 Two-Dimensional Complex Tori

Let X be a complex two-dimensional torus. As usual we have

Large complex structure limit: Going to a cusp at the boundary of the moduli of complex tori leads to a Gromov-Hausdorff colapse of X to a real

2 dimensional torus: $X \quad \rightsquigarrow \quad T := \mathbb{R}^2/\Gamma, \quad$ where Γ

- rank 2 lattice.

Mirror symplectic torus (Y, ω): is obtained as the quotient of T^*T by the fiberwise action of the lattice $\widehat{\Gamma} \subset (\mathbb{R}^2)^*$ of all linear functionals on $\mathbb{R}^2$ that take integral values on Γ; ω is induced from the standard form on T^*T.

Note:

- T can be viewed as a tropical degeneration of X.

- Objects in $Fuk(Y, \omega)$ can be constructed as images of conormal bundles to piecewise linear submanifolds in T.

The mirror map between the period matrices for X and the complexified symplectic forms $B + \sqrt{-1}\omega$ on Y, can be given explicitly:

$$\begin{pmatrix} a & b \\ c & d \end{pmatrix} \quad \leftrightarrow \quad a \cdot d_1 \wedge dx_2 + b \cdot dx_3 \wedge dx_2 + c \cdot dx_1 \wedge dx_4 + d \cdot dx_3 \wedge dx_4$$

We can use this mirror map to run some consistency checks for HMS in this case. The main idea here is that the mirror map should transform Noether–Lefschetz loci of periods (= loci of periods with extra objects in $D^b(X)$) into loci of complexified symplectic forms with additional objects in the Fukaya category. We will use conormal bundles to tropical subvarieties in T to construct the additional objects in $Fuk(Y, B + \sqrt{-1})$. We have the following matches:

- $\left\{\begin{matrix} \text{period matrices} \\ \text{with } b = -c \end{matrix}\right\} \leftrightarrow \left\{ \begin{matrix} \text{images of} \\ \text{conormal} \\ \text{bundles of} \end{matrix} \right\}$.

Note that in this case the Lagrangian in Y correspomding to the above tropical scheme has $m_0 \neq 0$. It does correspond to Hodge but non-algebraic cycle – well known counterexample to the Hodge conjecture for complex tori in dimension 2 [37].

- $\left\{\begin{matrix} \text{period matrices with} \\ b = c \end{matrix}\right\} \leftrightarrow \left\{ \begin{matrix} \text{images of} \\ \text{conormal} \\ \text{bundles of} \end{matrix} \right\}$.

The corresponding Lagrangian in Y is the mirror image for the theta divisor of the genus two Jacobian.

- $\left\{\begin{matrix} \text{period matrices with} \\ b = c = 0 \end{matrix}\right\} \leftrightarrow \left\{ \begin{matrix} \text{conormal bundles of tropical} \\ \text{elliptic curves} \end{matrix} \right\}$.

The corresponding Lagrangians in Y are the mirrors of genus one surfaces.

- $\left\{\begin{matrix} \text{period matrices} \\ \text{with } b = c = 0, \\ a = d = 1 \end{matrix}\right\} \leftrightarrow \left\{ \begin{matrix} \text{image of the} \\ \text{conormal} \\ \text{bundle of} \end{matrix} \right\}$.

In this case a new A brane appears – a coisotropic brane. The last one is a bit lost in the tropical interpretation. We loose some information by going to a tropical limit. Still, as we attempt to show in [23] where we work out the classical theory of Prym varieties and connic bundles symplectically, enough information survives. We combine the ideas of the previous two subsections in [22] where we study the Hodge conjecture for some Weil type of abelian varieties in dimension 4,6,8.

The considerations above suggest the following conjectures for L a la-. grangian 4 cycle in a four-dimensional abelian variety A.

Conjecture 7.2 ([22]) If a Lefschetz thimble (cycle) L obtained via step 5 of the construction satisfies the following conditions:

(1) L satisfies balance condition and tropically looks locally as a 4 or 6 graph.

(2) The deformations of L are unobstructed.

then it does correspond to an algebraic cycle in the mirror of A.

Examples we have considered seem to suggest that in the case of four-dimensional abelian varieties after adding big enough power of the tropicalization of the lagrangian corresponding to Θ^2 the above conditions can be achieved. Similarly we have:

Conjecture 7.3 The Hodge conjecture holds for four-dimensional rationally connected varieties.

Remark 7.4 This part of the paper came after some illuminating discussions with P. Deligne, I, Sol, R. Donagi.

Acknowledgments

We are grateful to D. Auroux, M. Gross, T. Pantev, B. Hassett. A. Kuznetsov P. Seidel, D. Orlov, M. Kontsevich for many useful conversations. Many thanks go to V. Boutchaktchiev without whom this paper would not have been written.

References

1. M. Abouzaid, On the Fukaya categories of higher genus surfaces. Adv. Math. 217 (2008), no. 3, 1192–1235.
2. D. Auroux, L. Katzarkov, S. Donaldson, and M. Yotov. Fundamental groups of complements of plane curves and symplectic invariants. Topology 43 (2004), no. 6, 1285–1318.
3. D. Auroux, L. Katzarkov, and D. Orlov. Mirror symmetry for Del Pezzo surfaces: Vanishing cycles and coherent sheaves 6. Invent. Math. 166 (2006), no. 3, 537–582.

4. D. Auroux, L. Katzarkov, T. Pantev and D. Orlov. Mirror symmetry for Del Pezzo surfaces II. in prep.
5. F. Bogomolov. The Brauer group of quotient spaces of linear representations., Izv. Akad. Nauk SSSR Ser. Mat. 51 (1987), no. 3, 485–516, 688.
6. A. Beauville and R. Donagi. La variété des droites d'une hypersurface cubique de dimension 4. C. R. Acad. Sci. Paris Sér. I Math. 301 (1985), no. 14, 703–706.
7. H. Clemens. Cohomology and obstructions II. (2002) AG 0206219.
8. H. Clemens and Ph. Griffiths. The intermediate Jacobian of the cubic threefold. Ann. of Math. 95 (1972), no. 2, 281–356.
9. M. Green and Ph. Griffiths. Algebraic Cycles I,II. Preprints (2006).
10. D. Cox and S. Katz. *Mirror symmetry and algebraic geometry*, volume 68 of *Mathematical Surveys and Monographs*. American Mathematical Society, Providence, RI, (1999).
11. T. deFernex. Adjunction beyond thresholds and birationally rigid hypersurfaces (2006). AG/0604213.
12. L. Ein and M. Musta. Mircea Invariants of singularities of pairs. International Congress of Mathematicians. Vol. II, 583–602, Eur. Math. Soc., Zrich, 2006. (Reviewer: Tommaso De Fernex) 14J17 (14E15).
13. M. Gross and B. Siebert. Mirror Symmetry via logarithmic degeneration data. I. J. Differential Geom. 72 (2006), no. 2, 169–338.
14. M. Gross and L. Katzarkov. Mirror Symmetry and vanishing cycles. in preparation.
15. B. Hassett. Some rational cubic fourfolds. J. Algebraic Geom. 8 (1999), no. 1, 103–114.
16. V.A. Iskovskikh, Y.I. Manin. Three-dimensional quartics and counterexamples to the *Lüroth* problem. , Mat. Sb. (N.S.) 86 (1971), no. 128 140–166.
17. V.A. Iskovskikh, On Rationality Criteria for Connic Bundles, Mat. Sb. 7, (1996), 75–92.
18. A. Iliev and L. Manivel. Cubic hypersurfaces and integrable systems., preprint 0606211.
19. K. Hori, S. Katz, A. Klemm, R. Pandharipande, R. Thomas, C. Vafa, R. Vakil, and E. Zaslow. *Mirror symmetry*, volume 1 of *Clay Mathematics Monographs*. American Mathematical Society, Providence, RI, 2003. With a preface by Vafa.
20. K. Hori and C. Vafa. Mirror symmetry, (2000), hep-th/0002222.
21. A. Kapustin, L. Katzarkov, D. Orlov, and M. Yotov. Homological mirror symmetry for manifolds of general type, 2004. preprint.
22. L. Katzarkov. Algebraic cycles and Homological mirror symmetry, in preparation.
23. L. Katzarkov. Mirror symmetry and nonrationality, in preparation.
24. L. Katzarkov. Tropical Geometry and Algebraic cycles, in preparation.
25. L. Katzarkov, M. Kontsevich, and T. Pantev. Hodge theoretic aspects of mirror symmetry. arXiv:0806.0107.
26. J. Kollár. Singularities of pairs, Algebraic geometry—Santa Cruz 1995, 221–287, Proc. Sympos. Pure Math., 62, Part 1, Amer. Math. Soc., Providence, RI, (1997).
27. A. Kuznetsov. Derived categories of cubic fourfolds. 0808.3351.
28. G. D. Kerr. Weighted Blowups and Mirror symmetry for Toric Surfaces, 0609162v1, To appear in Adv in Math.
29. G. Mikhalkin. Tropical geometry and its applications., 0601041, preprint.

30. D. Orlov. Derived categories of icoherent sheaves and equivalences between them. Uspekhi Mat. Nauk 58 (2003), no. 3(351), 89–172; translation in Russian Math. Surveys 58 (2003), no. 3, 511–591.

31. D. Orlov. Triangulated categories of singularities and D-branes in Landau-Ginzburg models. *Tr. Mat. Inst. Steklova*, 246(Algebr. Geom. Metody, Svyazi i Prilozh.):240–262, 2004.

32. A. Pukhlikov. Birationally rigid Fano varieties. The Fano Conference, 659–681, Univ. Torino, Turin, 2004.

33. P. Seidel. Vanishing cycles and mutation., European Congress of Mathematics, Vol. II (Barcelona, 2000), 65–85, Progr. Math., 202, Birkhäuser, Basel, 2001.

34. P. Seidel. Fukaya categories and deformations., Proceedings of the International Congress of Mathematicians, Vol. II (Beijing, 2002), 351–360, Higher Ed. Press, Beijing, (2002).

35. J. Wlodarczyk. Birational cobordisms and factorization of birational maps, J. Algebraic Geom. 9 (2000), no. 3, 425–449.

36. C. Voisin. Torelli theorems for cubics in $\mathbb{CP}^5$, Inv. Math. 86, (1986), no. 3, 577–601.

37. S. Zucker. The Hodge conjecture for four dimensional cubic. Comp. Math. 34, (1977), 199–209.

Notes on A_∞-Algebras, A_∞-Categories and Non-Commutative Geometry

M. Kontsevich[1] and Y. Soibelman[2]

[1] IHES, 35 route de Chartres, F-91440, France
   maxim@ihes.fr
[2] Department of Mathematics, KSU, Manhattan, KS 66506, USA
   soibel@math.ksu.edu

Abstract. We develop a geometric approach to A-infinity algebras and A-infinity categories based on the notion of formal scheme in the category of graded vector spaces. The geometric approach clarifies several questions, e.g. the notion of homological unit or A-infinity structure on A-infinity functors. We discuss Hochschild complexes of A-infinity algebras from geometric point of view. The chapter contains homological versions of the notions of properness and smoothness of projective varieties as well as the non-commutative version of the Hodge-to-de Rham degeneration conjecture. We also discuss a generalization of Deligne's conjecture which includes both Hochschild chains and cochains. We conclude the chapter with the description of an action of the PROP of singular chains of the topological PROP of two-dimensional surfaces on the Hochschild chain complex of an A-infinity algebra with scalar product (this action is more or less equivalent to the structure of two-dimensional Topological Field Theory associated with an "abstract" Calabi–Yau manifold).

1 Introduction

1.1 A_∞-Algebras as Spaces

The notion of A_∞-algebra introduced by Stasheff (or the notion of A_∞-category introduced by Fukaya) has two different interpretations. First one is operadic: an A_∞-algebra is an algebra over the A_∞-operad (one of its versions is the operad of singular chains of the operad of intervals in the real line). Second one is geometric: an A_∞-algebra is the same as a non-commutative formal graded manifold X over, say, field k, having a marked k-point pt and equipped with a vector field d of degree $+1$ such that $d|_{pt} = 0$ and $[d, d] = 0$ (such vector fields are called *homological*). By definition the algebra of functions on the non-commutative formal pointed graded manifold is isomorphic to the algebra of formal series $\sum_{n \geq 0} \sum_{i_1, i_2, \ldots, i_n \in I} a_{i_1 \ldots i_n} x_{i_1} \ldots x_{i_n} := \sum_M a_M x^M$ of free graded variables $x_i, i \in I$ (the set I can be infinite). Here

Kontsevich, M., Soibelman Y.: *Notes on A_∞-Algebras, A_∞-Categories and Non-Commutative Geometry*. Lect. Notes Phys. **757**, 153–219 (2009)
DOI 10.1007/978-3-540-68030-7_6

$M = (i_1, ..., i_n), n \geq 0$ is a non-commutative multi-index, i.e. an element of the free monoid generated by I. Homological vector field makes the above graded algebra into a complex of vector spaces. The triple (X, pt, d) is called a *non-commutative formal pointed differential-graded (or simply dg-) manifold.*

It is an interesting problem to make a dictionary from the pure algebraic language of A_∞-algebras and A_∞-categories to the language of non-commutative geometry.[3] One purpose of these notes is to make few steps in this direction.

From the point of view of Grothendieck's approach to the notion of "space," our formal pointed manifolds are given by functors on graded associative Artin algebras commuting with finite projective limits. It is easy to see that such functors are represented by graded coalgebras. These coalgebras can be thought of as coalgebras of distributions on formal pointed manifolds. The above-mentioned algebras of formal power series are dual to the coalgebras of distributions.

In the case of (small) A_∞-categories considered in the subsequent paper we will slightly modify the above definitions. Instead of one marked point one will have a closed subscheme of disjoint points (objects) in a formal graded manifold and the homological vector field d must be compatible with the embedding of this subscheme as well as with the projection onto it.

1.2 Some Applications of Geometric Language

Geometric approach to A_∞-algebras and A_∞-categories clarifies several long-standing questions. In particular one can obtain an explicit description of the A_∞-structure on A_∞-functors. This will be explained in detail in the subsequent paper. Here we make few remarks. In geometric terms A_∞-functors are interpreted as maps between non-commutative formal dg-manifolds commuting with homological vector fields. We will introduce a non-commutative formal dg-manifold of maps between two such spaces. Functors are just "commutative" points of the latter. The case of A_∞-categories with one object (i.e., A_∞-algebras) is considered in this chapter. The general case reflects the difference between quivers with one vertex and quivers with many vertices (vertices correspond to objects).[4] As a result of the above considerations one can describe explicitly the A_∞-structure on functors in terms of sums over sets of trees. Among other applications of our geometric language we mention an interpretation of the Hochschild chain complex of an A_∞-algebra in terms of cyclic differential forms on the corresponding formal pointed dg-manifold (Sect. 7.2).

Geometric language simplifies some proofs as well. For example, Hochschild cohomology of an A_∞-category $\mathcal{C}$ is isomorphic to $Ext^\bullet(Id_\mathcal{C}, Id_\mathcal{C})$ taken in the

---

[3] We use "formal" non-commutative geometry in tensor categories, which is different from the non-commutative geometry in the sense of Alain Connes.

[4] Another, purely algebraic approach to the A_∞-structure on functors was suggested in [39].

A_∞-category of endofunctors $\mathcal{C} \to \mathcal{C}$. This result admits an easy proof, if one interprets Hochschild cochains as vector fields and functors as maps (the idea to treat $Ext^\bullet(Id_\mathcal{C}, Id_\mathcal{C})$ as the tangent space to deformations of the derived category $D^b(\mathcal{C})$ goes back to A.Bondal).

1.3 Content of the Paper

Present paper contains two parts out of three (the last one is devoted to A_∞-categories and will appear later). Here we discuss A_∞-algebras (=non-commutative formal pointed dg-manifolds with fixed affine coordinates). We have tried to be precise and provide details of most of the proofs.

Part I is devoted to the geometric description of A_∞-algebras. We start with basics on formal graded affine schemes, then add a homological vector field, thus arriving to the geometric definition of A_∞-algebras as formal pointed dg-manifolds. Most of the material is well-known in algebraic language. We cannot completely avoid A_∞-categories (subject of the subsequent paper). They appear in the form of categories of A_∞-modules and A_∞-bimodules, which can be defined directly.

Since in the A_∞-world many notions are defined "up to quasi-isomorphism", their geometric meaning is not obvious. As an example we mention the notion of *weak unit*. Basically, this means that the unit exists at the level of cohomology only. In Sect. 4 we discuss the relationship of weak units with the "differential-graded" version of the affine line.

We start Part II with the definition of the Hochschild complexes of A_∞-algebras. As we already mentioned, Hochschild cochain complex is interpreted in terms of graded vector fields on the non-commutative formal affine space. Dualizing, Hochschild chain complex is interpreted in terms of degree one cyclic differential forms. This interpretation is motivated by [30]. It differs from the traditional picture (see e.g. [7, 11]) where one assigns to a Hochschild chain $a_0 \otimes a_1 \otimes ... \otimes a_n$ the differential form $a_0 da_1...da_n$. In our approach we intererpet a_i as the dual to an affine coordinate x_i and the above expression is dual to the cyclic differential 1-form $x_1...x_n dx_0$. We also discuss graphical description of Hochschild chains, the differential, etc.

After that we discuss homologically smooth compact A_∞-algebras. Those are analogs of smooth projective varieties in algebraic geometry. Indeed, the derived category $D^b(X)$ of coherent sheaves on a smooth projective variety X is A_∞-equivalent to the category of perfect modules over a homologically smooth compact A_∞-algebra (this can be obtained using the results of [5]). The algebra contains as much information about the geometry of X as the category $D^b(X)$ does. A good illustration of this idea is given by the "abstract" version of Hodge theory presented in Sect. 9. It is largely conjectural topic, which eventually should be incorporated in the theory of "non-commutative motives." Encoding smooth proper varieties by homologically smooth compact A_∞-algebras we can forget about the underlying commutative geometry and try to develop a theory of "non-commutative smooth projective varieties" in an abstract form. Let us briefly explain what does it mean for the Hodge theory. Let $(C_\bullet(A, A), b)$ be the Hochschild chain complex of a (weakly unital) homologically smooth compact A_∞-algebra A. The corresponding negative cyclic complex $(C_\bullet(A, A)[[u]], b + uB)$ gives rise to a family of complexes over the formal affine line $\mathbf{A}^1_{form}[+2]$ (shift of the grading reflects the fact that the variable u has degree $+2$, cf. [7, 11]). We conjecture that the corresponding family of cohomology groups gives rise to a vector bundle over the formal line. The generic fiber of this vector bundle is isomorphic to periodic cyclic homology, while the fiber over $u = 0$ is isomorphic to the Hochschild homology. If compact homologically smooth A_∞-algebra A corresponds to a smooth projective variety as explained above, then the generic fiber is just the algebraic de Rham cohomology of the variety, while the fiber over $u = 0$ is the Hodge cohomology. Then our conjecture becomes the classical theorem which claims degeneration of the spectral sequence Hodge-to-de Rham.[5]

Last section of Part II is devoted to the relationship between moduli spaces of points on a cylinder and algebraic structures on the Hochschild complexes. In Sect. 11.3 we formulate a generalization of Deligne's conjecture. Recall that Deligne's conjecture says (see e.g., [35]) that the Hochschild cochain complex of an A_∞-algebra is an algebra over the operad of chains on the topological operad of little discs. In the conventional approach to non-commutative geometry Hochschild cochains correspond to polyvector fields, while Hochschild chains correspond to de Rham differential forms. One can contract a form with a polyvector field or take a Lie derivative of a form with respect to a polyvector field. This geometric point of view leads to a generalization of Deligne's conjecture which includes Hochschild chains equipped with the structure of (homotopy) module over cochains and to the "Cartan type" calculus which involves both chains and cochains (cf. [11, 48]). We unify both approaches under one roof formulating a theorem which says that the pair consisting of the Hochschild chain and Hochschild cochain complexes of the same A_∞-algebra is an algebra over the colored operad of singular chains on configurations of discs on a cylinder with marked points on each of the boundary circles.[6]

Sections 10 and 11.6 are devoted to A_∞-algebras with scalar product, which is the same as non-commutative formal symplectic manifolds. In Sect. 10 we also discuss a homological version of this notion and explain that it corresponds to the notion of Calabi–Yau structure on a manifold. In Sect. 11.6 we define an action of the PROP of singular chains of the topological PROP of smooth oriented two-dimensional surfaces with boundaries on the Hochschild chain complex of an A_∞-algebra with scalar product. If in addition A is homologically smooth and the spectral sequence Hodge-to-de Rham degenerates, then the above action extends to the action of the PROP of singular chains

---

[5] In a recent preprint [24], D. Kaledin claims the proof of our conjecture. He uses a different approach.

[6] After our paper was finished we received the paper [49] where the authors proved an equivalent result.

on the topological PROP of stable two-dimensional surfaces. This is essentially equivalent to a structure of two-dimensional Cohomological TFT (similar ideas have been developed by Kevin Costello, see [8]). More details and an application of this approach to the calculation of Gromov–Witten invariants will be given in [22].

1.4 Generalization to A_∞-Categories

Let us say few words about the subsequent paper which is devoted to A_∞-categories. The formalism of present paper admits a straightforward generalization to the case of A_∞-categories. The latter should be viewed as non-commutative formal dg-manifolds with a closed marked subscheme of objects. Although some parts of the theory of A_∞-categories admit nice interpretation in terms of non-commutative geometry, some other still wait for it. This includes e.g. triangulated A_∞-categories. We will present the theory of triangulated A_∞-categories from the point of view of A_∞-functors from "elementary" categories to a given A_∞-category (see a summary in [33, 46, 47]). Those "elementary" categories are, roughly speaking, derived categories of representations of quivers with small number of vertices. Our approach has certain advantages over the traditional one. For example the complicated "octahedron axiom" admits a natural interpretation in terms of functors from the A_∞-category associated with the quiver of the Dynkin diagram A_2 (there are six indecomposable objects in the category $D^b(A_2 - mod)$ corresponding to six vertices of the octahedron). In some sections of the paper on A_∞-categories we have not been able to provide pure geometric proofs of the results, thus relying on less flexible approach which uses differential-graded categories (see [14]). As a compromise, we will present only part of the theory of A_∞-categories, with sketches of proofs, which are half-geomeric and half-algebraic, postponing more coherent exposition for future publications.

In the present and subsequent studies we mostly consider A_∞-algebras and categories over a field of characteristic zero. This assumption simplifies many results, but also makes some other less general. We refer the reader to [39, 40] for a theory over a ground ring instead of ground field (the approach of [39, 40] is pure algebraic and different from ours). Most of the results of present paper are valid for an A_∞-algebra A over the unital commutative associative ring k, as long as the graded module A is flat over k. More precisely, the results of Part I remain true except of the results of Sect. 3.2 (the minimal model theorem). In these two cases we assume that k is a field of characteristic zero. *Constructions* of Part II work over a commutative ring k. The results of Sect. 10 are valid (and the conjectures are expected to be valid) over a field of characteristic zero. Algebraic version of Hodge theory from Sect. 9 and the results of Sect. 11 are formulated for an A_∞-algebra over the field of characteristic zero, although the Conjecture 2 is expected to be true for any $\mathbf{Z}$-flat A_∞-algebra.

Part I: A_∞-Algebras and Non-commutative dg-Manifolds

2 Coalgebras and Non-commutative Schemes

Geometric description of A_∞-algebras will be given in terms of geometry of non-commutative ind-affine schemes in the tensor category of graded vector spaces (we will use $\mathbf{Z}$-grading or $\mathbf{Z}/2$-grading). In this section we are going to describe these ind-schemes as functors from finite-dimensional algebras to sets (cf. with the description of formal schemes in [20]). More precisely, such functors are represented by counital coalgebras. Corresponding geometric objects are called *non-commutative thin schemes*.

2.1 Coalgebras as Functors

Let k be a field and $\mathcal{C}$ be a k-linear Abelian symmetric monoidal category (we will call such categories *tensor*), which admits infinite sums and products (we refer to [13] about all necessary terminology of tensor categories). Then we can do simple linear algebra in $\mathcal{C}$, in particular, speak about associative algebras or coassociative coalgebras. For the rest of the paper, unless we say otherwise, we will assume that either $\mathcal{C} = Vect_k^{\mathbf{Z}}$, which is the tensor category of $\mathbf{Z}$-graded vector spaces $V = \oplus_{n \in \mathbf{Z}} V_n$, or $\mathcal{C} = Vect_k^{\mathbf{Z}/2}$, which is the tensor category of $\mathbf{Z}/2$-graded vector spaces (then $V = V_0 \oplus V_1$), or $\mathcal{C} = Vect_k$, which is the tensor category of k-vector spaces. Associativity morphisms in $Vect_k^{\mathbf{Z}}$ or $Vect_k^{\mathbf{Z}/2}$ are identity maps and commutativity morphisms are given by the Koszul rule of signs: $c(v_i \otimes v_j) = (-1)^{ij} v_j \otimes v_i$, where v_n denotes an element of degree n.

We will denote by $\mathcal{C}^f$ the Artinian category of finite-dimensional objects in $\mathcal{C}$ (i.e. objects of finite length). The category $Alg_{\mathcal{C}^f}$ of unital finite-dimensional algebras is closed with respect to finite projective limits. In particular, finite products and finite fiber products exist in $Alg_{\mathcal{C}^f}$. One has also the categories $Coalg_{\mathcal{C}}$ (resp. $Coalg_{\mathcal{C}^f}$) of coassociative counital (resp. coassociative counital finite-dimensional) coalgebras. In the case $\mathcal{C} = Vect_k$ we will also use the notation Alg_k, Alg_k^f, $Coalg_k$ and $Coalg_k^f$ for these categories. The category $Coalg_{\mathcal{C}^f} = Alg_{\mathcal{C}^f}^{op}$ admits finite inductive limits.

We will need simple facts about coalgebras. We will present proofs in the Appendix for completness.

Theorem 2.1 *Let $F : Alg_{\mathcal{C}^f} \to Sets$ be a covariant functor commuting with finite projective limits. Then it is isomorphic to a functor of the type $A \mapsto Hom_{Coalg_{\mathcal{C}}}(A^*, B)$ for some counital coalgebra B. Moreover, the category of such functors is equivalent to the category of counital coalgebras.*

Proposition 2.2 If $B \in Ob(Coalg_{\mathcal{C}})$, then B is a union of finite-dimensional counital coalgebras.

Objects of the category $Coalg_{\mathcal{C}f} = Alg_{\mathcal{C}f}^{op}$ can be interpreted as "very thin" non-commutative affine schemes (cf. with finite schemes in algebraic geometry). Proposition 1 implies that the category $Coalg_{\mathcal{C}}$ is naturally equivalent to the category of ind-objects in $Coalg_{\mathcal{C}f}$.

For a counital coalgebra B we denote by $Spc(B)$ (the "spectrum" of the coalgebra B) the corresponding functor on the category of finite-dimensional algebras. A functor isomorphic to $Spc(B)$ for some B is called a *non-commutative thin scheme*. The category of non-commutative thin schemes is equivalent to the category of counital coalgebras. For a non-commutative scheme X we denote by B_X the corresponding coalgebra. We will call it the coalgebra of *distributions* on X. The *algebra of functions* on X is by definition $\mathcal{O}(X) = B_X^*$.

Non-commutative thin schemes form a full monoidal subcategory $NAff_{\mathcal{C}}^{th}$ $\subset Ind(NAff_{\mathcal{C}})$ of the category of non-commutative ind-affine schemes (see Appendix). Tensor product corresponds to the tensor product of coalgebras.

Let us consider few examples.

Example 2.3 Let $V \in Ob(\mathcal{C})$. Then $T(V) = \oplus_{n \geq 0} V^{\otimes n}$ carries a structure of counital cofree coalgebra in $\mathcal{C}$ with the coproduct $\Delta(v_0 \otimes ... \otimes v_n) = \sum_{0 \leq i \leq n} (v_0 \otimes ... \otimes v_i) \otimes (v_{i+1} \otimes ... \otimes v_n)$. The corresponding non-commutative thin scheme is called non-commutative formal affine space V_{form} (or formal neighborhood of zero in V).

Definition 2.4 A non-commutative formal manifold X is a non-commutative thin scheme isomorphic to some $Spc(T(V))$ from the example above. The dimension of X is defined as $dim_k V$.

The algebra $\mathcal{O}(X)$ of functions on a non-commutative formal manifold X of dimension n is isomorphic to the topological algebra $k\langle\langle x_1, ..., x_n \rangle\rangle$ of formal power series in free graded variables $x_1, ..., x_n$.

Let X be a non-commutative formal manifold and $pt: k \to B_X$ a k-point in X,

Definition 2.5 The pair (X, pt) is called a non-commutative formal pointed manifold. If $\mathcal{C} = Vect_k^{\mathbf{Z}}$ it will be called non-commutative formal pointed graded manifold. If $\mathcal{C} = Vect_k^{\mathbf{Z}/2}$ it will be called non-commutative formal pointed supermanifold.

The following example is a generalization of the Example 1 (which corresponds to a quiver with one vertex).

Example 2.6 Let I be a set and $B_I = \oplus_{i \in I} 1_i$ be the direct sum of trivial coalgebras. We denote by $\mathcal{O}(I)$ the dual topological algebra. It can be thought of as the algebra of functions on a discrete non-commutative thin scheme I.

A quiver Q in C with the set of vertices I is given by a collection of objects $E_{ij} \in \mathcal{C}, i, j \in I$ called spaces of arrows from i to j. The coalgebra of Q is the coalgebra B_Q generated by the $\mathcal{O}(I) - \mathcal{O}(I)$-bimodule $E_Q = \oplus_{i,j \in I} E_{ij}$, i.e. $B_Q \simeq \oplus_{n \geq 0} \oplus_{i_0, i_1, \ldots, i_n \in I} E_{i_0 i_1} \otimes \ldots \otimes E_{i_{n-1} i_n} := \oplus_{n \geq 0} B_Q^n$, $B_Q^0 := B_I$. Elements of B_Q^0 are called *trivial paths*. Elements of B_Q^n are called paths of the length n. Coproduct is given by the formula

$$\Delta(e_{i_0 i_1} \otimes \ldots \otimes e_{i_{n-1} i_n}) = \oplus_{0 \leq m \leq n} (e_{i_0 i_1} \otimes \ldots \otimes e_{i_{m-1} i_m}) \otimes (e_{i_m i_{m+1}} \ldots \otimes \ldots \otimes e_{i_{n-1} i_n}),$$

where for $m = 0$ (resp. $m = n$) we set $e_{i_{-1} i_0} = 1_{i_0}$ (resp. $e_{i_n i_{n+1}} = 1_{i_n}$).

In particular, $\Delta(1_i) = 1_i \otimes 1_i, i \in I$ and $\Delta(e_{ij}) = 1_i \otimes e_{ij} + e_{ij} \otimes 1_j$, where $e_{ij} \in E_{ij}$ and $1_m \in B_I$ corresponds to the image of $1 \in \mathbf{1}$ under the natural embedding into $\oplus_{m \in I} \mathbf{1}$.

The coalgebra B_Q has a counit ε such that $\varepsilon(1_i) = 1_i$ and $\varepsilon(x) = 0$ for $x \in B_Q^n, n \geq 1$.

Example 2.7 (Generalized quivers). Here we replace $\mathbf{1}_i$ by a unital simple algebra A_i (e.g. $A_i = Mat(n_i, D_i)$, where D_i is a division algebra). Then E_{ij} are $A_i - mod - A_j$-bimodules. We leave as an exercise to the reader to write down the coproduct (one uses the tensor product of bimodules) and to check that we indeed obtain a coalgebra.

Example 2.8 Let I be a set. Then the coalgebra $B_I = \oplus_{i \in I} \mathbf{1}_i$ is a direct sum of trivial coalgebras, isomorphic to the unit object in $\mathcal{C}$. This is a special case of Example 2. Note that in general B_Q is a $\mathcal{O}(I) - \mathcal{O}(I)$-bimodule.

Example 2.9 Let A be an associative unital algebra. It gives rise to the functor $F_A : Coalg_{\mathcal{C}f} \to Sets$ such that $F_A(B) = Hom_{Alg_{\mathcal{C}}}(A, B^*)$. This functor describes finite-dimensional representations of A. It commutes with finite direct limits, hence it is representable by a coalgebra. If $A = \mathcal{O}(X)$ is the algebra of regular functions on the affine scheme X, then in the case of algebraically closed field k the coalgebra representing F_A is isomorphic to $\oplus_{x \in X(k)} \mathcal{O}_{x,X}^*$, where $\mathcal{O}_{x,X}^*$ denotes the topological dual to the completion of the local ring $\mathcal{O}_{x,X}$. If X is smooth of dimension n, then each summand is isomorphic to the topological dual to the algebra of formal power series $k[[t_1, \ldots, t_n]]$. In other words, this coalgebra corresponds to the disjoint union of formal neighborhoods of all points of X.

Remark 2.10 One can describe non-commutative thin schemes more precisely by using structure theorems about finite-dimensional algebras in $\mathcal{C}$. For example, in the case $\mathcal{C} = Vect_k$ any finite-dimensional algebra A is isomorphic to a sum $A_0 \oplus r$, where A_0 is a finite sum of matrix algebras $\oplus_i Mat(n_i, D_i)$, D_i are division algebras and r is the radical. In $\mathbf{Z}$-graded case a similar decomposition holds, with A_0 being a sum of algebras of the type $End(V_i) \otimes D_i$, where V_i are some graded vector spaces and D_i are division algebras of degree zero. In $\mathbf{Z}/2$-graded case the description is slightly more complicated. In particular A_0 can contain summands isomorphic to $(End(V_i) \otimes D_i) \otimes D_\lambda$, where V_i and D_i are $\mathbf{Z}/2$-graded analogs of the above-described objects and D_λ is a $1|1$-dimensional superalgebra isomorphic to $k[\xi]/(\xi^2 = \lambda)$, $deg\, \xi = 1, \lambda \in k^*/(k^*)^2$.

2.2 Smooth Thin Schemes

Recall that the notion of an ideal has meaning in any abelian tensor category. A two-sided ideal J is called *nilpotent* if the multiplication map $J^{\otimes n} \to J$ has zero image for a sufficiently large n.

Definition 2.11 Counital coalgebra B in a tensor category $\mathcal{C}$ is called smooth if the corresponding functor $F_B \colon Alg_{\mathcal{C}^f} \to Sets, F_B(A) = Hom_{Coalgc}(A^*, B)$ satisfies the following lifting property: for any two-sided nilpotent ideal $J \subset A$ the map $F_B(A) \to F_B(A/J)$ induced by the natural projection $A \to A/J$ is surjective. Non-commutative thin scheme X is called smooth if the corresponding counital coalgebra $B = B_X$ is smooth.

Proposition 2.12 For any quiver Q in $\mathcal{C}$ the corresponding coalgebra B_Q is smooth.

Proof. First let us assume that the result holds for all finite quivers. We remark that if A is finite-dimensional and Q is an infinite quiver then for any morphism $f \colon A^* \to B_Q$ we have: $f(A^*)$ belongs to the coalgebra of a finite sub-quiver of Q. Since the lifting property holds for the latter, the result follows. Finally, we need to prove the Proposition for a finite quiver Q. Let us choose a basis $\{e_{ij,\alpha}\}$ of each space of arrows E_{ij}. Then for a finite-dimensional algebra A the set $F_{B_Q}(A)$ is isomorphic to the set $\{((\pi_i), x_{ij,\alpha})_{i,j \in I}\}$, where $\pi_i \in A, \pi_i^2 = \pi_i, \pi_i \pi_j = \pi_j \pi_i$, if $i \neq j, \sum_{i \in I} \pi_i = 1_A$ and $x_{ij,\alpha} \in \pi_i A \pi_j$ satisfy the condition: there exists $N \geq 1$ such that $x_{i_1 j_1, \alpha_1} ... x_{i_m j_m, \alpha_m} = 0$ for all $m \geq N$. Let now $J \subset A$ be the nilpotent ideal from the definition of smooth coalgebra and $(\pi_i', x_{ij,\alpha}')$ be elements of A/J satisfying the above constraints. Our goal is to lift them to A. We can lift the them to the projectors π_i and elements $x_{ij,\alpha}$ for A in such a way that the above constraints are satisfied except of the last one, which becomes an inclusion $x_{i_1 j_1, \alpha_1} ... x_{i_m j_m, \alpha_m} \in J$ for $m \geq N$. Since $J^n = 0$ in A for some n we see that $x_{i_1 j_1, \alpha_1} ... x_{i_m j_m, \alpha_m} = 0$ in A for $m \geq nN$. This proves the result. ∎

Remark 2.13 (a) According to Cuntz and Quillen [10] a non-commutative *algebra* R in $Vect_k$ is called *smooth* if the functor $Alg_k \to Sets, F_R(A) = Hom_{Alg_k}(R, A)$ satisfies the lifting property from the Definition 3 applied to all (not only finite-dimensional) algebras. We remark that if R is smooth in the sense of Cuntz and Quillen then the coalgebra R_{dual} representing the functor $Coalg_k^f \to Sets, B \mapsto Hom_{Alg_k^f}(R, B^*)$ is smooth. One can prove that any smooth coalgebra in $Vect_k$ is isomorphic to a coalgebra of a generalized quiver (see Example 3).

(b) Almost all examples of non-commutative smooth thin schemes considered in this paper are formal pointed manifolds, i.e. they are isomorphic to $Spc(T(V))$ for some $V \in Ob(\mathcal{C})$. It is natural to try to "globalize" our results to the case of non-commutative "smooth" schemes X which satisfy the property that the completion of X at a "commutative" point gives rise to a formal pointed manifold in our sense. An example of the space of maps is considered in the next subsection.

(c) The tensor product of non-commutative smooth thin schemes is typically non-smooth, since it corresponds to the *tensor product* of coalgebras (the latter is not a categorical product).

Let now x be a k-point of a non-commutative smooth thin scheme X. By definition x is a homomorphism of counital coalgebras $x \colon k \to B_X$ (here $k = 1$ is the trivial coalgebra corresponding to the unit object). The completion $\widehat{X}_x$ of X at x is a formal pointed manifold which can be described such as follows. As a functor $F_{\widehat{X}_x} \colon Alg_C^f \to Sets$ it assigns to a finite-dimensional algebra A the set of such homomorphisms of counital colagebras $f \colon A^* \to B_X$ which are compositions $A^* \to A_1^* \to B_X$, where $A_1^* \subset B_X$ is a conilpotent extension of x (i.e., A_1 is a finite-dimensional unital nilpotent algebra such that the natural embedding $k \to A_1^* \to B_X$ coinsides with $x \colon k \to B_X$).

Description of the coalgebra $B_{\widehat{X}_x}$ is given in the following Proposition.

Proposition 2.14 The formal neighborhood $\widehat{X}_x$ corresponds to the counital sub-coalgebra $B_{\widehat{X}_x} \subset B_X$ which is the preimage under the natural projection $B_X \to B_X/x(k)$ of the sub-coalgebra consisting of conilpotent elements in the non-counital coalgebra $B/x(k)$. Moreover, $\widehat{X}_x$ is universal for all morphisms from nilpotent extensions of x to X.

We discuss in Appendix a more general construction of the completion along a non-commutative thin subscheme.

We leave as an exercise to the reader to prove the following result.

Proposition 2.15 Let Q be a quiver and $pt_i \in X = X_{B_Q}$ corresponds to a vertex $i \in I$. Then the formal neighborhood $\widehat{X}_{pt_i}$ is a formal pointed manifold corresponding to the tensor coalgebra $T(E_{ii}) = \oplus_{n \geq 0} E_{ii}^{\otimes n}$, where E_{ii} is the space of loops at i.

2.3 Inner Hom

Let X, Y be non-commutative thin schemes and B_X, B_Y the corresponding coalgebras.

Theorem 2.16 *The functor $Alg_{C^f} \to Sets$ such that*

$$A \mapsto Hom_{Coalg_C}(A^* \otimes B_X, B_Y)$$

is representable. The corresponding non-commutative thin scheme is denoted by $Maps(X, Y)$.

Proof. It is easy to see that the functor under consideration commutes with finite projective limits. Hence it is of the type $A \mapsto Hom_{Coalg_C}(A^*, B)$, where

B is a counital coalgebra (Theorem 1). The corresponding non-commutative thin scheme is the desired $Maps(X, Y)$. ∎

It follows from the definition that $Maps(X, Y) = \underline{Hom}(X, Y)$, where the inner Hom is taken in the symmetric monoidal category of non-commutative thin schemes. By definition $\underline{Hom}(X, Y)$ is a non-commutative thin scheme, which satisfies the following functorial isomorphism for any $Z \in Ob(NAff_C^{th})$:

$$Hom_{NAff_C^{th}}(Z, \underline{Hom}(X, Y)) \simeq Hom_{NAff_C^{th}}(Z \otimes X, Y).$$

Note that the monoidal category $NAff_C$ of all non-commutative affine schemes does not have inner $Hom's$ even in the case $C = Vect_k$. If $C = Vect_k$ then one can define $\underline{Hom}(X, Y)$ for $X = Spec(A)$, where A is a finite-dimensional unital algebra and Y is arbitrary. The situation is similar to the case of "commutative" algebraic geometry, where one can define an affine scheme of maps from a scheme of finite length to an arbitrary affine scheme. On the other hand, one can show that the category of non-commutative ind-affine schemes admit inner Hom's (the corresponding result for commutative ind-affine schemes is known).

Remark 2.17 The non-commutative thin scheme $Maps(X, Y)$ gives rise to a quiver, such that its vertices are k-points of $Maps(X, Y)$. In other words, vertices correspond to homomorphisms $B_X \to B_Y$ of the coalgebras of distributions. Taking the completion at a k-point we obtain a formal pointed manifold. More generally, one can take a completion along a subscheme of k-points, thus arriving to a non-commutative formal manifold with a marked closed subscheme (rather than one point). This construction will be used in the subsequent paper for the desription of the A_∞-structure on A_∞-functors. We also remark that the space of arrows E_{ij} of a quiver is an example of the geometric notion of bitangent space at a pair of k-points i, j. It will be discussed in the subsequent paper.

Example 2.18 Let $Q_1 = \{i_1\}$ and $Q_2 = \{i_2\}$ be quivers with one vertex such that $E_{i_1 i_1} = V_1, E_{i_2 i_2} = V_2, \dim V_i < \infty, i = 1, 2$. Then $B_{Q_i} = T(V_i), i = 1, 2$ and $Maps(X_{B_{Q_1}}, X_{B_{Q_2}})$ corresponds to the quiver Q such that the set of vertices $I_Q = Hom_{Coalg_C}(B_{Q_1}, B_{Q_2}) = \prod_{n \geq 1} \underline{Hom}(V_1^{\otimes n}, V_2)$ and for any two vertices $f, g \in I_Q$ the space of arrows is isomorphic to $E_{f,g} = \prod_{n \geq 0} \underline{Hom}(V_1^{\otimes n}, V_2)$.

Definition 2.19 Homomorphism $f: B_1 \to B_2$ of counital coalgebras is called a minimal conilpotent extension if it is an inclusion and the induced coproduct on the non-counital coalgebra $B_2/f(B_1)$ is trivial.

Composition of minimal conilpotent extensions is simply called a conilpotent extension. Definition 2.2.1 can be reformulated in terms of finite-dimensional coalgebras. Coalgebra B is smooth if the functor $C \mapsto Hom_{Coalg_C}(C, B)$ satisfies the lifting property with respect to conilpotent extensions of finite-dimensional counital coalgebras. The following proposition shows that we can drop the condition of finite-dimensionality.

Proposition 2.20 If B is a smooth coalgebra then the functor $Coalg_C \to Sets$ such that $C \mapsto Hom_{Coalg_C}(C, B)$ satisfies the lifting property for conilpotent extensions.

Proof. Let $f\colon B_1 \to B_2$ be a conilpotent extension, and $g\colon B_1 \to B$ and be an arbitrary homomorphism of counital algebras. It can be thought of as homomorphism of $f(B_1) \to B$. We need to show that g can be extended to B_2. Let us consider the set of pairs (C, g_C) such $f(B_1) \subset C \subset B_2$ and $g_C\colon C \to B$ defines an extension of counital coalgebras, which coincides with g on $f(B_1)$. We apply Zorn lemma to the partially ordered set of such pairs and see that there exists a maximal element (B_{max}, g_{max}) in this set. We claim that $B_{max} = B_2$. Indeed, let $x \in B_2 \setminus B_{max}$. Then there exists a finite-dimensional coalgebra $B_x \subset B_2$ which contains x. Clearly B_x is a conilpotent extension of $f(B_1) \cap B_x$. Since B is smooth we can extend $g_{max}\colon f(B_1) \cap B_x \to B$ to $g_x\colon B_x \to B$ and, finally to $g_{x,max}\colon B_x + B_{max} \to B$. This contradicts to maximality of (B_{max}, g_{max}). Proposition is proved. ∎

Proposition 2.21 If X, Y are non-commutative thin schemes and Y is smooth then $Maps(X, Y)$ is also smooth.

Proof. Let $A \to A/J$ be a nilpotent extension of finite-dimensional unital algebras. Then $(A/J)^* \otimes B_X \to A^* \otimes B_X$ is a conilpotent extension of counital coalgebras. Since B_Y is smooth then the previous Proposition implies that the induced map $Hom_{Coalg_C}(A^* \otimes B_X, B_Y) \to Hom_{Coalg_C}((A/J)^* \otimes B_X, B_Y)$ is surjective. This concludes the proof. ∎

Let us consider the case when (X, pt_X) and (Y, pt_Y) are non-commutative formal pointed manifolds in the category $\mathcal{C} = Vect_k^{\mathbf{Z}}$. One can describe "in coordinates" the non-commutative formal pointed manifold, which is the formal neighborhood of a k-point of $Maps(X, Y)$. Namely, let $X = Spc(B)$ and $Y = Spc(C)$, and let $f \in Hom_{NAff_C^{fth}}(X, Y)$ be a morphism preserving marked points. Then f gives rise to a k-point of $Z = Maps(X, Y)$. Since $\mathcal{O}(X)$ and $\mathcal{O}(Y)$ are isomorphic to the topological algebras of formal power series in free graded variables, we can choose sets of free topological generators $(x_i)_{i \in I}$ and $(y_j)_{j \in J}$ for these algebras. Then we can write for the corresponding homomorphism of algebras $f^*\colon \mathcal{O}(Y) \to \mathcal{O}(X)$:

$$f^*(y_j) = \sum_I c_{j,M}^0 x^M,$$

where $c_{j,M}^0 \in k$ and $M = (i_1, ..., i_n), i_s \in I$ is a non-commutative multi-index (all the coefficients depend on f, hence a better notation should be $c_{j,M}^{f,0}$). Notice that for $M = 0$ one gets $c_{j,0}^0 = 0$ since f is a morphism of pointed schemes. Then we can consider an "infinitesimal deformation" f_{def} of f

$$f_{def}^*(y_j) = \sum_M (c_{j,M}^0 + \delta c_{j,M}^0) x^M,$$

where $\delta c^0_{j,M}$ are *new variables* commuting with all x_i. Then $\delta c^0_{j,M}$ can be thought of as coordinates in the formal neighborhood of f. More pedantically it can be spelled out such as follows. Let $A = k \oplus m$ be a finite-dimensional graded unital algebra, where m is a graded nilpotent ideal of A. Then an A-point of the formal neighborhood U_f of f is a morphism $\phi \in Hom_{NAff^{th}_\mathcal{C}}(Spec(A) \otimes X, Y)$, such that it reduces to f modulo the nilpotent ideal m. We have for the corresponding homomorphism of algebras:

$$\phi^*(y_j) = \sum_M c_{j,M} x^M,$$

where M is a non-commutative multi-index, $c_{j,M} \in A$, and $c_{j,M} \mapsto c^0_{j,M}$ under the natural homomorphism $A \to k = A/m$. In particular $c_{j,0} \in m$. We can treat coefficients $c_{j,M}$ as A-points of the formal neighborhood U_f of $f \in Maps(X, Y)$.

Remark 2.22 The above definitions will play an important role in the subsequent paper, where the non-commutative smooth thin scheme $Spc(B_Q)$ will be assigned to a (small) A_∞-category, the non-commutative smooth thin scheme $Maps(Spc(B_{Q_1}), Spc(B_{Q_2}))$ will be used for the description of the category of A_∞-functors between A_∞-categories and the formal neighborhood of a point in the space $Maps(Spc(B_{Q_1}), Spc(B_{Q_2}))$ will correspond to natural transformations between A_∞-functors.

3 A_∞-Algebras

3.1 Main Definitions

From now on assume that $\mathcal{C} = Vect^\mathbf{Z}_k$ unless we say otherwise. If X is a thin scheme then a vector field on X is, by definition, a derivation of the coalgebra B_X. Vector fields form a graded Lie algebra $Vect(X)$.

Definition 3.1 A non-commutative thin differential-graded (dg for short) scheme is a pair (X, d) where X is a non-commutative thin scheme and d is a vector field on X of degree $+1$ such that $[d, d] = 0$.

We will call the vector field d *homological vector field*.

Let X be a formal pointed manifold and x_0 be its unique k-point. Such a point corresponds to a homomorphism of counital coalgebras $k \to B_X$. We say that the vector field d vanishes at x_0 if the corresponding derivation kills the image of k.

Definition 3.2 A non-commutative formal pointed dg-manifold is a pair $((X, x_0), d)$ such that (X, x_0) is a non-commutative formal pointed graded manifold and $d = d_X$ is a homological vector field on X such that $d|_{x_0} = 0$.

Homological vector field d has an infinite Taylor decomposition at x_0. More precisely, let $T_{x_0}X$ be the tangent space at x_0. It is canonically isomorphic to the graded vector space of primitive elements of the coalgebra B_X, i.e. the set of $a \in B_X$ such that $\Delta(a) = 1 \otimes a + a \otimes 1$ where $1 \in B_X$ is the image of $1 \in k$ under the homomorphism of coalgebras $x_0 : k \to B_X$ (see Appendix for the general definition of the tangent space). Then $d := d_X$ gives rise to a (non-canonically defined) collection of linear maps $d_X^{(n)} := m_n$: $T_{x_0}X^{\otimes n} \to T_{x_0}X[1], n \geq 1$ called *Taylor coefficients of d* which satisfy a system of quadratic relations arising from the condition $[d,d] = 0$. Indeed, our non-commutative formal pointed manifold is isomorphic to the formal neighborhood of zero in $T_{x_0}X$, hence the corresponding non-commutative thin scheme is isomorphic to the cofree tensor coalgebra $T(T_{x_0}X)$ generated by $T_{x_0}X$. Homological vector field d is a derivation of a cofree coalgebra, hence it is uniquely determined by a sequence of linear maps m_n.

Definition 3.3 Non-unital A_∞-algebra over k is given by a non-commutative formal pointed dg-manifold (X, x_0, d) together with an isomorphism of counital coalgebras $B_X \simeq T(T_{x_0}X)$.

Choice of an isomorphism with the tensor coalgebra generated by the tangent space is a non-commutative analog of a choice of affine structure in the formal neighborhood of x_0.

From the above definitions one can recover the traditional one. We present it below for convenience of the reader.

Definition 3.4 A structure of an A_∞-algebra on $V \in Ob(Vect_k^{\mathbf{Z}})$ is given by a derivation d of degree $+1$ of the non-counital cofree coalgebra $T_+(V[1]) = \oplus_{n \geq 1} V^{\otimes n}$ such that $[d,d] = 0$ in the differential-graded Lie algebra of coalgebra derivations.

Traditionally the Taylor coefficients of $d = m_1 + m_2 + \cdots$ are called (higher) multiplications for V. The pair (V, m_1) is a complex of k-vector spaces called the *tangent complex*. If $X = Spc(T(V))$ then $V[1] = T_0X$ and $m_1 = d_X^{(1)}$ is the first Taylor coefficient of the homological vector field d_X. The tangent cohomology groups $H^i(V, m_1)$ will be denoted by $H^i(V)$. Clearly $H^\bullet(V) = \oplus_{i \in \mathbf{Z}} H^i(V)$ is an associative (non-unital) algebra with the product induced by m_2.

An important class of A_∞-algebras consists of *unital* (or strictly unital) and *weakly unital* (or homologically unital) ones. We are going to discuss the definition and the geometric meaning of unitality later.

Homomorphism of A_∞-algebras can be described geometrically as a morphism of the corresponding non-commutative formal pointed dg-manifolds. In the algebraic form one recovers the following traditional definition.

Definition 3.5 A homomorphism of non-unital A_∞-algebras (A_∞-morphism for short) $(V, d_V) \to (W, d_W)$ is a homomorphism of differential-graded coalgebras $T_+(V[1]) \to T_+(W[1])$.

A homomorphism f of non-unital A_∞-algebras is determined by its Taylor coefficients $f_n \colon V^{\otimes n} \to W[1-n], n \geq 1$ satisfying the system of equations

$$\sum_{1 \leq l_1 < \ldots < l_i = n} (-1)^{\gamma_i} m_i^W (f_{l_1}(a_1, \ldots, a_{l_1}),$$
$$f_{l_2-l_1}(a_{l_1+1}, \ldots, a_{l_2}), \ldots, f_{n-l_{i-1}}(a_{n-l_{i-1}+1}, \ldots, a_n)) =$$
$$\sum_{s+r=n+1} \sum_{1 \leq j \leq s} (-1)^{\epsilon_s} f_s(a_1, \ldots, a_{j-1}, m_r^V(a_j, \ldots, a_{j+r-1}), a_{j+r}, \ldots, a_n).$$

Here $\epsilon_s = r \sum_{1 \leq p \leq j-1} deg(a_p) + j - 1 + r(s-j)$, $\gamma_i = \sum_{1 \leq p \leq i-1} (i-p)(l_p - l_{p-1} - 1) + \sum_{1 \leq p \leq i-1} \nu(l_p) \sum_{l_{p-1}+1 \leq q \leq l_p} deg(a_q)$, where we use the notation $\nu(l_p) = \sum_{p+1 \leq m \leq i} (1 - l_m + l_{m-1})$ and set $l_0 = 0$.

Remark 3.6 All the above definitions and results are valid for $\mathbf{Z}/2$-graded A_∞-algebras as well. In this case we consider formal manifolds in the category $Vect_k^{\mathbf{Z}/2}$ of $\mathbf{Z}/2$-graded vector spaces. We will use the correspodning results without further comments. In this case one denotes by ΠA the $\mathbf{Z}/2$-graded vector space $A[1]$.

3.2 Minimal Models of A_∞-Algebras

One can do simple differential geometry in the symmetric monoidal category of non-commutative formal pointed dg-manifolds. New phenomenon is the possibility to define some structures up to a quasi-isomorphism.

Definition 3.7 Let $f \colon (X, d_X, x_0) \to (Y, d_Y, y_0)$ be a morphism of non-commutative formal pointed dg-manifolds. We say that f is a quasi-isomorphism if the induced morphism of the tangent complexes $f_1 \colon (T_{x_0}X, d_X^{(1)}) \to (T_{y_0}Y, d_Y^{(1)})$ is a quasi-isomorphism. We will use the same terminology for the corresponding A_∞-algebras.

Definition 3.8 An A_∞-algebra A (or the corresponding non-commutative formal pointed dg-manifold) is called minimal if $m_1 = 0$. It is called contractible if $m_n = 0$ for all $n \geq 2$ and $H^\bullet(A, m_1) = 0$.

The notion of minimality is coordinate independent, while the notion of contractibility is not.

It is easy to prove that any A_∞-algebra A has a *minimal model* M_A, i.e. M_A is minimal and there is a quasi-isomorphism $M_A \to A$ (the proof is similar to the one from [29, 36]). The minimal model is unique up to an A_∞-isomorphism. We will use the same terminology for non-commutative formal pointed dg-manifolds. In geometric language a non-commutative formal pointed dg-manifold X is isomorphic to a *categorical* product (i.e. corresponding to the completed free product of algebras of functions) $X_m \times X_{lc}$, where X_m is minimal and X_{lc} is linear contractible. The above-mentioned quasi-isomorphism corresponds to the projection $X \to X_m$.

The following result (homological inverse function theorem) can be easily deduced from the above product decomposition.

Proposition 3.9 If $f : A \to B$ is a quasi-isomorphism of A_∞-algebras then there is a (non-canonical) quasi-isomorphism $g : B \to A$ such that fg and gf induce identity maps on zero cohomologies $H^0(B)$ and $H^0(A)$ respectively.

3.3 Centralizer of an A_∞-Morphism

Let A and B be two A_∞-algebras, and (X, d_X, x_0) and (Y, d_Y, y_0) be the corresponding non-commutative formal pointed dg-manifolds. Let $f : A \to B$ be a morphism of A_∞-algebras. Then the corresponding k-point $f \in Maps(Spc(A), Spc(B))$ gives rise to the formal pointed manifold $U_f = \widehat{Maps(X,Y)}_f$ (completion at the point f). Functoriality of the construction of $Maps(X,Y)$ gives rise to a homomorphism of graded Lie algebras of vector fields $Vect(X) \oplus Vect(Y) \to Vect(Maps(X,Y))$. Since $[d_X, d_Y] = 0$ on $X \otimes Y$, we have a well-defined homological vector field d_Z on $Z = Maps(X,Y)$. It corresponds to $d_X \otimes 1_Y - 1_X \otimes d_Y$ under the above homomorphism. It is easy to see that $d_Z|_f = 0$ and in fact morphisms $f : A \to B$ of A_∞-algebras are exactly zeros of d_Z. We are going to describe below the A_∞-algebra $Centr(f)$ (centralizer of f) which corresponds to the formal neighborhood U_f of the point $f \in Maps(X,Y)$. We can write (see Sect. 2.3 for the notation)

$$c_{j,M} = c^0_{j,M} + r_{j,M},$$

where $c^0_{j,M} \in k$ and $r_{j,M}$ are formal non-commutative coordinates in the neighborhood of f. Then the A_∞-algebra $Centr(f)$ has a basis $(r_{j,M})_{j,M}$ and the A_∞-structure is defined by the restriction of the homological vector d_Z to U_f.

As a $\mathbf{Z}$-graded vector space $Centr(f) = \prod_{n \geq 0} Hom_{Vect^{\mathbf{Z}}_k}(A^{\otimes n}, B)[-n]$. Let $\phi_1, ..., \phi_n \in Centr(f)$ and $a_1, ..., a_N \in A$. Then we have $m_n(\phi_1, ..., \phi_n)(a_1, ..., a_N) = I + R$. Here I corresponds to the term $= 1_X \otimes d_Y$ and is given by the following expression

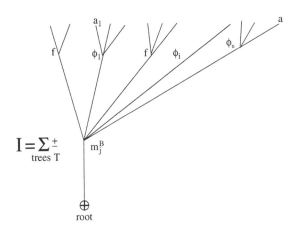

Similarly R corresponds to the term $d_X \otimes 1_Y$ and is described by the following figure

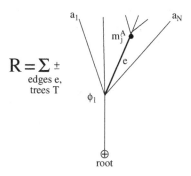

Comments on the figure describing I.

(1) We partition a sequence $(a_1, ..., a_N)$ into $l \geq n$ non-empty subsequences.
(2) We mark n of these subsequences counting from the left (the set can be empty).
(3) We apply multilinear map $\phi_i, 1 \leq i \leq n$ to the ith marked group of elements a_l.
(4) We apply Taylor coefficients of f to the remaining subsequences.

Notice that the term R appear only for m_1 (i.e. $n = 1$). For all subsequences we have $n \geq 1$.

From geometric point of view the term I corresponds to the vector field d_Y, while the term R corresponds to the vector field d_X.

Proposition 3.10 Let $d_{Centr(f)}$ be the derivation corresponding to the image of $d_X \oplus d_Y$ in $Maps(X, Y)$.
One has $[d_{Centr(f)}, d_{Centr(f)}] = 0$.

Proof. Clear. ∎

Remark 3.11 The A_∞-algebra $Centr(f)$ and its generalization to the case of A_∞-categories discussed in the subsequent paper provide geometric description of the notion of natural transformaion in the A_∞-case (see [39, 40] for a pure algebraic approach to this notion).

4 Non-Commutative dg-line L and Weak Unit

4.1 Main Definition

Definition 4.1 An A_∞-algebra is called *unital* (or strictly unital) if there exists an element $1 \in V$ of degree zero, such that $m_2(1, v) = m_2(v, 1)$ and

$m_n(v_1, ..., 1, ..., v_n) = 0$ for all $n \neq 2$ and $v, v_1, ..., v_n \in V$. It is called *weakly unital* (or homologically unital) if the graded associative unital algebra $H^\bullet(V)$ has a unit $1 \in H^0(V)$.

The notion of strict unit depends on a choice of affine coordinates on $Spc(T(V))$, while the notion of weak unit is "coordinate free." Moreover, one can show that a weakly unital A_∞-algebra becomes strictly unital after an appropriate change of coordinates.

The category of unital or weakly unital A_∞-algebras are defined in the natural way by the requirement that morphisms preserve the unit (or weak unit) structure.

In this section we are going to discuss a non-commutative dg-version of the odd one-dimensional supervector space $\mathbf{A}^{0|1}$ and its relationship to weakly unital A_∞-algebras. The results are valid for both $\mathbf{Z}$-graded and $\mathbf{Z}/2$-graded A_∞-algebras.

Definition 4.2 Non-commutative formal dg-line $\mathbf{L}$ is a non-commutative formal pointed dg-manifold corresponding to the one-dimensional A_∞-algebra $A \simeq k$ such that $m_2 = id, m_{n \neq 2} = 0$.

The algebra of functions $\mathcal{O}(\mathbf{L})$ is isomorphic to the topological algebra of formal series $k\langle\langle\xi\rangle\rangle$, where $deg\,\xi = 1$. The differential is given by $\partial(\xi) = \xi^2$.

4.2 Adding a Weak Unit

Let (X, d_X, x_0) be a non-commutative formal pointed dg-manifold correspodning to a non-unital A_∞-algebra A. We would like to describe geometrically the procedure of adding a weak unit to A.

Let us consider the non-commutative formal pointed graded manifold $X_1 = \mathbf{L} \times X$ corresponding to the free product of the coalgebras $B_\mathbf{L} * B_X$. Clearly one can lift vector fields d_X and $d_\mathbf{L} := \partial/\partial\xi$ to X_1.

Lemma 4.3 *The vector field*

$$d := d_{X_1} = d_X + ad(\xi) - \xi^2 \partial/\partial\xi$$

satisfies the condition $[d, d] = 0$.

Proof. Straightforward check. ∎

It follows from the formulas given in the proof that ξ appears in the expansion of d_X in quadratic expressions only. Let A_1 be an A_∞-algebras corresponding to X_1 and $1 \in T_{pt}X_1 = A_1[1]$ be the element of $A_1[1]$ dual to ξ (it corresponds to the tangent vector $\partial/\partial\xi$). Thus we see that $m_2^{A_1}(1, a) = m_2^{A_1}(a, 1) = a, m_2^{A_1}(1, 1) = 1$ for any $a \in A$ and $m_n^{A_1}(a_1, ..., 1, ..., a_n) = 0$ for all $n \geq 2, a_1, ..., a_n \in A$. This proves the following result.

Proposition 4.4 *The A_∞-algebra A_1 has a strict unit.*

Notice that we have a canonical morphism of non-commutative formal pointed dg-manifolds $e\colon X \to X_1$ such that $e^*|_X = id, e^*(\xi) = 0$.

Definition 4.5 Weak unit in X is given by a morphism of non-commutative formal pointed dg-manifolds $p : X_1 \to X$ such that $p \circ e = id$.

It follows from the definition that if X has a weak unit then the associative algebra $H^\bullet(A, m_1^A)$ is unital. Hence our geometric definition agrees with the pure algebraic one (explicit algebraic description of the notion of weak unit can be found, e.g., in [15], Sect. 20^7).

5 Modules and Bimodules

5.1 Modules and Vector Bundles

Recall that a topological vector space is called linearly compact if it is a projective limit of finite-dimensional vector spaces. The duality functor $V \mapsto V^*$ establishes an anti-equivalence between the category of vector spaces (equipped with the discrete topology) and the category of linearly compact vector spaces. All that can be extended in the obvious way to the category of graded vector spaces.

Let X be a non-commutative thin scheme in $Vect_k^{\mathbf{Z}}$.

Definition 5.1 Linearly compact vector bundle $\mathcal{E}$ over X is given by a linearly compact topologically free $\mathcal{O}(X)$-module $\Gamma(\mathcal{E})$, where $\mathcal{O}(X)$ is the algebra of function on X. Module $\Gamma(\mathcal{E})$ is called the module of *sections* of the linearly compact vector bundle $\mathcal{E}$.

Suppose that (X, x_0) is formal graded manifold. *The fiber* of $\mathcal{E}$ over x_0 is given by the quotient space $\mathcal{E}_{x_0} = \Gamma(\mathcal{E})/\overline{m_{x_0}\Gamma(\mathcal{E})}$ where $m_{x_0} \subset \mathcal{O}(X)$ is the two-sided maximal ideal of functions vanishing at x_0 and the bar means the closure.

Definition 5.2 A dg-vector bundle over a formal pointed dg-manifold (X, d_X, x_0) is given by a linearly compact vector bundle $\mathcal{E}$ over (X, x_0) such that the corresponding module $\Gamma(\mathcal{E})$ carries a differential $d_{\mathcal{E}} \colon \Gamma(\mathcal{E}) \to \Gamma(\mathcal{E})[1], d_{\mathcal{E}}^2 = 0$ so that $(\Gamma(\mathcal{E}), d_{\mathcal{E}})$ becomes a dg-module over the dg-algebra $(\mathcal{O}(X), d_X)$ and $d_{\mathcal{E}}$ vanishes on $\mathcal{E}_{x_0}$.

Definition 5.3 Let A be a non-unital A_∞-algebra. A left A-module M is given by a dg-bundle E over the formal pointed dg-manifold $X = Spc(T(A[1]))$ together with an isomorphism of vector bundles $\Gamma(\mathcal{E}) \simeq \mathcal{O}(X)\widehat{\otimes}M^*$ called a trivialization of $\mathcal{E}$.

---

[7] V. Lyubashenko has informed us that the equivalence of two descriptions also follows from his results with Yu. Bespalov and O. Manzyuk.

Passing to dual spaces we obtain the following algebraic definition.

Definition 5.4 Let A be an A_∞-algebra and M be a **Z**-graded vector space. A structure of a left A_∞-module on M over A (or simply a structure of a left A-module on M) is given by a differential d_M of degree $+1$ on $T(A[1]) \otimes M$ which makes it into a dg-comodule over the dg-coalgebra $T(A[1])$.

The notion of *right* A_∞-module is similar. Right A-module is the same as left A^{op}-module. Here A^{op} is the *opposite* A_∞-algebra, which coincides with A as a **Z**-graded vector space and for the higher multiplications one has: $m_n^{op}(a_1, ..., a_n) = (-1)^{n(n-1)/2} m_n(a_n, ..., a_1)$. The A_∞-algebra A carries the natural structures of the left and right A-modules. If we simply say "A-module" it will always mean "left A-module."

Taking the Taylor series of d_M we obtain a collection of k-linear maps (higher action morphisms) for any $n \geq 1$

$$m_n^M : A^{\otimes(n-1)} \otimes M \to M[2-n],$$

satisfying the compatibility conditions which can be written in exactly the same form as compatibility conditions for the higher products m_n^A (see e.g., [27]). All those conditions can be derived from just one property that the cofree $T_+(A[1])$-comodule $T_+(A[1], M) = \oplus_{n \geq 0} A[1]^{\otimes n} \otimes M$ carries a derivation $m^M = (m_n^M)_{n \geq 0}$ such that $[m^M, m^M] = 0$. In particular (M, m_1^M) is a complex of vector spaces.

Definition 5.5 Let A be a weakly unital A_∞-algebra. An A-module M is called weakly unital if the cohomology $H^\bullet(M, m_1^M)$ is a unital $H^\bullet(A)$-module.

It is easy to see that left A_∞-modules over A form a dg-category $A - mod$ with morphisms being homomorphisms of the corresponding comodules. As a graded vector space

$$Hom_{A-mod}(M, N) = \oplus_{n \geq 0} \underline{Hom}_{Vect_k^{\mathbf{Z}}}(A[1]^{\otimes n} \otimes M, N).$$

It easy to see that $Hom_{A-mod}(M, N)$ is a complex.

If M is a right A-module and N is a left A-module then one has a naturally defined structure of a complex on $M \otimes_A N := \oplus_{n \geq 0} M \otimes A[1]^{\otimes n} \otimes N$. The differential is given by the formula:

$$d(x \otimes a_1 \otimes ... \otimes a_n \otimes y) = \sum \pm m_i^M(x \otimes a_1 \otimes ... \otimes a_i) \otimes a_{i+1} \otimes ... \otimes a_n \otimes y)$$

$$+ \sum \pm x \otimes a_1 \otimes ... \otimes a_{i-1} \otimes m_k^A(a_i \otimes ... \otimes a_{i+k-1}) \otimes a_{i+k} \otimes ... \otimes a_n \otimes y$$

$$+ \sum \pm x \otimes a_1 \otimes ... \otimes a_{i-1} \otimes m_j^N(a_i \otimes ... \otimes a_n \otimes y).$$

We call this complex the *derived tensor product* of M and N.

For any A_∞-algebras A and B we define an $A - B$-bimodule as a $\mathbf{Z}$-graded vector space M together with linear maps

$$c^M_{n_1,n_2} : A[1]^{\otimes n_1} \otimes M \otimes B[1]^{\otimes n_2} \to M[1]$$

satisfying the natural compatibility conditions (see e.g. [27]). If X and Y are formal pointed dg-manifolds corresponding to A and B respectively then an $A - B$-bimodule is the same as a dg-bundle $\mathcal{E}$ over $X \otimes Y$ equipped with a homological vector field $d_\mathcal{E}$ which is a lift of the vector field $d_X \otimes 1 + 1 \otimes d_Y$.

Example 5.6 Let $A = B = M$. We define a structure of diagonal bimodule on A by setting $c^A_{n_1,n_2} = m^A_{n_1+n_2+1}$.

Proposition 5.7 (1) To have a structure of an A_∞-module on the complex M is the same as to have a homomorphism of A_∞-algebras $\phi\colon A \to \underline{End}_{\mathbf{K}}(M)$, where $\mathbf{K}$ is a category of complexes of k-vector spaces.

(2) To have a structure of an $A - B$-bimodule on a graded vector space M is the same as to have a structure of left A-module on M and to have a morphism of A_∞-algebras $\varphi_{A,B}\colon B^{op} \to Hom_{A-mod}(M, M)$.

Let A be an A_∞-algebra, M be an A-module and $\varphi_{A,A}\colon A^{op} \to Hom_{A-mod}$ (M, M) be the corresponding morphism of A_∞-algebras. Then the dg-algebra $Centr(\varphi)$ is isomorphic to the dg-algebra $Hom_{A-mod}(M, M)$.

If $M =_A M_B$ is an $A - B$-bimodule and $N =_B N_C$ is a $B - C$-bimodule then the complex $_A M_B \otimes_B {}_B N_C$ carries an $A - C$-bimodule structure. It is called the *tensor product* of M and N.

Let $f\colon X \to Y$ be a homomorphism of formal pointed dg-manifolds corresponding to a homomorphism of A_∞-algebras $A \to B$. Recall that in Sect. 4 we constructed the formal neighborhood U_f of f in $Maps(X, Y)$ and the A_∞-algebra $Centr(f)$. On the other hand, we have an $A - mod - B$ bimodule structure on B induced by f. Let us denote this bimodule by $M(f)$. We leave the proof of the following result as an exercise to the reader. It will not be used in the paper.

Proposition 5.8 If B is weakly unital then the dg-algebra $End_{A-mod-B}$ $(M(f))$ is quasi-isomorphic to $Centr(f)$.

A_∞-bimodules will be used in Part II for study of homologically smooth A_∞-algebras. In the subsequent paper devoted to A_∞-categories we will explain that bimodules give rise to A_∞-functors between the corresponding categories of modules. Tensor product of bimodules corresponds to the composition of A_∞-functors.

5.2 On the Tensor Product of A_∞-Algebras

The tensor product of two dg-algebras A_1 and A_2 is a dg-algebra. For A_∞-algebras there is no canonical simple formula for the A_∞-structure on $A_1 \otimes_k A_2$

which generalizes the one in the dg-algebras case. Some complicated formulas were proposed in [44]. They are not symmetric with respect to the permutation $(A_1, A_2) \mapsto (A_2, A_1)$. We will give below the definition of the dg-algebra which is quasi-isomorphic to the one from [44] in the case when both A_1 and A_2 are weakly unital. Namely, we define the A_∞-tensor product

$$A_1 \text{``} \otimes'' A_2 = End_{A_1-mod-A_2}(A_1 \otimes A_2).$$

Note that it is a unital dg-algebra. One can show that the dg-category $A - mod - B$ is equivalent (as a dg-category) to $A_1 \text{``} \otimes'' A_2^{op} - mod$.

6 Yoneda Lemma

6.1 Explicit Formulas for the Product and Differential on $Centr(f)$

Let A be an A_∞-algebra and $B = End_{\mathbf{K}}(A)$ be the dg-algebra of endomorphisms of A in the category $\mathbf{K}$ of complexes of k-vector spaces. Let $f = f_A : A \to B$ be the natural A_∞-morphism coming from the left action of A on itself. Notice that B is always a unital dg-algebra, while A can be non-unital. The aim of this Section was to discuss the relationship between A and $Centr(f_A)$. This is a simplest case of the A_∞-version of Yoneda lemma (the general case easily follows from this one. See also [39, 40]).

As a graded vector space $Centr(f_A)$ is isomorphic to $\prod_{n \geq 0} \underline{Hom}(A^{\otimes(n+1)}, A)[-n]$.

Let us describe the product in $Centr(f)$ for $f = f_A$. Let ϕ, ψ be two homogeneous elements of $Centr(f)$. Then

$$(\phi \cdot \psi)(a_1, a_2, \ldots, a_N) = \pm \phi(a_1, \ldots, a_{p-1}, \psi(a_p, \ldots, a_N)).$$

Here ψ acts on the last group of variables $a_p, \ldots, a_N$ and we use the Koszul sign convention for A_∞-algebras in order to determine the sign.

Similarly one has the following formula for the differential (see Sect. 3.3):

$$(d\phi)(a_1, \ldots, a_N) = \sum \pm \phi(a_1, \ldots, a_s, m_i(a_{s+1}, \ldots, a_{s+i}), a_{s+i+1} \cdots, a_N)$$

$$+ \sum \pm m_i(a_1, \ldots, a_{s-1}, \phi(a_s, \ldots, a_j, \ldots, a_N)).$$

6.2 Yoneda Homomorphism

If M is an $A - B$-bimodule then one has a homomorphism of A_∞-algebras $B^{op} \to Centr(\phi_{A,M})$ (see Propositions 5.1.7 and 5.1.8). We would like to apply this general observation in the case of the diagonal bimodule structure on A. Explicitly, we have the A_∞-morphism $A^{op} \to End_{mod-A}(A)$ or, equivalently, the collection of maps $A^{\otimes m} \to Hom(A^{\otimes n}, A)$. By conjugation it gives us a collection of maps

$$A^{\otimes m} \otimes Hom(A^{\otimes n}, A) \to Hom(A^{\otimes(m+n)}, A).$$

In this way we get a natural A_∞-morphism $Yo\colon A^{op} \to Centr(f_A)$ called the *Yoneda homomorphism.*

Proposition 6.1 The A_∞-algebra A is weakly unital if and only if the Yoneda homomorphism is a quasi-isomorphism.

Proof. Since $Centr(f_A)$ is weakly unital, then A must be weakly unital as long as Yoneda morphism is a quasi-isomorphism.

Let us prove the opposite statement. We assume that A is weakly unital. It suffices to prove that the cone $Cone(Yo)$ of the Yoneda homomorphism has trivial cohomology. Thus we need to prove that the cone of the morphism of complexes

$$(A^{op}, m_1) \to (\oplus_{n \geq 1} Hom(A^{\otimes n}, A), m_1^{Centr(f_A)}).$$

is contractible. In order to see this, one considers the extended complex $A \oplus Centr(f_A)$. It has natural filtration arising from the tensor powers of A. The corresponding spectral sequence collapses, which gives an explicit homotopy of the extended complex to the trivial one. This implies the desired quasi-isomorphism of $H^0(A^{op})$ and $H^0(Centr(f_A))$. ∎

Remark 6.2 It look like the construction of $Centr(f_A)$ is the first known canonical construction of a unital dg-algebra quasi-isomorphic to a given A_∞-algebra (canonical but not functorial). This is true even in the case of strictly unital A_∞-algebras. Standard construction via bar and cobar resolutions gives a non-unital dg-algebra.

Part II: Smoothness and Compactness

7 Hochschild Cochain and Chain Complexes of an A_∞-Algebra

7.1 Hochschild Cochain Complex

We change the notation for the homological vector field to Q, since the letter d will be used for the differential.[8] Let $((X, pt), Q)$ be a non-commutative

---

[8] We recall that the super version of the notion of formal dg-manifold was introduced by A. Schwarz under the name "Q-manifold." Here letter Q refers to the supercharge notation from Quantum Field Theory.

formal pointed dg-manifold corresponding to a non-unital A_∞-algebra A and $Vect(X)$ the graded Lie algebra of vector fields on X (i.e., continuous derivations of $\mathcal{O}(X)$).

We denote by $C^\bullet(A, A) := C^\bullet(X, X) := Vect(X)[-1]$ the *Hochschild cochain complex* of A. As a $\mathbf{Z}$-graded vector space

$$C^\bullet(A, A) = \prod_{n \geq 0} \underline{Hom}_C(A[1]^{\otimes n}, A).$$

The differential on $C^\bullet(A, A)$ is given by $[Q, \bullet]$. Algebraically, $C^\bullet(A, A)[1]$ is a DGLA of derivations of the coalgebra $T(A[1])$ (see Sect. 3).

Theorem 7.1 *Let X be a non-commutative formal pointed dg-manifold and $C^\bullet(X, X)$ be the Hochschild cochain complex. Then one has the following quasi-isomorphism of complexes*

$$C^\bullet(X, X)[1] \simeq T_{id_X}(Maps(X, X)),$$

where T_{id_X} denotes the tangent complex at the identity map.

Proof. Notice that $Maps(Spec(k[\varepsilon]/(\varepsilon^2)) \otimes X, X)$ is the non-commutative dg ind-manifold of vector fields on X. The tangent space T_{id_X} from the theorem can be identified with the set of such $f \in Maps(Spec(k[\varepsilon]/(\varepsilon^2)) \otimes X, X)$ that $f|_{\{pt\} \otimes X} = id_X$. On the other hand the DGLA $C^\bullet(X, X)[1]$ is the DGLA of vector fields on X. The theorem follows. ∎

The Hochschild complex admits a couple of other interpretations. We leave to the reader to check the equivalence of all of them. First, $C^\bullet(A, A) \simeq Centr(id_A)$. Finally, for a weakly unital A one has $C^\bullet(A, A) \simeq Hom_{A-mod-A}(A, A)$. Both are quasi-isomorphisms of complexes.

Remark 7.2 Interpretation of $C^\bullet(A, A)[1]$ as vector fields gives a DGLA structure on this space. It is a Lie algebra of the "commutative" formal group in $Vect_k^{\mathbf{Z}}$, which is an abelianization of the non-commutative formal group of inner (in the sense of tensor categories) automorphisms $\underline{Aut}(X) \subset Maps(X, X)$. Because of this non-commutative structure underlying the Hochschild cochain complex, it is natural to expect that $C^\bullet(A, A)[1]$ carries more structures than just DGLA. Indeed, Deligne's conjecture (see e.g., [35] and the last section of this paper) claims that the DGLA algebra structure on $C^\bullet(A, A)[1]$ can be extended to a structure of an algebra over the operad of singular chains of the topological operad of little discs. Graded Lie algebra structure can be recovered from cells of highest dimension in the cell decomposition of the topological operad.

7.2 Hochschild Chain Complex

In this subsection we are going to construct a complex of k-vector spaces which is dual to the Hochschild chain complex of a non-unital A_∞-algebra.

Cyclic Differential Forms of Order Zero

Let (X, pt) be a non-commutative formal pointed manifold over k and $\mathcal{O}(X)$ the algebra of functions on X. For simplicity we will assume that X is finite-dimensional, i.e., $dim_k T_{pt}X < \infty$. If $B = B_X$ is a counital coalgebra corresponding to X (coalgebra of distributions on X) then $\mathcal{O}(X) \simeq B^*$. Let us choose affine coordinates $x_1, x_2, ..., x_n$ at the marked point pt. Then we have an isomorphism of $\mathcal{O}(X)$ with the topological algebra $k\langle\langle x_1, ..., x_n\rangle\rangle$ of formal series in free graded variables $x_1, ..., x_n$.

We define the space of *cyclic differential degree zero forms on X* as

$$\Omega^0_{cycl}(X) = \mathcal{O}(X)/[\mathcal{O}(X), \mathcal{O}(X)]_{top},$$

where $[\mathcal{O}(X), \mathcal{O}(X)]_{top}$ denotes the topological commutator (the closure of the algebraic commutator in the adic topology of the space of non-commutative formal power series).

Equivalently, we can start with the graded k-vector space $\Omega^0_{cycl,dual}(X)$ defined as the kernel of the composition $B \to B \otimes B \to \bigwedge^2 B$ (first map is the coproduct $\Delta: B \to B \otimes B$, while the second one is the natural projection to the skew-symmetric tensors). Then $\Omega^0_{cycl}(X) \simeq (\Omega^0_{cycl,dual}(X))^*$ (dual vector space).

Higher Order Cyclic Differential Forms

We start with the definition of the *odd tangent bundle* $T[1]X$. This is the dg-analog of the total space of the tangent supervector bundle with the changed parity of fibers. It is more convenient to describe this formal manifold in terms of algebras rather than coalgebras. Namely, the algebra of functions $\mathcal{O}(T[1]X)$ is a unital topological algebra isomorphic to the algebra of formal power series $k\langle\langle x_i, dx_i\rangle\rangle, 1 \leq i \leq n$, where $deg\, dx_i = deg\, x_i + 1$ (we do not impose any commutativity relations between generators). More invariant description involves the odd line. Namely, let $t_1 := Spc(B_1)$, where $(B_1)^* = k\langle\langle \xi \rangle\rangle/(\xi^2), deg\, \xi = +1$. Then we define $T[1]X$ as the formal neighborhood in $Maps(t_1, X)$ of the point p which is the composition of pt with the trivial map of t_1 into the point $Spc(k)$.

Definition 7.3 (a) The graded vector space

$$\mathcal{O}(T[1]X) = \Omega^\bullet(X) = \prod_{m \geq 0} \Omega^m(X)$$

is called the space of de Rham differential forms on X.

(b) The graded space

$$\Omega^0_{cycl}(T[1]X) = \prod_{m \geq 0} \Omega^m_{cycl}(X)$$

is called the space of cyclic differential forms on X.

In coordinate description the grading is given by the total number of dx_i. Clearly each space $\Omega^n_{cycl}(X), n \geq 0$ is dual to some vector space $\Omega^n_{cycl,dual}(X)$ equipped with the discrete topology (since this is true for $\Omega^0(T[1]X)$).

The *de Rham differential on* $\Omega^\bullet(X)$ corresponds to the vector field $\partial/\partial\xi$ (see description which uses the odd line, it is the same variable ξ). Since Ω^0_{cycl} is given by the natural (functorial) construction, the de Rham differential descends to the subspace of cyclic differential forms. We will denote the former by d_{DR} and the latter by d_{cycl}.

The space of *cyclic 1-forms* $\Omega^1_{cycl}(X)$ is a (topological) span of expressions $x_1 x_2 ... x_l \, dx_j, x_i \in \mathcal{O}(X)$. Equivalently, the space of cyclic 1-forms consists of expressions $\sum_{1 \leq i \leq n} f_i(x_1, ..., x_n) \, dx_i$ where $f_i \in k\langle\langle x_1, ..., x_n\rangle\rangle$.

There is a map $\varphi : \Omega^1_{cycl}(X) \to \mathcal{O}(X)_{red} := \mathcal{O}(X)/k$, which is defined on $\Omega^1(X)$ by the formula $a\,db \mapsto [a,b]$ (check that the induced map on the cyclic 1-forms is well-defined). This map does not have an analog in the commutative case.[9]

Non-commutative Cartan Calculus

Let X be a formal graded manifold over a field k. We denote by $g := g_X$ the graded Lie algebra of continuous linear maps $\mathcal{O}(T[1]X) \to \mathcal{O}(T[1]X)$ generated by de Rham differential $d = d_{dR}$ and contraction maps $i_\xi, \xi \in Vect(X)$ which are defined by the formulas $i_\xi(f) = 0, i_\xi(df) = \xi(f)$ for all $f \in \mathcal{O}(T[1]X)$. Let us define the Lie derivative $Lie_\xi = [d, i_\xi]$ (graded commutator). Then one can easily checks the usual formulas of the Cartan calculus

$$[d,d] = 0, Lie_\xi = [d, i_\xi], [d, Lie_\xi] = 0,$$

$$[Lie_\xi, i_\eta] = i_{[\xi,\eta]}, [Lie_\xi, Lie_\eta] = Lie_{[\xi,\eta]}, [i_\xi, i_\eta] = 0,$$

for any $\xi, \eta \in Vect(X)$.

By naturality, the graded Lie algebra g_X acts on the space $\Omega^\bullet_{cycl}(X)$ as well as one the dual space $(\Omega^\bullet_{cycl}(X))^*$.

Differential on the Hochschild Chain Complex

Let Q be a homological vector field on (X, pt). Then $A = T_{pt}X[-1]$ is a non-unital A_∞-algebra.

We define the *dual Hochschild chain complex* $(C_\bullet(A, A))^*$ as $\Omega^1_{cycl}(X)[2]$ with the differential Lie_Q. Our terminology is explained by the observation that $\Omega^1_{cycl}(X)[2]$ is dual to the conventional Hochschild chain complex

---

[9] V. Ginzburg pointed out that the geometric meaning of the map φ as a "contraction with double derivation" was suggested in Sect. 5.4 of [19].

$$C_\bullet(A, A) = \oplus_{n \geq 0}(A[1])^{\otimes n} \otimes A.$$

Note that we use the cohomological grading on $C_\bullet(A, A)$, i.e. chains of degree n in conventional (homological) grading have degree $-n$ in our grading. The differential has degree $+1$.

In coordinates the isomorphism identifies an element $f_i(x_1, ..., x_n) \otimes x_i \in (A[1]^{\otimes n} \otimes A)^*$ with the homogeneous element $f_i(x_1, ..., x_n) \, dx_i \in \Omega^1_{cycl}(X)$. Here $x_i \in (A[1])^*, 1 \leq i \leq n$ are affine coordinates.

The graded Lie algebra $Vect(X)$ of vector fields of all degrees acts on any functorially defined space, in particular, on all spaces $\Omega^j(X), \Omega^j_{cycl}(X)$, etc. Then we have a differential on $\Omega^j_{cycl}(X)$ given by $b = Lie_Q$ of degree $+1$. There is an explicit formula for the differential b on $C_\bullet(A, A)$ (cf. [T]):

$$b(a_0 \otimes ... \otimes a_n) = \sum \pm a_0 \otimes ... \otimes m_l(a_i \otimes ... \otimes a_j) \otimes ... \otimes a_n$$

$$+ \sum \pm m_l(a_j \otimes ... \otimes a_n \otimes a_0 \otimes ... \otimes a_i) \otimes a_{i+1} \otimes ... \otimes a_{j-1}.$$

It is convenient to depict a cyclic monomial $a_0 \otimes ... \otimes a_n$ in the following way. We draw a clockwise oriented circle with $n+1$ points labeled from 0 to n such that one point is marked We assign the elements $a_0, a_1, ..., a_n$ to the points with the corresponding labels, putting a_0 at the marked point.

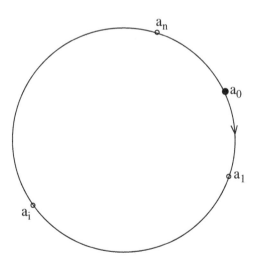

Then we can write $b = b_1 + b_2$ where b_1 is the sum (with appropriate signs) of the expressions depicted below:

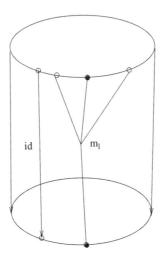

Similarly, b_2 is the sum (with appropriate signs) of the expressions depicted below:

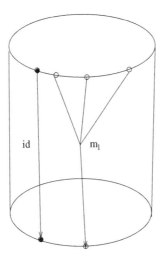

In both cases maps m_l are applied to a consequitive cyclically ordered sequence of elements of A assigned to the points on the top circle. The identity map is applied to the remaining elements. Marked point on the top circle is the position of the element of a_0. Marked point on the bottom circle depicts the first tensor factor of the corresponding summand of b. In both the cases we start cyclic count of tensor factors clockwise from the marked point.

7.3 The Case of Strictly Unital A_∞-Algebras

Let A be a strictly unital A_∞-algebra. There is a *reduced* Hochschild chain complex

$$C_\bullet^{red}(A, A) = \oplus_{n \geq 0} A \otimes ((A/k \cdot 1)[1])^{\otimes n},$$

which is the quotient of $C_\bullet(A, A)$. Similarly there is a reduced Hochschild cochain complex

$$C_{red}^\bullet(A, A) = \prod_{n \geq 0} \underline{Hom}_C((A/k \cdot 1)[1]^{\otimes n}, A),$$

which is a subcomplex of the Hochschild cochain complex $C^\bullet(A, A)$.

Also, $C_\bullet(A, A)$ carries also the "Connes's differential" B of degree -1 (called sometimes "de Rham differential") given by the formula (see [7], [T])

$$B(a_0 \otimes \ldots \otimes a_n) = \sum_i \pm 1 \otimes a_i \otimes \ldots \otimes a_n \otimes a_0 \otimes \ldots \otimes a_{i-1}, B^2 = 0, Bb + bB = 0.$$

Here is a graphical description of B (it will receive an explanation in the section devoted to generalized Deligne's conjecture)

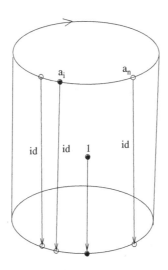

Let u be an independent variable of degree $+2$. It follows that for a strictly unital A_∞-algebra A one has a differential $b + uB$ of degree $+1$ on the graded vector space $C_\bullet(A, A)[[u]]$ which makes the latter into a complex called *negative cyclic complex* (see [7, T]). In fact $b + uB$ is a differential on a smaller complex $C_\bullet(A, A)[u]$. In the non-unital case one can use Cuntz–Quillen complex instead of a negative cyclic complex (see next subsection).

7.4 Non-unital Case: Cuntz–Quillen Complex

In this subsection we are going to present a formal dg-version of the mixed complex introduced by Cuntz and Quillen [9]. In the previous subsection we introduced the Connes differential B in the case of strictly unital A_∞-algebras. In the non-unital case the construction has to be modified. Let $X = A[1]_{form}$ be the corresponding non-commutative formal pointed dg-manifold. The algebra of functions $\mathcal{O}(X) \simeq \prod_{n \geq 0} (A[1]^{\otimes n})^*$ is a complex with the differential Lie_Q.

Proposition 7.4 If A is weakly unital then all non-zero cohomology of the complex $\mathcal{O}(X)$ are trivial and $H^0(\mathcal{O}(X)) \simeq k$.

Proof. Let us calculate the cohomology using the spectral sequence associated with the filtration $\prod_{n \geq n_0} (A[1]^{\otimes n})^*$. The term E_1 of the spectral sequence is isomorphic to the complex $\prod_{n \geq 0} ((H^\bullet(A[1], m_1))^{\otimes n})^*$ with the differential induced by the multiplication m_2^A on $H^\bullet(A, m_1^A)$. By assumption $H^\bullet(A, m_1^A)$ is a unital algebra, hence all the cohomology groups vanish except of the zeroth one, which is isomorphic to k. This concludes the proof. ∎

It follows from the above Proposition that the complex $\mathcal{O}(X)/k$ is acyclic. We have the following two morphisms of complexes

$$d_{cycl} \colon (\mathcal{O}(X)/k \cdot 1, Lie_Q) \to (\Omega^1_{cycl}(X), Lie_Q)$$

and

$$\varphi \colon (\Omega^1_{cycl}(X), Lie_Q) \to (\mathcal{O}(X)/k \cdot 1, Lie_Q).$$

Here d_{cycl} and φ were introduced in the Sect. 7.2. We have: $deg(d_{cycl}) = +1$, $deg(\varphi) = -1, d_{cycl} \circ \varphi = 0, \varphi \circ d_{cycl} = 0..$

Let us consider a *modified* Hochschild chain complex

$$C^{mod}_\bullet(A, A) := (\Omega^1_{cycl}(X)[2])^* \oplus (\mathcal{O}(X)/k \cdot 1)^*$$

with the differential

$$b = \begin{pmatrix} (Lie_Q)^* & \varphi^* \\ 0 & (Lie_Q)^* \end{pmatrix}$$

Let

$B = \begin{pmatrix} 0 & 0 \\ d^*_{cycl} & 0 \end{pmatrix}$ be an endomorphism of $C^{mod}_\bullet(A, A)$ of degree -1. Then $B^2 = 0$. Let u be a formal variable of degree $+2$. We define modified negative cyclic, periodic cyclic and cyclic chain complexes such as follows

$$CC^{-,mod}_\bullet(A) = (C^{mod}_\bullet(A, A)[[u]], b + uB),$$

$$CP^{mod}_\bullet(A) = (C^{mod}_\bullet(A, A)((u)), b + uB),$$

$$CC^{mod}_\bullet(A) = (CP^{mod}_\bullet(A)/CC^{-,mod}_\bullet(A))[-2].$$

For unital dg-algebras these complexes are quasi-isomorphic to the standard ones. If $char\, k = 0$ and A is weakly unital then $CC_\bullet^{-,mod}(A)$ is quasi-isomorphic to the complex $(\Omega_{cycl}^0(X), Lie_Q)^*$. Note that the $k[[u]]$-module structure on the cohomology $H^\bullet((\Omega_{cycl}^0(X), Lie_Q)^*)$ is not visible from the definition.

8 Homologically Smooth and Compact A_∞-Algebras

From now on we will assume that all A_∞-algebras are weakly unital unless we say otherwise.

8.1 Homological Smoothness

Let A be an A_∞-algebra over k and $E_1, E_2, ..., E_n$ be a sequence of A-modules. Let us consider a sequence $(E_{\leq i})_{1 \leq i \leq n}$ of A-modules together with exact triangles

$$E_i \to E_{\leq i} \to E_{i+1} \to E_i[1],$$

such that $E_{\leq 1} = E_1$.

We will call $E_{\leq n}$ an *extension* of the sequence $E_1, ..., E_n$.

The reader also notices that the above definition can be given also for the category of $A - A$-bimodules.

Definition 8.1 (1) A perfect A-module is the one which is quasi-isomorphic to a direct summand of an extension of a sequence of modules each of which is quasi-isomorphic to $A[n], n \in \mathbf{Z}$.

(2) A perfect $A - A$-bimodule is the one which is quasi-isomorphic to a direct summand of an extension of a sequence consisting of bimodules each of which is quasi-isomorphic to $(A \otimes A)[n], n \in \mathbf{Z}$.

Perfect A-modules form a full subcategory $Perf_A$ of the dg-category $A - mod$. Perfect $A - A$-bimodules form a full subcategory $Perf_{A-mod-A}$ of the category of $A - A$-bimodules.[10]

Definition 8.2 We say that an A_∞-algebra A is homologically smooth if it is a perfect $A - A$-bimodule (equivalently, A is a perfect module over the A_∞-algebra A "$\otimes$" A^{op}).

Remark 8.3 An $A - B$-bimodule M gives rise to a dg-functor $B - mod \to A - mod$ such that $V \mapsto M \otimes_B V$. The diagonal bimodule A corresponds to the identity functor $Id_{A-mod} : A - mod \to A - mod$. The notion of homological smoothness can be generalized to the framework of A_∞-categories. The corresponding notion of saturated A_∞-category can be spelled out entirely in terms of the identity functor.

---

[10] Sometimes $Perf_A$ is called a thick triangulated subcategory of $A-mod$ generated by A. Then it is denoted by $\langle A \rangle$. In the case of $A - A$-bimodules we have a thik triangulated subcategory generated by $A \otimes A$, which is denoted by $\langle A \otimes A \rangle$.

Let us list few examples of homologically smooth A_∞-algebras.

Example 8.4 (a) Algebra of functions on a smooth affine scheme.

(b) $A = k[x_1, ..., x_n]_q$, which is the algebra of polynomials in variables $x_i, 1 \le i \le n$ subject to the relations $x_i x_j = q_{ij} x_j x_i$, where $q_{ij} \in k^*$ satisfy the properties $q_{ii} = 1, q_{ij} q_{ji} = 1$. More generally, all quadratic Koszul algebras, which are deformations of polynomial algebras are homologically smooth.

(c) Algebras of regular functions on quantum groups (see [37]).

(d) Free algebras $k\langle x_1, ..., x_n \rangle$.

(e) Finite-dimensional associative algebras of finite homological dimension.

(f) If X is a smooth scheme over k then the bounded derived category $D^b(Perf(X))$ of the category of perfect complexes (it is equivalent to $D^b(Coh(X))$) has a generator P (see [5]). Then the dg-algebra $A = End(P)$ (here we understand endomorphisms in the "derived sense", see [28]) is a homologically smooth algebra.

Let us introduce an $A - A$-bimodule $A^! = Hom_{A-mod-A}(A, A \otimes A)$ (cf. [18]). The structure of an $A - A$-bimodule is defined similarly to the case of associative algebras.

Proposition 8.5 If A is homologically smooth then $A^!$ is a perfect $A - A$-bimodule.

Proof. We observe that $Hom_{C-mod}(C, C)$ is a dg-algebra for any A_∞-algebra C. The Yoneda embedding $C \to Hom_{C-mod}(C, C)$ is a quasi-isomorphism of A_∞-algebras. Let us apply this observation to $C = A \otimes A^{op}$. Then using the A_∞-algebra $A " \otimes " A^{op}$ (see Sect. 5.2) we obtain a quasi-isomorphism of $A - A$-bimodules $Hom_{A-mod-A}(A \otimes A, A \otimes A) \simeq A \otimes A$. By assumption A is quasi-isomorphic (as an A_∞-bimodule) to a direct summand in an extension of a sequence $(A \otimes A)[n_i]$ for $n_i \in \mathbf{Z}$. Hence $Hom_{A-mod-A}(A \otimes A, A \otimes A)$ is quasi-isomorphic to a direct summand in an extension of a sequence $(A \otimes A)[m_i]$ for $m_i \in \mathbf{Z}$. The result follows. ∎

Definition 8.6 The bimodule $A^!$ is called the inverse dualizing bimodule.

The terminology is explained by an observation that if $A = End(P)$ where P is a generator of of $Perf(X)$ (see example 8f)) then the bimodule $A^!$ corresponds to the functor $F \mapsto F \otimes K_X^{-1}[-dim X]$, where K_X is the canonical class of X.[11]

Remark 8.7 In [50] the authors introduced a stronger notion of fibrant dg-algebra. Informally it corresponds to "non-commutative homologically smooth affine schemes of finite type." In the compact case (see the next section) both notions are equivalent.

---

[11] We thank Amnon Yekutieli for pointing out that the inverse dualizing module was first mentioned in the paper by M. van den Bergh "Existence theorems for dualizing complexes over non-commutative graded and filtered rings," *J. Algebra*, 195:2, 1997, 662–679.

8.2 Compact A_∞-Algebras

Definition 8.8 We say that an A_∞-algebra A is compact if the cohomology $H^\bullet(A, m_1)$ is finite-dimensional.

Example 8.9 (a) If $dim_k A < \infty$ then A is compact.

(b) Let X/k be a proper scheme of finite type. According to [5] there exists a compact dg-algebra A such that $Perf_A$ is equivalent to $D^b(Coh(X))$.

(c) If $Y \subset X$ is a proper subscheme (possibly singular) of a smooth scheme X then the bounded derived category $D_Y^b(Perf(X))$ of the category of perfect complexes on X, which are supported on Y has a generator P such that $A = End(P)$ is compact. In general it is not homologically smooth for $Y \neq X$. More generally, one can replace X by a formal smooth scheme containing Y, e.g., by the formal neighborhood of Y in the ambient smooth scheme. In particular, for $Y = \{pt\} \subset X = \mathbf{A}^1$ and the generator $\mathcal{O}_Y$ of $D^b(Perf(X))$ the corresponding graded algebra is isomorphic to $k\langle\xi\rangle/(\xi^2)$, where $deg\,\xi = 1$.

Proposition 8.10 If A is compact and homologically smooth then the Hochschild homology and cohomology of A are finite-dimensional.

Proof. (a) Let us start with Hochschild cohomology. We have an isomorphism of complexes $C^\bullet(A, A) \simeq Hom_{A-mod-A}(A, A)$. Since A is homologically smooth the latter complex is quasi-isomorphic to a direct summand of an extension of the bimodule $Hom_{A-mod-A}(A \otimes A, A \otimes A)$. The latter complex is quasi-isomorphic to $A \otimes A$ (see the proof of the Proposition 8.1.5). Since A is compact, the complex $A \otimes A$ has finite-dimensional cohomology. Therefore any perfect $A - A$-bimodule enjoys the same property. We conclude that the Hochschild cohomology groups are finite-dimensional vector spaces.

(b) Let us consider the case of Hochschild homology. With any $A - A$-bimodule E we associate a complex of vector spaces $E^\sharp = \oplus_{n \geq 0} A[1]^{\otimes n} \otimes E$ (cf. [18]). The differential on $E^\sharp$ is given by the same formulas as the Hochschild differential for $C_\bullet(A, A)$ with the only change: we place an element $e \in E$ instead of an element of A at the marked vertex (see Sect. 7). Taking $E = A$ with the structure of the diagonal $A - A$-bimodule we obtain $A^\sharp = C_\bullet(A, A)$. On the other hand, it is easy to see that the complex $(A \otimes A)^\sharp$ is quasi-isomorphic to (A, m_1), since $(A \otimes A)^\sharp$ is the quotient of the canonical free resolution (bar resolution) for A by a subcomplex A. The construction of $E^\sharp$ is functorial, hence $A^\sharp$ is quasi-isomorphic to a direct summand of an extension (in the category of complexes) of a shift of $(A \otimes A)^\sharp$, because A is smooth. Since $A^\sharp = C_\bullet(A, A)$ we see that the Hochschild homology $H_\bullet(A, A)$ is isomorphic to a direct summand of the cohomology of an extension of a sequence of k-modules $(A[n_i], m_1)$. Since the vector space $H^\bullet(A, m_1)$ is finite-dimensional the result follows. ∎

Remark 8.11 For a homologically smooth compact A_∞-algebra A one has a quasi-isomorphism of complexes $C_\bullet(A, A) \simeq Hom_{A-mod-A}(A^!, A)$ Also, the complex $Hom_{A-mod-A}(M^!, N)$ is quasi-isomorphic to $(M \otimes_A N)^\sharp$ for two $A - A$-bimodules M, N, such that M is perfect. Here $M^! := Hom_{A-mod-A}$ $(M, A \otimes A)$ Having this in mind one can offer a version of the above proof which uses the isomorphism

$$Hom_{A-mod-A}(A^!, A) \simeq Hom_{A-mod-A}(Hom_{A-mod-A}(A, A \otimes A), A).$$

Indeed, since A is homologically smooth the bimodule $Hom_{A-mod-A}(A, A \otimes A)$ is quasi-isomorphic to a direct summand P of an extension of a shift of $Hom_{A-mod-A}(A \otimes A, A \otimes A) \simeq A \otimes A$. Similarly, $Hom_{A-mod-A}(P, A)$ is quasi-isomorphic to a direct summand of an extension of a shift of $Hom_{A-mod-A}(A \otimes A, A \otimes A) \simeq A \otimes A$. Combining the above computations we see that the complex $C_\bullet(A, A)$ is quasi-isomorphic to a direct summand of an extension of a shift of the complex $A \otimes A$. The latter has finite-dimensional cohomology, since A enjoys this property.

Besides algebras of finite quivers there are two main sources of homologically smooth compact $\mathbf{Z}$-graded A_∞-algebras.

Example 8.12 (a) Combining Examples 8.1.4(f) and 8.2.2(b) we see that the derived category $D^b(Coh(X))$ is equivalent to the category $Perf_A$ for a homologically smooth compact A_∞-algebra A.

(b) According to [45] the derived category $D^b(F(X))$ of the Fukaya category of a K3 surface X is equivalent to $Perf_A$ for a homologically smooth compact A_∞-algebra A. The latter is generated by Lagrangian spheres, which are vanishing cycles at the critical points for a fibration of X over $\mathbf{CP}^1$. This result can be generalized to other Calabi–Yau manifolds.

In $\mathbf{Z}/2$-graded case examples of homologically smooth compact A_∞-algebras come from Landau–Ginzburg categories (see [42, 43]) and from Fukaya categories for Fano varieties.

Remark 8.13 Formal deformation theory of smooth compact A_∞-algebras gives a finite-dimensional formal pointed (commutative) dg-manifold. The global moduli stack can be constructed using methods of [50]). It can be thought of as a moduli stack of non-commutative smooth proper varieties.

9 Degeneration Hodge-to-de Rham

9.1 Main Conjecture

Let us assume that $char \, k = 0$ and A is a weakly unital A_∞-algebra, which can be $\mathbf{Z}$-graded or $\mathbf{Z}/2$-graded.

For any $n \geq 0$ we define the *truncated modified negative cyclic* complex $C_\bullet^{mod,(n)}(A, A) = (C_\bullet^{mod}(A, A) \otimes k[u]/(u^n), b + uB)$, where $deg \, u = +2$. Its cohomology will be denoted by $H^\bullet(C_\bullet^{mod,(n)}(A, A))$.

Definition 9.1 We say that an A_∞-algebra A satisfies the degeneration property if for any $n \geq 1$ one has: $H^\bullet(C^{mod,(n)}_\bullet(A, A))$ is a flat $k[u]/(u^n)$-module.

Conjecture 9.2 (Degeneration Hodge-to-de Rham). Let A be a weakly unital compact homologically smooth A_∞-algebra. Then it satisfies the degeneration property.

We will call the above statement the *degeneration conjecture*.

Corollary 9.3 If the A satisfies the degeneration property then the negative cyclic homology coincides with $\varprojlim_n H^\bullet(C^{mod,(n)}_\bullet(A, A))$ and it is a flat $k[[u]]$-module.

Remark 9.4 One can speak about degeneration property (modulo u^n) for A_∞-algebras which are flat over unital commutative k-algebras. For example, let R be an Artinian local k-algebra with the maximal ideal m and A be a flat R-algebra such that A/m is weakly unital, homologically smooth and compact. Then, assuming the degeneration property for A/m, one can easily see that it holds for A as well. In particular, the Hochschild homology of A gives rise to a vector bundle over $Spec(R) \times \mathbf{A}^1_{form}[-2]$.

Assuming the degeneration property for A we see that there is a $\mathbf{Z}$-graded vector bundle ξ_A over $\mathbf{A}^1_{form}[-2] = Spf(k[[u]])$ with the space of sections isomorphic to

$$\varprojlim_n H^\bullet(C^{mod,(n)}_\bullet(A, A)) = HC^{-,mod}_\bullet(A),$$

which is the negative cyclic homology of A. The fiber of ξ_A at $u = 0$ is isomorphic to the Hochschild homology $H^{mod}_\bullet(A, A) := H_\bullet(C_\bullet(A, A))$.

Note that $\mathbf{Z}$-graded $k((u))$-module $HP^{mod}_\bullet(A)$ of periodic cyclic homology can be described in terms of just one $\mathbf{Z}/2$-graded vector space $HP^{mod}_{even}(A) \oplus \Pi HP^{mod}_{odd}(A)$, where $HP^{mod}_{even}(A)$ (resp. $HP^{mod}_{odd}(A)$) consists of elements of degree zero (resp. degree $+1$) of $HP^{mod}_\bullet(A)$ and Π is the functor of changing the parity. We can interpret ξ_A in terms of ($\mathbf{Z}/2$-graded) supergeometry as a $\mathbf{G}_m$-equivariant supervector bundle over the even formal line $\mathbf{A}^1_{form}$. The structure of a $\mathbf{G}_m$-equivariant supervector bundle ξ_A is equivalent to a filtration F (called Hodge filtration) by even numbers on $HP^{mod}_{even}(A)$ and by odd numbers on $HP^{mod}_{odd}(A)$. The associated $\mathbf{Z}$-graded vector space coincides with $H_\bullet(A, A)$.

We can say few words in support of the degeneration conjecture. One is, of course, the classical Hodge-to-de Rham degeneration theorem (see Sect. 9.2 below). It is an interesting question to express the classical Hodge theory algebraically, in terms of a generator $\mathcal{E}$ of the derived category of coherent sheaves and the corresponding A_∞-algebra $A = RHom(\mathcal{E}, \mathcal{E})$. The degeneration conjecture also trivially holds for algebras of finite quivers without relations.

In classical algebraic geometry there are basically two approaches to the proof of degeneration conjecture. One is analytic and uses Kähler metric,

Hodge decomposition, etc. Another one is pure algebraic and uses the technique of reduction to finite characteristic (see [12]). Recently Kaledin (see [24]) suggested a proof of a version of the degeneration conjecture based on the reduction to finite characterstic.

Below we will formulate a conjecture which could lead to the definition of crystalline cohomology for A_∞-algebras. Notice that one can define homologically smooth and compact A_∞-algebras over any commutative ring, in particular, over the ring of integers $\mathbf{Z}$. We assume that A is a flat $\mathbf{Z}$-module.

Conjecture 9.5 Suppose that A is a weakly unital A_∞-algebra over $\mathbf{Z}$, such that it is homologically smooth (but not necessarily compact). Truncated negative cyclic complexes $(C_\bullet(A, A) \otimes \mathbf{Z}[[u, p]]/(u^n, p^m), b + uB)$ and $(C_\bullet(A, A) \otimes \mathbf{Z}[[u, p]]/(u^n, p^m), b - puB)$ are quasi-isomorphic for all $n, m \geq 1$ and all prime numbers p.

If, in addition, A is compact then the homology of either of the above complexes is a flat module over $\mathbf{Z}[[u, p]]/(u^n, p^m)$.

If the above conjecture is true then the degeneration conjecture, probably, can be deduced along the lines of [12]. One can also make some conjectures about Hochschild complex of an arbitrary A_∞-algebra, not assuming that it is compact or homologically smooth. More precisely, let A be a unital A_∞-algebra over the ring of p-adic numbers $\mathbf{Z}_p$. We assume that A is topologically free $\mathbf{Z}_p$-module. Let $A_0 = A \otimes_{\mathbf{Z}_p} \mathbf{Z}/p$ be the reduction modulo p. Then we have the Hochschild complex $(C_\bullet(A_0, A_0), b)$ and the $\mathbf{Z}/2$-graded complex $(C_\bullet(A_0, A_0), b + B)$.

Conjecture 9.6 For any i there is natural isomorphism of $\mathbf{Z}/2$-graded vector spaces over the field $\mathbf{Z}/p$:

$$H^\bullet(C_\bullet(A_0, A_0), b) \simeq H^\bullet(C_\bullet(A_0, A_0), b + B).$$

There are similar isomorphisms for weakly unital and non-unital A_∞-algebras, if one replaces $C_\bullet(A_0, A_0)$ by $C_\bullet^{mod}(A_0, A_0)$. Also one has similar isomorphisms for $\mathbf{Z}/2$-graded A_∞-algebras.

The last conjecture presumably gives an isomorphism used in [12], but does not imply the degeneration conjecture.

Remark 9.7 As we will explain elsewhere there are similar conjectures for saturated A_∞-categories (recall that they are generalizations of homologically smooth compact A_∞-algebras). This observation supports the idea of introducing the category $NCMot$ of non-commutative pure motives. Objects of the latter will be saturated A_∞-categories over a field and $Hom_{NCMot}(\mathcal{C}_1, \mathcal{C}_2) = K_0(Funct(\mathcal{C}_1, \mathcal{C}_2)) \otimes \mathbf{Q}/equiv$ where K_0 means the K_0-group of the A_∞-category of functors and *equiv* means numerical equivalence (i.e., the quivalence relation generated by the kernel of the Euler form $\langle E, F \rangle := \chi(RHom(E, F))$, where χ is the Euler characteristic). The above category is worth of consideration and will be discussed elsewhere (see [32]). In particular, one can formulate non-commutative analogs of Weil and Beilinson conjectures for the category $NCMot$.

9.2 Relationship with the Classical Hodge Theory

Let X be a quasi-projective scheme of finite type over a field k of characteristic zero. Then the category $Perf(X)$ of perfect sheaves on X is equivalent to $H^0(A - mod)$, where $A - mod$ is the category of A_∞-modules over a dg-algebra A. Let us recall a construction of A. Consider a complex E of vector bundles which generates the bounded derived category $D^b(Perf(X))$ (see [5]). Then A is quasi-isomorphic to $RHom(E, E)$. More explicitly, let us fix an affine covering $X = \cup_i U_i$. Then the complex $A := \oplus_{i_0, i_1, \ldots, i_n} \Gamma(U_{i_0} \cap \ldots \cap U_{i_n}, E^* \otimes E)[-n]$, $n = \dim X$ computes $RHom(E, E)$ and carries a structure of dg-algebra. Different choices of A give rise to equivalent categories $H^0(A - mod)$ (derived Morita equivalence).

Properties of X are encoded in the properties of A. In particular:
(a) X is smooth iff A is homologically smooth;
(b) X is compact iff A is compact.
Moreover, if X is smooth then

$$H^\bullet(A, A) \simeq Ext^\bullet_{D^b(Coh(X \times X))}(\mathcal{O}_\Delta, \mathcal{O}_\Delta) \simeq$$

$$\oplus_{i,j \geq 0} H^i(X, \wedge^j T_X)[-(i+j)]]$$

where $\mathcal{O}_\Delta$ is the structure sheaf of the diagonal $\Delta \subset X \times X$.

Similarly

$$H_\bullet(A, A) \simeq \oplus_{i,j \geq 0} H^i(X, \wedge^j T_X^*)[j - i].$$

The RHS of the last formula is the Hodge cohomology of X. One can consider the hypercohomology $\mathbf{H}^\bullet(X, \Omega_X^\bullet[[u]]/u^n \Omega_X^\bullet[[u]])$ equipped with the differential $u d_{dR}$. Then the classical Hodge theory ensures degeneration of the corresponding spectral sequence, which means that the hypercohomology is a flat $k[u]/(u^n)$-module for any $n \geq 1$. Usual de Rham cohomology $H_{dR}^\bullet(X)$ is isomorphic to the generic fiber of the corresponding flat vector bundle over the formal line $\mathbf{A}^1_{form}[-2]$, while the fiber at $u = 0$ is isomorphic to the Hodge cohomology $H_{Hodge}^\bullet(X) = \oplus_{i,j \geq 0} H^i(X, \wedge^j T_X^*)[j - i]$. In order to make a connection with the "abstract" theory of the previous subsection we remark that $H_{dR}^\bullet(X)$ is isomorphic to the periodic cyclic homology $HP_\bullet(A)$ while $H_\bullet(A, A)$ is isomorphic to $H_{Hodge}^\bullet(X)$.

10 A_∞-Algebras with Scalar Product

10.1 Main Definitions

Let (X, pt, Q) be a finite-dimensional formal pointed dg-manifold over a field k of characteristic zero.

Definition 10.1 A symplectic structure of degree $N \in \mathbf{Z}$ on X is given by a cyclic closed 2-form ω of degree N such that its restriction to the tangent space $T_{pt}X$ is non-degenerate.

One has the following non-commutative analog of the Darboux lemma.

Proposition 10.2 Symplectic form ω has constant coefficients in some affine coordinates at the point pt.

Proof. Let us choose an affine structure at the marked point and write down $\omega = \omega_0 + \omega_1 + \omega_2 + ...$, where $\omega_l = \sum_{i,j} c_{ij}(x)dx_i \otimes dx_j$ and $c_{ij}(x)$ is homogeneous of degree l (in particular, ω_0 has constant coefficients). Next we observe that the following lemma holds.

Lemma 10.3 *Let $\omega = \omega_0 + r$, where $r = \omega_l + \omega_{l+1} + ..., l \geq 1$. Then there is a change of affine coordinates $x_i \mapsto x_i + O(x^{l+1})$ which transforms ω into $\omega_0 + \omega_{l+1} + ...$*

Lemma implies the Proposition, since we can make an infinite product of the above changes of variables (it is a well-defined infinite series). The resulting automorphism of the formal neighborhood of x_0 transforms ω into ω_0.

Proof of the lemma. We have $d_{cycl}\omega_j = 0$ for all $j \geq l$ (see Sect. 7.2 for the notation). The change of variables is determined by a vector field $v = (v_1, ..., v_n)$ such that $v(x_0) = 0$. Namely, $x_i \mapsto x_i - v_i, 1 \leq i \leq n$. Moreover, we will be looking for a vector field such that $v_i = O(x^{l+1})$ for all i.

We have $Lie_v(\omega) = d(i_v\omega_0) + d(i_v r)$. Since $d\omega_l = 0$ we have $\omega_l = d\alpha_{l+1}$ for some form $\alpha_{l+1} = O(x^{l+1})$ in the obvious notation (formal Poincare lemma). Therefore in order to kill the term with ω_l we need to solve the equation $d\alpha_{l+1} = d(i_v\omega_0)$. It suffices to solve the equation $\alpha_{l+1} = i_v\omega_0$. Since ω_0 is non-degenerate, there exists a unique vector field $v = O(x^{l+1})$ solving last equation. This proves the lemma. ∎

Definition 10.4 Let (X, pt, Q, ω) be a non-commutative formal pointed symplectic dg-manifold. A scalar product of degree N on the A_∞-algebra $A = T_{pt}X[-1]$ is given by a choice of affine coordinates at pt such that the ω becomes constant and gives rise to a non-degenerate bilinear form $A \otimes A \to k[-N]$.

Remark 10.5 Note that since $Lie_Q(\omega) = 0$ there exists a cyclic function $S \in \Omega^0_{cycl}(X)$ such that $i_Q\omega = dS$ and $\{S, S\} = 0$ (here the Poisson bracket corresponds to the symplectic form ω). It follows that the deformation theory of a non-unital A_∞-algebra A with the scalar product is controlled by the DGLA $\Omega^0_{cycl}(X)$ equipped with the differential $\{S, \bullet\}$.

We can restate the above definition in algebraic terms. Let A be a finite-dimensional A_∞-algebra, which carries a non-degenerate symmetric bilinear form $(,)$ of degree N. This means that for any two elements $a, b \in A$ such that $deg(a) + deg(b) = N$ we are given a number $(a, b) \in k$ such that:

(1) for any collection of elements $a_1, ..., a_{n+1} \in A$ the expression $(m_n(a_1, ..., a_n), a_{n+1})$ is cyclically symmetric in the graded sense (i.e., it satisfies the Koszul rule of signs with respect to the cyclic permutation of arguments);

(2) bilinear form $(\bullet, \bullet)$ is non-degenerate.

In this case we will say that A is an A_∞-algebra with the scalar product of degree N.

10.2 Calabi–Yau Structure

The above definition requires A to be finite-dimensional. We can relax this condition requesting that A is compact. As a result we will arrive to a homological version of the notion of scalar product. More precisely, assume that A is weakly unital compact A_∞-algebra. Let $CC_\bullet^{mod}(A) = (CC_\bullet^{mod}(A, A)[u^{-1}], b + uB)$ be the cyclic complex of A. Let us choose a cohomology class $[\varphi] \in H^\bullet(CC_\bullet^{mod}(A))^*$ of degree N. Since the complex (A, m_1) is a subcomplex of $C_\bullet^{mod}(A, A) \subset CC_\bullet^{mod}(A)$ we see that $[\varphi]$ defines a linear functional $Tr_{[\varphi]} : H^\bullet(A) \to k[-N]$.

Definition 10.6 We say that $[\varphi]$ is homologically non-degenerate if the bilinear form of degree N on $H^\bullet(A)$ given by $(a, b) \mapsto Tr_{[\varphi]}(ab)$ is non-degenerate.

Note that the above bilinear form defines a symmetric scalar product of degree N on $H^\bullet(A)$.

Theorem 10.7 For a weakly unital compact A_∞-algebra A a homologically non-degenerate cohomology class $[\varphi]$ gives rise to a class of isomorphisms of non-degenerate scalar products on a minimal model of A.

Proof. Since $char\, k = 0$ the complex $(CC_\bullet^{mod}(A))^*$ is quasi-isomorphic to $(\Omega_{cycl}^0(X)/k, Lie_Q)$.

Lemma 10.8 Complex $(\Omega_{cycl}^{2,cl}(X), Lie_Q)$ is quasi-isomorphic to the complex $(\Omega_{cycl}^0(X)/k, Lie_Q)$.[12]

Proof. Notice that as a complex $(\Omega_{cycl}^{2,cl}(X), Lie_Q)$ is isomorphic to the complex $\Omega_{cycl}^1(X)/d_{cycl}\,\Omega_{cycl}^0(X)$. The latter is quasi-isomorphic to $[\mathcal{O}(X), \mathcal{O}(X)]_{top}$ via $a\, db \mapsto [a, b]$ (recall that $[\mathcal{O}(X), \mathcal{O}(X)]_{top}$ denotes the topological closure of the commutator).

By definition $\Omega_{cycl}^0(X) = \mathcal{O}(X)/[\mathcal{O}(X), \mathcal{O}(X)]_{top}$. We know that $\mathcal{O}(X)/k$ is acyclic, hence $\Omega_{cycl}^0(X)/k$ is quasi-isomorphic to $[\mathcal{O}(X), \mathcal{O}(X)]_{top}$. Hence the complex $(\Omega_{cycl}^{2,cl}(X), Lie_Q)$ is quasi-isomorphic to $(\Omega_{cycl}^0(X)/k, Lie_Q)$. $\blacksquare$

---

[12] See also Proposition 5.5.1 from [19].

As a corollary we obtain an isomorphism of cohomology groups $H^\bullet(\Omega^{2,cl}_{cycl}(X)) \simeq H^\bullet(\Omega^0_{cycl}(X)/k)$. Having a non-degenerate cohomology class $[\varphi] \in H^\bullet(CC^{mod}_\bullet(A))^* \simeq H^\bullet(\Omega^{2,cl}_{cycl}(X), Lie_Q)$ as above, we can choose its representative $\omega \in \Omega^{2,cl}_{cycl}(X)$, $Lie_Q\omega = 0$. Let us consider $\omega(x_0)$. It can be described pure algebraically such as follows. Notice that there is a natural projection $H^\bullet(\Omega^0_{cycl}(X)/k) \to (A/[A,A])^*$ which corresponds to the taking the first Taylor coefficient of the cyclic function. Then the above evaluation $\omega(x_0)$ is the image of $\varphi(x_0)$ under the natural map $(A/[A,A])^* \to (Sym^2(A))^*$ which assigns to a linear functional l the bilinear form $l(ab)$.

We claim that the total map $H^\bullet(\Omega^{2,cl}_{cycl}(X)) \to (Sym^2(A))^*$ is the same as the evaluation at x_0 of the closed cyclic 2-form. Equivalently, we claim that $\omega(x_0)(a,b) = Tr_\varphi(ab)$. Indeed, if $f \in \Omega^0_{cycl}(X)/k$ is the cyclic function corresponding to ω then we can write $f = \sum_i a_i x_i + O(x^2)$. Therefore $Lie_Q(f) = \sum_{l,i,j} a_i c^{ij}_l [x_i, x_j] + O(x^3)$, where c^{ij}_l are structure constants of $\mathcal{O}(X)$. Dualizing we obtain the claim.

Proposition 10.9 Let ω_1 and ω_2 be two symplectic structures on the finite-dimensional formal pointed minimal dg-manifold (X, pt, Q) such that $[\omega_1] = [\omega_2]$ in the cohomology of the complex $(\Omega^{2,cl}_{cycl}(X), Lie_Q)$ consisting of closed cyclic 2-forms. Then there exists a change of coordinates at x_0 preserving Q which transforms ω_1 into ω_2.

Corollary 10.10 Let (X, pt, Q) be a (possibly infinite-dimensional) formal pointed dg-manifold endowed with a (possibly degenerate) closed cyclic 2-form ω. Assume that the tangent cohomology $H^0(T_{pt}X)$ is finite-dimensional and ω induces a non-degenerate pairing on it. Then on the minimal model of (X, pt, Q) we have a canonical isomorphism class of symplectic forms modulo the action of the group $Aut(X, pt, Q)$.

Proof. Let M be a (finite-dimensional) minimal model of A. Choosing a cohomology class $[\varphi]$ as above we obtain a non-degenerate bilinear form on M, which is the restriction $\omega(x_0)$ of a representative $\omega \in \Omega^{2,cl}(X)$. By construction this scalar product depends on ω. We would like to show that in fact it depends on the cohomology class of ω, i.e., on φ only. This is the corollary of the following result.

Lemma 10.11 Let $\omega_1 = \omega + Lie_Q(d\alpha)$. Then there exists a vector field v such that $v(x_0) = 0$, $[v, Q] = 0$ and $Lie_v(\omega) = Lie_Q(d\alpha)$.

Proof. As in the proof of Darboux lemma we need to find a vector field v, satisfying the condition $di_v(\omega) = Lie_Q(d\alpha)$. Let $\beta = Lie_Q(\alpha)$. Then $d\beta = dLie_Q(\alpha) = 0$. Since ω is non-degenerate we can find v satisfying the conditions of the Proposition and such that $di_v(\omega) = Lie_Q(d\alpha)$. Using this v we can change affine coordinates transforming $\omega + Lie_Q(d\alpha)$ back to ω. This concludes the proof of the Proposition and the Theorem.■

Presumably the above construction is equivalent to the one given in [23]. We will sometimes call the cohomology class $[\varphi]$ a *Calabi–Yau structure* on A (or on the corresponding non-commutative formal pointed dg-manifold X). The following example illustrates the relation to geometry.

Example 10.12 Let X be a complex Calabi–Yau manifold of dimension n. Then it carries a nowhere vanishing holomorphic n-form *vol*. Let us fix a holomorphic vector bundle E and consider a dg-algebra $A = \Omega^{0,*}(X, End(E))$ of Dolbeault $(0, p)$-forms with values in $End(E)$. This dg-algebra carries a linear functional $a \mapsto \int_X Tr(a) \wedge vol$. One can check that this is a cyclic cocycle which defines a non-degenerate pairing on $H^\bullet(A)$ in the way described above.

There is another approach to Calabi–Yau structures in the case when A is homologically smooth. Namely, we say that A carries a Calabi–Yau structure of dimension N if $A^! \simeq A[N]$ (recall that $A^!$ is the $A - A$-bimodule $Hom_{A-mod-A}(A, A \otimes A)$ introduced in Sect. 8.1. Then we expect the following conjecture to be true.

Conjecture 10.13 If A is a homologically smooth compact finite-dimensional A_∞-algebra then the existence of a non-degenerate cohomology class $[\varphi]$ of degree $dim\, A$ is equivalent to the condition $A^! \simeq A[dim\, A]$.

If A is the dg-algebra of endomorphisms of a generator of $D^b(Coh(X))$ where X is Calabi–Yau then the above conjecture holds trivially.

Finally, we would like to illustrate the relationship of the non-commutative symplectic geometry discussed above with the commutative symplectic geometry of certain spaces of representations.[13] More generally we would like to associate with $X = Spc(T(A[1]))$ a collection of formal algebraic varieties, so that some "non-commutative" geometric structure on X becomes a collection of compatible "commutative" structures on formal manifolds $\mathcal{M}(X, n) := \widehat{Rep}_0(\mathcal{O}(X), Mat_n(k))$, where $Mat_n(k)$ is the associative algebra of $n \times n$ matrices over k, $\mathcal{O}(X)$ is the algebra of functions on X and $\widehat{Rep}_0(...)$ means the formal completion at the trivial representation. In other words, we would like to define a collection of compatible geometric structure on "$Mat_n(k)$-points" of the formal manifold X. In the case of symplectic structure this philosophy is illustrated by the following result.

Theorem 10.14 *Let X be a non-commutative formal symplectic manifold in $Vect_k$. Then it defines a collection of symplectic structures on all manifolds $\mathcal{M}(X, n), n \geq 1$.*

Proof. Let $\mathcal{O}(X) = A, \mathcal{O}(\mathcal{M}(X, n)) = B$. Then we can choose isomorphisms $A \simeq k\langle\langle x_1, ..., x_m\rangle\rangle$ and $B \simeq \langle\langle x_1^{\alpha,\beta}, ..., x_m^{\alpha,\beta}\rangle\rangle$, where $1 \leq \alpha, \beta \leq n$. To any $a \in A$ we can assign $\widehat{a} \in B \otimes Mat_n(k)$ such that:

---

[13] It goes back to [30] and since that time has been discussed in many papers, see e.g. [18].

$$\hat{x}_i = \sum_{\alpha,\beta} x_i^{\alpha,\beta} \otimes e_{\alpha,\beta},$$

where $e_{\alpha,\beta}$ is the $n \times n$ matrix with the only non-trivial element equal to 1 on the intersection of α-th line and β-th column. The above formulas define an algebra homomorphism. Composing it with the map $id_B \otimes Tr_{Mat_n(k)}$ we get a linear map $\mathcal{O}_{cycl}(X) \to \mathcal{O}(\mathcal{M}(X,n))$. Indeed the closure of the commutator $[A, A]$ is mapped to zero. Similarly, we have a morphism of complexes $\Omega_{cycl}^\bullet(X) \to \Omega^\bullet(\mathcal{M}(X,n))$, such that

$$dx_i \mapsto \sum_{\alpha,\beta} dx_i^{\alpha,\beta} e_{\alpha,\beta}.$$

Clearly, continuous derivations of A (i.e., vector fields on X) give rise to the vector fields on $\mathcal{M}(X,n)$.

Finally, one can see that a non-degenerate cyclic 2-form ω is mapped to the tensor product of a non-degenerate 2-form on $\mathcal{M}(X,n)$ and a nondegenerate 2-form $Tr(XY)$ on $Mat_n(k)$. Therefore a symplectic form on X gives rise to a symplectic form on $\mathcal{M}(X,n), n \geq 1$. ∎

11 Hochschild Complexes as Algebras Over Operads and PROPs

Let A be a strictly unital A_∞-algebra over a field k of characteristic zero. In this section we are going to describe a colored dg-operad P such that the pair $(C^\bullet(A, A), C_\bullet(A, A))$ is an algebra over this operad. More precisely, we are going to describe $\mathbf{Z}$-graded k-vector spaces $A(n, m)$ and $B(n, m)$, $n, m \geq 0$ which are components of the colored operad such that $B(n, m) \neq 0$ for $m = 1$ only and $A(n, m) \neq 0$ for $m = 0$ only together with the colored operad structure and the action

(a) $A(n, 0) \otimes (C^\bullet(A, A))^{\otimes n} \to C^\bullet(A, A)$,

(b) $B(n, 1) \otimes (C^\bullet(A, A))^{\otimes n} \otimes C_\bullet(A, A) \to C_\bullet(A, A)$.

Then, assuming that A carries a non-degenerate scalar product, we are going to describe a PROP R associated with moduli spaces of Riemannian surfaces and a structure of R-algebra on $C_\bullet(A, A)$.

11.1 Configuration Spaces of Discs

We start with the spaces $A(n, 0)$. They are chain complexes. The complex $A(n, 0)$ coincides with the complex M_n of the minimal operad $M = (M_n)_{n \geq 0}$ described in [35], Sect. 5. Without going into details which can be found in loc. cit. we recall main facts about the operad M. A basis of M_n as a k-vector space is formed by n-labeled planar trees (such trees have internal vertices labeled by the set $\{1, ..., n\}$ as well as other internal vertices which are non-labeled and each has the valency at least 3).

We can depict n-labeled trees such as follows

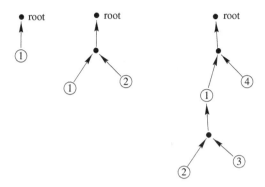

Labeled vertices are depicted as circles with numbers inscribed, non-labeled vertices are depicted as black vertices. In this way we obtain a graded operad M with the total degree of the basis element corresponding to a tree T equal to

$$deg(T) = \sum_{v \in V_{lab}(T)} (1 - |v|) + \sum_{v \in V_{nonl}(T)} (3 - |v|)$$

where $V_{lab}(T)$ and $V_{nonl}(T)$ denote the sets of labeled and non-labeled vertices respectively , and $|v|$ is the valency of the vertex v, i.e., the cardinality of the set of edges attached to v.

The notion of an *angle* between two edges incoming in a vertex is illustrated in the following figure (angles are marked by asteriscs).

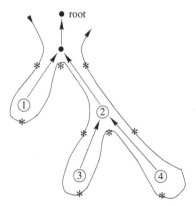

Operadic composition and the differential are described in [35], sects. 5.2, 5.3. We borrow from there the following figure which illustrates the operadic composition of generators corresponding to labeled trees T_1 and T_2.

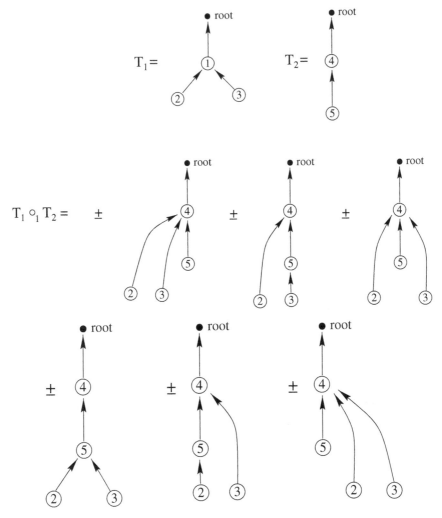

Informally speaking, the operadic gluing of T_2 to T_1 at an internal vertex v of T_1 is obtained by:

(a) Removing from T_1 the vertex v together with all incoming edges and vertices.

(b) Gluing T_2 to v (with the root vertex removed from T_2). Then

(c) Inserting removed vertices and edges of T_1 in all angles between incoming edges to the new vertex v_{new}.

(d) Taking the sum (with appropriate signs) over all possible inserting of edges in (c).

The differential d_M is a sum of the "local" differentials d_v, where v runs through the set of all internal vertices. Each d_v inserts a new edge into the set of edges attached to v. The following figure borrowed from [35] illustrates the difference between labeled (white) and non-labeled (black) vertices.

In this way we make M into a dg-operad. It was proved in [35], that M is quasi-isomorphic to the dg-operad $Chains(FM_2)$ of singular chains on the Fulton–Macpherson operad FM_2. The latter consists of the compactified moduli spaces of configurations of points in $\mathbf{R}^2$ (see e.g. [35], Sect. 7.2 for a description). It was also proved in [35] that $C^\bullet(A, A)$ is an algebra over the operad M (Deligne's conjecture follows from this fact). The operad FM_2 is homotopy equivalent to the famous operad $C_2 = (C_2(n))_{n\geq 0}$ of two-dimensional discs (little disc operad). Thus $C^\bullet(A, A)$ is an algebra (in the homotopy sense) over the operad $Chains(C_2)$.

11.2 Configurations of Points on the Cylinder

Let $\Sigma = S^1 \times [0, 1]$ denotes the standard cylinder.

Let us denote by $S(n)$ the set of isotopy classes of the following graphs $\Gamma \subset \Sigma$:

(a) every graph Γ is a forest (i.e., disjoint union of finitely many trees $\Gamma = \sqcup_i T_i$);

(b) the set of vertices $V(\Gamma)$ is decomposed into the union $V_{\partial\Sigma} \sqcup V_{lab} \sqcup V_{nonl} \sqcup V_1$ of four sets with the following properties:

(b1) the set $V_{\partial\Sigma}$ is the union $\{in\} \cup \{out\} \cup V_{out}$ of three sets of points which belong to the boundary $\partial\Sigma$ of the cylinder. The set $\{in\}$ consists of one marked point which belongs to the boundary circle $S^1 \times \{1\}$ while the set $\{out\}$ consists of one marked point which belongs to the boundary circle $S^1 \times \{0\}$. The set V_{out} consists of a finitely many unlableled points on the boundary circle $S^1 \times \{0\}$;

(b2) the set V_{lab} consists of n labeled points which belong to the surface $S^1 \times (0, 1)$ of the cylinder;

(b3) the set V_{nonl} consists of a finitely many non-labeled points which belong to the surface $S^1 \times (0, 1)$ of the cylinder;

(b4) the set V_1 is either empty or consists of only one element denoted by $\mathbf{1} \in S^1 \times (0,1)$ and called *special* vertex;

(c) the following conditions on the valencies of vertices are imposed:

(c1) the valency of the vertex *out* is ≤ 1;

(c2) the valency of each vertex from the set $V_{\partial \Sigma} \setminus V_{out}$ is equal to 1;

(c3) the valency of each vertex from V_{lab} is at least 1;

(c4) the valency of each vertex from V_{nonl} is at least 3;

(c5) if the set V_1 is non-empty then the valency of the special vertex is equal to 1. In this case the only outcoming edge connects $\mathbf{1}$ with the vertex *out*.

(d) Every tree T_i from the forest Γ has its root vertex in the set $V_{\partial \Sigma}$.

(e) We orient each tree T_i down to its root vertex.

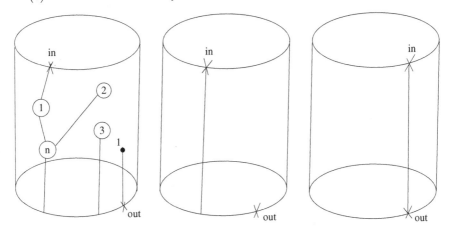

Remark 11.1 Let us consider the configuration space $X_n, n \geq 0$ which consists of (modulo $\mathbf{C}^*$-dilation) equivalence classes of n points on $\mathbf{CP}^1 \setminus \{0, \infty\}$ together with two direction lines at the tangent spaces at the points 0 and ∞. One-point compactification $\widehat{X}_n$ admits a cell decomposition with cells (except of the point $\widehat{X}_n \setminus X_n$) parametrized by elements of the set $S(n)$. This can be proved with the help of Strebel differentials (cf. [35], Sect. 5.5).

Previous remark is related to the following description of the sets $S(n)$ (it will be used later in the chapter). Let us contract both circles of the boundary $\partial \Sigma$ into points. In this way we obtain a tree on the sphere. Points become vertices of the tree and lines outcoming from the points become edges. There are two vertices marked by *in* and *out* (placed at the north and south poles respectively). We orient the tree towards to the vertex *out*. An additional structure consists of:

(a) Marked edge outcoming from *in* (it corresponds to the edge outcoming from *in*).

(b) Either a marked edge incoming to *out* (there was an edge incoming to *out* which connected it with a vertex not marked by $\mathbf{1}$) or an angle between two edges incoming to *out* (all edges which have one of the endpoint vertices on the bottom circle become after contracting it to a point the edges incoming to *out*, and if there was an edge connecting a point marked by **1** with *out*, we mark the angle between edges containing this line).

The reader notices that the star of the vertex *out* can be identified with a regular k-gon, where k is the number of incoming to *out* edges. For this k-gon we have either a marked point on an edge (case (a) above) or a marked angle with the vertex in *out* (case (b) above).

11.3 Generalization of Deligne's Conjecture

The definition of the operadic space $B(n,1)$ will be clear from the description of its action on the Hochschild chain complex. The space $B(n,1)$ will have a basis parametrized by elements of the set $S(n)$ described in the previous subsection. Let us describe the action of a generator of $B(n,1)$ on a pair $(\gamma_1 \otimes \ldots \otimes \gamma_n, \beta)$, where $\gamma_1 \otimes \ldots \otimes \gamma_n \in C^\bullet(A,A)^{\otimes n}$ and $\beta = a_0 \otimes a_1 \otimes \ldots \otimes a_l \in C_l(A,A)$. We attach elements $a_0, a_1, \ldots, a_l$ to points on Σ_h^{in}, in a cyclic order, such that a_0 is attached to the point *in*. We attach γ_i to the ith numbered point on the surface of Σ_h. Then we draw disjoint continuous segments (in all possible ways, considering pictures up to an isotopy) starting from each point marked by some element a_i and oriented downstairs, with the requirements (a–c) as above, with the only modification that we allow an arbitrary number of points on $S^1 \times \{1\}$. We attach higher multiplications m_j to all non-numbered vertices, so that j is equal to the incoming valency of the vertex. Reading from the top to the bottom and composing γ_i and m_j we obtain (on the bottom circle) an element $b_0 \otimes \ldots \otimes b_m \in C_\bullet(A,A)$ with b_0 attached to the vertex *out*. If the special vertex **1** is present then we set $b_0 = 1$. This gives the desired action.

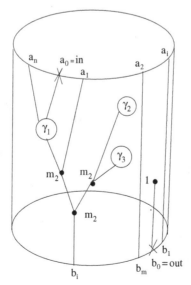

Composition of the operations in $B(n, 1)$ corresponds to the gluing of the cylinders such that the point *out* of the top cylinder is identified with the point *in* of the bottom cylinder. If after the gluing there is a line from the point marked **1** on the top cylinder which does not end at the point *out* of the bottom cylinder, we will declare such a composition to be equal to zero.

Let us now consider a topological colored operad $C_2^{col} = (C_2^{col}(n, m))_{n, m \geq 0}$ with two colors such that $C_2^{col}(n, m) \neq \emptyset$ only if $m = 0, 1$ and

(a) In the case $m = 0$ it is the little disc operad.

(b) In the case $m = 1$ $C_2^{col}(n, 1)$ is the moduli space (modulo rotations) of the configurations of $n \geq 1$ discs on the cyliner $S^1 \times [0, h]$ $h \geq 0$ and two marked points on the boundary of the cylinder. We also add the degenerate circle of configurations $n = 0, h = 0$. The topological space $C_2^{col}(n, 1)$ is homotopically equivalent to the configuration space X_n described in the previous subsection.

Let $Chains(C_2^{col})$ be the colored operad of singular chains on C_2^{col}. Then, similarly to [35], Sect. 7, one proves (using the explicit action of the colored operad $P = (A(n, m), B(n, m))_{n, m \geq 0}$ described above) the following result.

Theorem 11.2 *Let A be a unital A_∞-algebra. Then the pair $(C^\bullet(A, A), C_\bullet (A, A))$ is an algebra over the colored operad $Chains(C_2^{col})$ (which is quasi-isomorphic to P) such that for $h = 0, n = 0$ and coinciding points $in = out$, the corresponding operation is the identity.*

Remark 11.3 The above Theorem generalizes Deligne's conjecture (see e.g. [35]). It is related to the abstract calculus associated with A (see [T, 48]). The reader also notices that for $h = 0, n = 0$ we have the moduli space of two points on the circle. It is homeomorphic to S^1. Thus we have an action of S^1 on $C_\bullet(A, A)$. This action gives rise to the Connes differential B.

Similarly to the case of little disc operad, one can prove the following result.

Proposition 11.4 *The colored operad C_2^{col} is formal, i.e., it is quasi-isomorphic to its homology colored operad.*

If A is non-unital we can consider the direct sum $A_1 = A \oplus k$ and make it into a unital A_∞-algebra. The *reduced* Hochschild chain complex of A_1 is defined as $C_\bullet^{red}(A_1, A_1) = \oplus_{n \geq 0} A_1 \otimes ((A_1/k)[1])^{\otimes n}$ with the same differential as in the unital case. One defines the reduced Hochschild cochain complex $C_{red}^\bullet(A_1, A_1)$ similarly. We define the *modified* Hochschild chain complex $C_\bullet^{mod}(A, A)$ from the following isomorphism of complexes $C_\bullet^{red}(A_1, A_1) \simeq C_\bullet^{mod}(A, A) \oplus k$. Similarly, we define the modified Hochschild cochain complex from the decomposition $C_{red}^\bullet(A_1, A_1) \simeq C_{mod}^\bullet(A, A) \oplus k$. Then, similarly to the Theorem 11.3.1 one proves the following result.

Proposition 11.5 *The pair $(C_\bullet^{mod}(A, A), C_{mod}^\bullet(A, A))$ is an algebra over the colored operad which is an extension of $Chains(C_2^{col})$ by null-ary operations* on Hochschild chain and cochain complexes, which correspond to the unit in A, and such that for $h = 0, n = 0$ and coinciding points $in = out$, the corresponding operation is the identity.

11.4 Remark About Gauss–Manin Connection

Let $R = k[[t_1, ..., t_n]]$ be the algebra of formal series and A be an R-flat A_∞-algebra. Then the (modified) negative cyclic complex $CC_\bullet^{-,mod}(A) = (C_\bullet(A, A)[[u]], b + uB)$ is an $R[[u]]$-module. It follows from the existense of Gauss-Manin connection (see [16]) that the cohomology $HC_\bullet^{-,mod}(A)$ is in fact a module over the ring

$$D_R(A) := k[[t_1, ..., t_n, u]][u\partial/\partial t_1, ..., u\partial/\partial t_n].$$

Inedeed, if ∇ is the Gauss–Manin connection from [16] then $u\partial/\partial t_i$ acts on the cohomology as $u\nabla_{\partial/\partial t_i}, 1 \leq i \leq n$.

The above considerations can be explained from the point of view of conjecture below. Let $g = C^\bullet(A, A)[1]$ be the DGLA associated with the Hochschild cochain complex and $M := (CC_\bullet^{-,mod}(A))$. We define a DGLA $\hat{g}$ which is the crossproduct $(g \otimes k\langle \xi \rangle) \rtimes k(\partial/\partial \xi)$, where $deg\, \xi = +1$.

Conjecture 11.6 There is a structure of an L_∞-module on M over $\hat{g}$ which extends the natural structure of a g-module and such that $\partial/\partial \xi$ acts as Connes differential B. Moreover this structure should follow from the P-algebra structure described in Sect. 11.3.

It looks plausible that the formulas for the Gauss–Manin connection from [16] can be derived from our generalization of Deilgne's conjecture. We will discuss flat connections on periodic cyclic homology later in the text.

11.5 Flat Connections and the Colored Operad

We start with $\mathbf{Z}$-graded case. Let us interpret the $\mathbf{Z}$-graded formal scheme $Spf(k[[u]])$ as even formal line equipped with the $\mathbf{G}_m$-action $u \mapsto \lambda^2 u$. The space $HC_\bullet^{-,mod}(A)$ can be interpreted as a space of sections of a $\mathbf{G}_m$-equivariant vector bundle ξ_A over $Spf(k[[u]])$ corresponding to the $k[[u]]$-flat module $\varprojlim_n H^\bullet(C_\bullet^{(n)}(A, A))$. The action of $\mathbf{G}_m$ identifies fibers of this vector bundle over $u \neq 0$. Thus we have a natural flat connection ∇ on the restriction of ξ_A to the complement of the point 0 which has the pole of order one at $u = 0$.

Here we are going to introduce a different construction of the connection ∇ which works also in $\mathbf{Z}/2$-graded case. This connection will have in general a pole of degree two at $u = 0$. In particular we have the following result.

Proposition 11.7 The space of section of the vector bundle ξ_A can be endowed with a structure of a $k[[u]][[u^2\partial/\partial u]]$-module.

In fact we are going to give an explicit construction of the connection, which is based on the action of the colored dg-operad P discussed in Sect. 11.3 (more precisely, an extension P^{new} of P, see below). Before presenting an explicit formula, we will make few comments.

1. For any $\mathbf{Z}/2$-graded A_∞-algebra A one can define canonically a 1-parameter family of A_∞-algebras $A_\lambda, \lambda \in \mathbf{G}_m$, such that $A_\lambda = A$ as a $\mathbf{Z}/2$-graded vector space and $m_n^{A_\lambda} = \lambda m_n^A$.

2. For simplicity we will assume that A is strictly unital. Otherwise we will work with the pair $(C_\bullet^{mod}(A, A), C_{mod}^\bullet(A, A))$ of modified Hochschild complexes.

3. We can consider an extension P^{new} of the dg-operad P allowing any non-zero valency for a non-labeled (black) vertex(in the definition of P we required that such a valency was at least three). All the formulas remain the same. But the dg-operad P^{new} is no longer formal. It contains a dg-suboperad generated by trees with all vertices being non-labeled. Action of this suboperad P_{nonl}^{new} is responsible for the flat connection discussed below.

4. In addition to the connection along the variable u one has the Gauss–Manin connection which acts along the fibers of ξ_A (see Sect. 11.4). Probably one can write down an explicit formula for this connection using the action of the colored operad P^{new}. In what follows are going to describe a connection which presumably coincides with the Gauss-Manin connection.

Let us now consider a dg-algebra $k[B, \gamma_0, \gamma_2]$ which is generated by the following operations of the colored dg-operad P^{new}:

(a) Connes differential B of degree -1. It can be depicted such as follows (cf. Sect. 7.3):

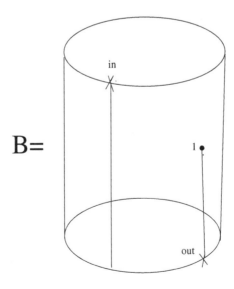

(b) Generator γ_2 of degree 2, corresponding to the following figure:

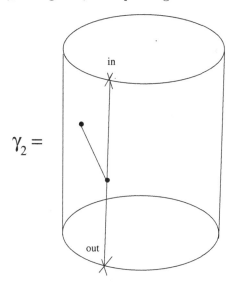

(c) Generator γ_0 of degree 0, where $2\gamma_0$ is depicted below:

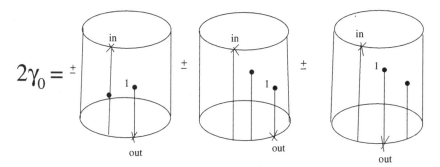

Proposition 11.8 The following identities hold in P^{new}:

$$B^2 = dB = d\gamma_2 = 0, d\gamma_0 = [B, \gamma_2],$$

$$B\gamma_0 + \gamma_0 B := [B, \gamma_0]_+ = -B.$$

Here by d we denote the Hochschild chain differential (previously it was denoted by b).

Proof. Let us prove that $[B, \gamma_0] = -B$, leaving the rest as an exercise to the reader. One has the following identities for the compositions of operations in P^{new}: $B\gamma_0 = 0$, $\gamma_0 B = B$. Let us check, for example, the last identity. Let us denote by W the first summand on the figure defining $2\gamma_0$. Then $\gamma_0 B = \frac{1}{2}WB$. The latter can be depicted in the following way:

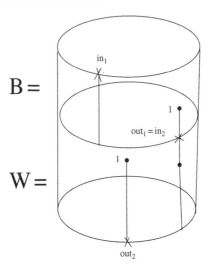

$$B =$$

$$W =$$

It is easily seen equals to $2 \cdot 1/2B = B$. ∎

Corollary 11.9 Hochschild chain complex $C_\bullet(A, A)$ is a dg-module over the dg-algebra $k[B, \gamma_0, \gamma_2]$.

Let us consider the truncated negative cyclic complex $(C_\bullet(A, A)[[u]]/(u^n), d_u = d + uB)$. We introduce a k-linear map ∇ of $C_\bullet(A, A)[[u]]/(u^n)$ into itself such that $\nabla_{u^2 \partial/\partial u} = u^2 \partial/\partial u - \gamma_2 + u\gamma_0$. Then we have:
(a) $[\nabla_{u^2 \partial/\partial u}, d_u] = 0$;
(b) $[\nabla_{u^2 \partial/\partial u}, u] = u^2$.
Let us denote by V the unital dg-algebra generated by $\nabla_{u^2 \partial/\partial u}$ and u, subject to the relations (a), (b) and the relation $u^n = 0$. From (a) and (b) one deduces the following result.

Proposition 11.10 The complex $(C_\bullet(A, A)[[u]]/(u^n), d_u = d + uB)$ is a V-module. Moreover, assuming the degeneration conjecture, we see that the operator $\nabla_{u^2 \partial/\partial u}$ defines a flat connection on the cohomology bundle

$$H^\bullet(C_\bullet(A, A)[[u]]/(u^n), d_u)$$

which has the only singularity at $u = 0$ which is a pole of second order.

Taking the inverse limit over n we see that $H^\bullet(C_\bullet(A, A)[[u]], d_u)$ gives rise to a vector bundle over $\mathbf{A}^1_{form}[-2]$ which carries a flat connection with the second order pole at $u = 0$. It is interesting to note the difference between $\mathbf{Z}$-graded and $\mathbf{Z}/2$-graded A_∞-algebras. It follows from the explicit formula for the connection ∇ that the coefficient of the second degree pole is represented by multiplication by a cocyle $(m_n)_{n \geq 1} \in C^\bullet(A, A)$. In cohomology it is trivial in $\mathbf{Z}$-graded case (because of the invariance with respect to the group action $m_n \mapsto \lambda \, m_n$), but non-trivial in $\mathbf{Z}/2$-graded case. Therefore the order of the pole of ∇ is equal to one for $\mathbf{Z}$-graded A_∞-algebras and is equal to two for

$\mathbf{Z}/2$-graded A_∞-algebras. We see that in $\mathbf{Z}$-graded case the connection along the variable u comes from the action of the group $\mathbf{G}_m$ on higher products m_n, while in $\mathbf{Z}/2$-graded case it is more complicated.

11.6 PROP of Marked Riemann Surfaces

In this section we will describe a PROP naturally acting on the Hochschild complexes of a finite-dimensional A_∞-algebra with the scalar product of degree N.

Since we have a quasi-isomorphism of complexes

$$C^\bullet(A, A) \simeq (C_\bullet(A, A))^*[-N]$$

it suffices to consider the chain complex only.

In this subsection we will assume that A is either $\mathbf{Z}$-graded (then N is an integer) or $\mathbf{Z}/2$-graded (then $N \in \mathbf{Z}/2$). We will present the results for non-unital A_∞-algebras. In this case we will consider the modified Hochschild chain complex

$$C_\bullet^{mod}(A, A) = \oplus_{n \geq 0} A \otimes (A[1])^{\otimes n} \bigoplus \oplus_{n \geq 1} (A[1])^{\otimes n},$$

equipped with the Hochschild chain differential (see Sect. 7.4).

Our construction is summarized in (i–ii) below.

(i) Let us consider the topological PROP $\mathcal{M} = (\mathcal{M}(n, m))_{n,m \geq 0}$ consisting of moduli spaces of metrics on compacts oriented surfaces with bondary consisting of $n + m$ circles and some additional marking (see precise definition below).

(ii) Let $Chains(\mathcal{M})$ be the corresponding PROP of singular chains. Then there is a structure of a $Chains(\mathcal{M})$-algebra on $C_\bullet^{mod}(A, A)$, which is encoded in a collection of morphisms of complexes

$$Chains(\mathcal{M}(n, m)) \otimes C_\bullet^{mod}(A, A)^{\otimes n} \to (C_\bullet^{mod}(A, A))^{\otimes m}.$$

In addition one has the following:

(iii) If A is homologically smooth and satisfies the degeneration property then the structure of $Chains(\mathcal{M})$-algebra extends to a structure of a $Chains(\overline{\mathcal{M}})$-algebra, where $\overline{\mathcal{M}}$ is the topological PROP of stable compactifications of $\mathcal{M}(n, m)$.

Definition 11.11 An element of $\mathcal{M}(n, m)$ is an isomorphism class of triples $(\Sigma, h, mark)$ where Σ is a compact oriented surface (not necessarily connected) with metric h and $mark$ is an orientation preserving isometry between a neighborhood of $\partial\Sigma$ and the disjoint union of $n + m$ flat semiannuli $\sqcup_{1 \leq i \leq n}(S^1 \times [0, \varepsilon)) \sqcup \sqcup_{1 \leq i \leq m}(S^1 \times [-\varepsilon, 0])$, where ε is a sufficiently small positive number. We will call n circle "inputs" and the rest m circles "outputs". We will assume that each connected component of Σ has at least one input and there are no discs among the connected components. Also we will add $\Sigma = S^1$ to $\mathcal{M}(1,1)$ as the identity morphism. It can be thought of as the limit of cylinders $S^1 \times [0, \varepsilon]$ as $\varepsilon \to 0$.

The composition is given by the natural gluing of surfaces.

Let us describe a construction of the action of $Chains(\mathcal{M})$ on the Hochschild chain complex. In fact, instead of $Chains(\mathcal{M})$ we will consider a quasi-isomorphic dg-PROP $R = (R(n,m)_{n,m\geq 0})$ generated by ribbon graphs with additional data. In what follows we will skip some technical details in the definition of the PROP R. They can be recovered in a more or less straightforward way.

It is well-known (and can be proved with the help of Strebel differentials) that $\mathcal{M}(n,m)$ admits a stratification with strata parametrized by graphs described below. More precisely, we consider the following class of graphs.

(1) Each graph Γ is a (not necessarily connected) ribbon graph (i.e., we are given a cyclic order on the set $Star(v)$ of edges attached to a vertex v of Γ). It is well-known that replacing an edge of a ribbon graph by a thin stripe (thus getting a "fat graph") and gluing stripes in the cyclic order one gets a Riemann surface with the boundary.

(2) The set $V(\Gamma)$ of vertices of Γ is the union of three sets: $V(\Gamma) = V_{in}(\Gamma) \cup V_{middle}(\Gamma) \cup V_{out}(\Gamma)$. Here $V_{in}(\Gamma)$ consists of n numbered vertices $in_1, ..., in_n$ of the valency 1 (the outcoming edges are called tails), $V_{middle}(\Gamma)$ consists of vertices of the valency ≥ 3, and $V_{out}(\Gamma)$ consists of m numbered vertices $out_1, ..., out_m$ of valency ≥ 1.

(3) We assume that the Riemann surface corresponding to Γ has n connected boundary components each of which has exactly one input vertex.

(4) For every vertex $out_j \in V_{out}(\Gamma), 1 \leq j \leq m$ we *mark* either an incoming edge or a pair of adjacent (we call such a pair of edges a *corner*).

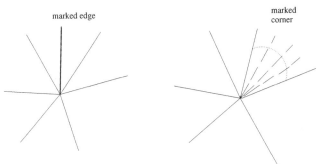

marked edge   marked corner

More pedantically, let $E(\Gamma)$ denotes the set of edges of Γ and $E^{or}(\Gamma)$ denotes the set of pairs (e, or) where $e \in E(\Gamma)$ and or is one of two possible orientations of e. There is an obvious map $E^{or}(\Gamma) \to V(\Gamma) \times V(\Gamma)$ which assigns to an oriented edge the pair of its endpoint vertices: source and target. The free involution σ acting on $E^{or}(\Gamma)$ (change of orientation) corresponds to the permutation map on $V(\Gamma) \times V(\Gamma)$. Cyclic order on each $Star(v)$ means that there is a bijection $\rho : E^{or}(\Gamma) \to E^{or}(\Gamma)$ such that orbits of iterations $\rho^n, n \geq 1$ are elements of $Star(v)$ for some $v \in V(\Gamma)$. In particular, the corner is given either by a pair of coinciding edges (e, e) such that $\rho(e) = e$ or by a pair edges $e, e' \in Star(v)$ such that $\rho(e) = e'$. Let us define a *face* as an orbit of $\rho \circ \sigma$. Then faces are oriented closed paths. It follows from the condition (2) that each face contains exactly one edge outcoming from some in_i.

We depict below two graphs in the case $g = 0, n = 2, m = 0$.

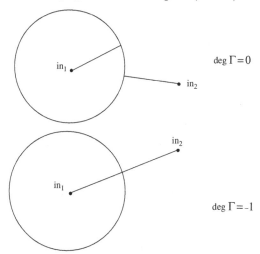

Here is a figure illustrating the notion of face

Two faces: one contains in_1, another contains in_2

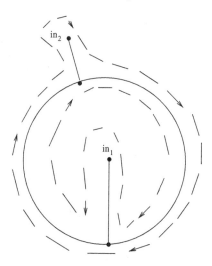

Remark 11.12 The above data (i.e., a ribbon graph with numerations of *in* and *out* vertices) have no automorphisms. Thus we can identify Γ with its isomorphism class.

The functional $(m_n(a_1, ..., a_n), a_{n+1})$ is depicted such as follows.

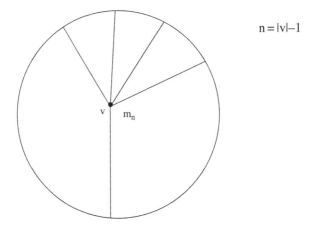

We define the *degree* of Γ by the formula

$$deg\,\Gamma = \sum_{v \in V_{middle}(\Gamma)} (3 - |v|) + \sum_{v \in V_{out}(\Gamma)} (1 - |v|) + \sum_{v \in V_{out}(\Gamma)} \epsilon_v - N\chi(\Gamma),$$

where $\epsilon_v = -1$, if v contains a marked corner and $\epsilon_v = 0$ otherwise. Here $\chi(\Gamma) = |V(\Gamma)| - |E(\Gamma)|$ denotes the Euler characteristic of Γ.

Definition 11.13 We define $R(n, m)$ as a graded vector space which is a direct sum $\oplus_\Gamma \psi_\Gamma$ of 1-dimensional graded vector spaces generated by graphs Γ as above, each summand has degree $deg\,\Gamma$.

One can see that ψ_Γ is naturally identified with the tensor product of one-dimensional vector spaces (determinants) corresponding to vertices of Γ.

Now, having a graph Γ which satisfies conditions (1–3) above and Hochschild chains $\gamma_1, ..., \gamma_n \in C^{mod}_\bullet(A, A)$ we would like to define an element of $C^{mod}_\bullet(A, A)^{\otimes m}$. Roughly speaking we are going to assign the above n elements of the Hochschild complex to n faces corresponding to vertices $in_i, 1 \le i \le n$, then assign tensors corresponding to higher products m_l to internal vertices $v \in V_{middle}(\Gamma)$, then using the convolution operation on tensors given by the scalar product on A to read off the resulting tensor from $out_j, 1 \le j \le m$. More precise algorithm is described below.

(a) We decompose the modified Hochschild complex such as follows:

$$C^{mod}_\bullet(A, A) = \oplus_{l \ge 0, \varepsilon \in \{0,1\}} C^{mod}_{l,\varepsilon}(A, A),$$

where $C^{mod}_{l,\varepsilon=0}(A, A) = A \otimes (A[1])^{\otimes l}$ and $C^{mod}_{l,\varepsilon=1}(A, A) = k \otimes (A[1])^{\otimes l}$ according to the definition of modified Hochschild chain complex. For any choice of $l_i \ge 0, \varepsilon_i \in \{0, 1\}, 1 \le i \le n$ we are going to construct a linear map of degree zero

$$f_\Gamma \colon \psi_\Gamma \otimes C^{mod}_{l_1,\varepsilon_1}(A,A) \otimes ... \otimes C^{mod}_{l_n,\varepsilon_1}(A,A) \to (C^{mod}_\bullet(A,A))^{\otimes m}.$$

The result will be a sum $f_\Gamma = \sum_{\Gamma'} f_{\Gamma'}$ of certain maps. The description of the collection of graphs Γ' is given below.

(b) Each new graph Γ' is obtained from Γ by adding new edges. More precisely one has $V(\Gamma') = V(\Gamma)$ and for each vertex $in_i \in V_{in}(\Gamma)$ we add l_i new outcoming edges. Then the valency of in_i becomes $l_i + 1$.

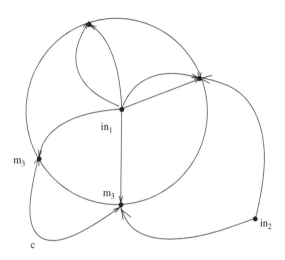

More pedantically, for every $i, 1 \leq i \leq n$ we have constructed a map from the set $\{1, ..., l_i\}$ to a cyclically ordered set which is an orbit of $\rho \circ \sigma$ with removed tail edge outcoming from in_i. Cyclic order on the edges of Γ' is induced by the cyclic order at every vertex and the cyclic order on the path forming the face corresponding to in_i.

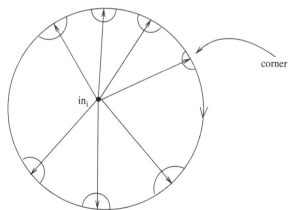

(c) We assign $\gamma_i \in C_{l_i,\varepsilon_i}$ to in_i. We depict γ_i as a "wheel" representing the Hochschild cocycle. It is formed by the endpoints of the $l_i + 1$ edges outcoming from $in_i \in V(\Gamma')$ and taken in the cyclic order of the corresponding face. If $\varepsilon_i = 1$ then (up to a scalar) $\gamma_i = 1 \otimes a_1 \otimes ... \otimes a_{l_i}$ and we require that the tensor factor 1 corresponds to zero in the cyclic order.

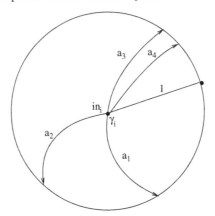

(d) We remove from considerations graphs Γ which do not obey the following property after the step (c):

the edge corresponding to the unit $1 \in k$ (see step c)) is of the type (in_i, v) where either $v \in V_{middle}(\Gamma')$ and $|v| = 3$ or $v = out_j$ for some $1 \leq j \leq m$ and the edge (in_i, out_j) was the marked edge for out_j.

Let us call *unit edge* the one which satisfies one of the above properties. We define a new graph Γ'' which is obtained from Γ by removing unit edges.

(e) Each vertex now has the valency $|v| \geq 2$. We attach to every such vertex either:

the tensor $c \in A \otimes A$ (inverse to the scalar product), if $|v| = 2$,

or

the tensor $(m_{|v|-1}(a_1, ..., a_{|v|-1}), a_{|v|})$ if $|v| \geq 3$. The latter can be identified with the element of $A^{\otimes |v|}$ (here we use the non-degenerate scalar product on A).

Let us illustrate this construction.

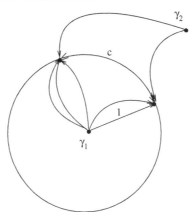

(f) Let us contract indices of tensors corresponding to $V_{in}(\Gamma'')\cup V_{middle}(\Gamma'')$ (see c, e) along the edges of Γ'' using the scalar product on A. The result will be an element a_{out} of the tensor product $\otimes_{1\leq j\leq m}A^{Star_{\Gamma''}(out_j)}$.

(g) Last thing we need to do is to interpret the element a_{out} as an element of $C_\bullet^{mod}(A,A)$. There are three cases.

Case 1. When we constructed Γ'' there was a unit edge incoming to some out_j. Then we reconstruct back the removed edge, attach $1 \in k$ to it, and interpret the resulting tensor as an element of $C_{|out_j|,\varepsilon_j=1}^{mod}(A,A)$.

Case 2. There was no removed unit edge incoming to out_j and we had a marked edge (not a marked corner) at the vertex out_j. Then we have an honest element of $C_{|out_j|,\varepsilon_j=0}^{mod}(A,A)$

Case 3. Same as in Case 2, but there was a marked corner at $out_j \in V_{out}(\Gamma)$. We have added and removed new edges when constructed Γ''. Therefore the marked corner gives rise to a new set of marked corners at out_j considered as a vertex of Γ''. Inside every such a corner we insert a new edge, attach the element $1 \in k$ to it and take the sum over all the corners. In this way we obtain an element of $C_{|out_j|,\varepsilon_j=1}^{mod}(A,A)$. This procedure is depicted below.

e_1 and e_2 are new edges.

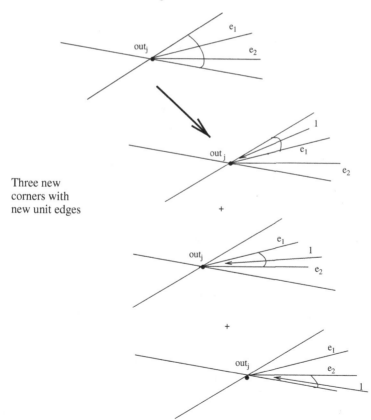

Three new
corners with
new unit edges

This concludes the construction of f_Γ. Notice that R is a dg-PROP with the differential given by the insertion of a new edge between two vertices from $V_{middle}(\Gamma)$.

Proof of the following Proposition will be given elsewhere.

Proposition 11.14 The above construction gives rise to a structure of a R-algebra on $C_\bullet^{mod}(A, A)$.

Remark 11.15 The above construction did not use homological smoothness of A.

Finally we would like to say few words about an extension of the R-action to the $Chains(\overline{\mathcal{M}})$-action. More details and application to Topological Field Theory will be given in [22].

If we assume the degeneration property for A, then the action of the PROP R can be extended to the action of the PROP $Chains(\overline{\mathcal{M}})$ of singular chains of the topological PROP of stable degenerations of $M_{g,n,m}^{marked}$. In order to see this, one introduces the PROP D freely generated by $R(2,0)$ and $R(1,1)$, i.e., by singular chains on the moduli space of cylinders with two inputs and zero outputs (they correspond to the scalar product on $C_\bullet(A, A)$) and by cylinders with one input and one output (they correspond to morphisms $C_\bullet(A, A) \to C_\bullet(A, A)$). In fact the (non-symmetric) bilinear form $h\colon H_\bullet(A, A) \otimes H_\bullet(A, A) \to k$ does exist for any compact A_∞-algebra A. It is described by the graph of degree zero on the figure in Sect. 11.6. This is a generalization of the bilinear form $(a, b) \in A/[A, A] \otimes A/[A, A] \mapsto Tr(axb) \in k$. It seems plausible that homological smoothness implies that h is non-degenerate. This allows us to extend the action of the dg sub-PROP $D \subset R$ to the action of the dg PROP $D' \subset R$ which contains also $R(0, 2)$ (i.e., the inverse to the above bilinear form). If we assume the degeneration property, then we can "shrink" the action of the homologically non-trivial circle of the cylinders (since the rotation around this circle corresponds to the differential B). Thus D' is quasi-isomorphic to the dg-PROP of chains on the (one-dimensional) retracts of the above cylinders (retraction contracts the circle). Let us denote the dg-PROP generated by singular chains on the retractions by D''. Thus, assuming the degeneration property, we see that the free product dg-PROP $R' = R *_D D''$ acts on $C_\bullet^{mod}(A, A)$. One can show that R' is quasi-isomorphic to the dg-PROP of chains on the topological PROP $\overline{M}_{g,n,m}^{marked}$ of stable compactifications of the surfaces from $M_{g,n,m}^{marked}$.

Remark 11.16 (a) The above construction is generalization of the construction from [31], which assigns cohomology classes of $M_{g,n}$ to a finite-dimensional A_∞-algebra with scalar product (trivalent graphs were used in [31]).

(b) Different approach to the action of the PROP R was suggested in [8]. The above Proposition gives rise to a structure of Topological Field Theory associated with a non-unital A_∞-algebra with scalar product. If the degeneration property holds for A then one can define a Cohomological Field Theory in the sense of [34].

(c) Homological smoothness of A is closely related to the existence of a non-commutative analog of the Chern class of the diagonal $\Delta \subset X \times X$ of a projective scheme X. This Chern class gives rise to the inverse to the scalar product on A. This topic will be discussed in the subsequent study devoted to A_∞-categories.

12 Appendix

12.1 Non-Commutative Schemes and Ind-Schemes

Let $\mathcal{C}$ be an Abelian k-linear tensor category. To simplify formulas we will assume that it is strict (see [41]). We will also assume that $\mathcal{C}$ admits infinite sums. To simplify the exposition we will assume below (and in the main body of the paper) that $\mathcal{C} = Vect_k^{\mathbf{Z}}$.

Definition 12.1 The category of non-commutative affine k-schemes in $\mathcal{C}$ (notation $NAff_\mathcal{C}$) is the one opposite to the category of associative unital k-algebras in $\mathcal{C}$.

The non-commutative scheme corresponding to the algebra A is denoted by $Spec(A)$. Conversely, if X is a non-commutative affine scheme then the corresponding algebra (algebra of regular functions on X) is denoted by $\mathcal{O}(X)$. By analogy with commutative case we call a morphism $f \colon X \to Y$ a *closed embedding* if the corresponding homomorphism $f^* \colon \mathcal{O}(Y) \to \mathcal{O}(X)$ is an epimorphism.

Let us recall some terminology of ind-objects (see e.g., [1, 20, 21]). For a covariant functor $\phi \colon I \to \mathcal{A}$ from a small filtering category I (called filtrant in [21]) there is a notion of an inductive limit "$\varinjlim$" $\phi \in \widehat{\mathcal{A}}$ and a projective limit "$\varprojlim$" $\phi \in \widehat{\mathcal{A}}$. By definition "$\varinjlim$" $\phi(X) = \varinjlim Hom_\mathcal{A}(X, \phi(i))$ and "$\varprojlim$" $\phi(X) = \varprojlim Hom_\mathcal{A}(\phi(i), X)$. All inductive limits form a full subcategory $Ind(\mathcal{A}) \subset \widehat{\mathcal{A}}$ of ind-objects in $\mathcal{A}$. Similarly all projective limits form a full subcategory $Pro(\mathcal{A}) \subset \widehat{\mathcal{A}}$ of pro-objects in $\mathcal{A}$.

Definition 12.2 Let I be a small filtering category and $F : I \to NAff_\mathcal{C}$ a covariant functor. We say that "$\varinjlim$" F is a non-commutative ind-affine scheme if for a morphism $i \to j$ in I the corresponding morphism $F(i) \to F(j)$ is a closed embedding.

In other words a non-commutative ind-affine scheme X is an object of $Ind(NAff_\mathcal{C})$, corresponding to the projective limit $\varprojlim A_\alpha, \alpha \in I$, where each A_α is a unital associative algebra in $\mathcal{C}$ and for a morphism $\alpha \to \beta$ in I the corresponding homomorphism $A_\beta \to A_\alpha$ is a surjective homomorphism of unital algebras (i.e., one has an exact sequence $0 \to J \to A_\beta \to A_\alpha \to 0$).

Remark 12.3 Not all categorical epimorphisms of algebras are surjective homomorphisms (although the converse is true). Nevertheless one can define closed embeddings of affine schemes for an arbitrary Abelian k-linear category, observing that a surjective homomorphism of algebras $f : A \to B$ is characterized categorically by the condition that B is the cokernel of the pair of the natural projections $f_{1,2} : A \times_B A \to A$ defined by f.

Morphisms between non-commutative ind-affine schemes are defined as morphisms between the corresponding projective systems of unital algebras. Thus we have

$$Hom_{NAffc}(\varinjlim_I X_i, \varinjlim_J Y_j) = \varprojlim_I \varinjlim_J Hom_{NAffc}(X_i, Y_j).$$

Let us recall that an algebra $M \in Ob(\mathcal{C})$ is called nilpotent if the natural morphism $M^{\otimes n} \to M$ is zero for all sufficiently large n.

Definition 12.4 A non-commutative ind-affine scheme $\hat{X}$ is called formal if it can be represented as $\hat{X} = \varinjlim Spec(A_i)$, where $(A_i)_{i \in I}$ is a projective system of associative unital algebras in $\mathcal{C}$ such that the homomorphisms $A_i \to A_j$ are surjective and have nilpotent kernels for all morphisms $j \to i$ in I.

Let us consider few examples in the case when $\mathcal{C} = Vect_k$.

Example 12.5 In order to define the non-commutative formal affine line $\hat{\mathbf{A}}^1_{NC}$ it suffices to define $Hom(Spec(A), \hat{\mathbf{A}}^1_{NC})$ for any associative unital algebra A. We define $Hom_{NAff_k}(Spec(A), \hat{\mathbf{A}}^1_{NC}) = \varinjlim Hom_{Alg_k}(k[[t]]/(t^n), A)$. Then the set of A-points of the non-commutative formal affine line consists of all nilpotent elements of A.

Example 12.6 For an arbitrary set I the non-commutative formal affine space $\hat{\mathbf{A}}^I_{NC}$ corresponds, by definition, to the topological free algebra $k\langle\langle t_i \rangle\rangle_{i \in I}$. If A is a unital k-algebra then any homomorphism $k\langle\langle t_i \rangle\rangle_{i \in I} \to A$ maps almost all t_i to zero and the remaining generators are mapped into nilpotent elements of A. In particular, if $I = \mathbf{N} = \{1, 2, ...\}$ then $\hat{\mathbf{A}}^{\mathbf{N}}_{NC} = \varinjlim Spec(k\langle\langle t_1, ..., t_n \rangle\rangle/(t_1, ..., t_n)^m)$, where $(t_1, ..., t_n)$ denotes the two-sided ideal generated by $t_i, 1 \le i \le n$ and the limit is taken over all $n, m \to \infty$.

By definition, a *closed subscheme* Y of a scheme X is defined by a two-sided ideal $J \subset \mathcal{O}(X)$. Then $\mathcal{O}(Y) = \mathcal{O}(X)/J$. If $Y \subset X$ is defined by a two-sided ideal $J \subset \mathcal{O}(X)$, then the completion of X along Y is a formal scheme corresponding to the projective limit of algebras $\varprojlim_n \mathcal{O}(X)/J^n$. This formal scheme will be denoted by $\hat{X}_Y$ or by $Spf(\mathcal{O}(X)/J)$.

Non-commutative affine schemes over a given field k form symmetric monoidal category. The tensor structure is given by the *ordinary tensor product* of unital algebras. The corresponding tensor product of non-commutative affine schemes will be denoted by $X \otimes Y$. It is not a categorical product, differently from the case of commutative affine schemes (where the tensor product of algebras corresponds to the Cartesian product $X \times Y$). For non-commutative affine schemes the analog of the Cartesian product is the *free product* of algebras.

Let A, B be free algebras. Then $Spec(A)$ and $Spec(B)$ are non-commutative manifolds. Since the tensor product $A \otimes B$ in general is not a smooth algebra, the non-commutative affine scheme $Spec(A \otimes B)$ is not a manifold.

Let X be a non-commutative ind-affine scheme in $\mathcal{C}$. A closed k-point $x \in X$ is by definition a homomorphism of $\mathcal{O}(X)$ to the tensor algebra generated by the unit object $\mathbf{1}$. Let m_x be the kernel of this homomorphism. We define the *tangent space* $T_x X$ in the usual way as $(m_x/m_x^2)^* \in Ob(\mathcal{C})$. Here m_x^2 is the image of the multiplication map $m_x^{\otimes 2} \to m_x$.

A non-commutative ind-affine scheme with a marked closed k-point will be called *pointed*. There is a natural generalization of this notion to the case of many points. Let $Y \subset X$ be a closed subscheme of disjoint closed k-points (it corresponds to the algebra homomorphism $\mathcal{O}(X) \to \mathbf{1} \oplus \mathbf{1} \oplus ...$). Then $\hat{X}_Y$ is a formal manifold. A pair $(\hat{X}_Y, Y)$ (often abbreviated by $\hat{X}_Y$) will be called (non-commutative) *formal manifold with marked points*. If Y consists of one such point then $(\hat{X}_Y, Y)$ will be called (non-commutative) *formal pointed manifold*.

12.2 Proof of Theorem 2.1.1

In the category $Alg_{\mathcal{C}f}$ every pair of morphisms has a kernel. Since the functor F is left exact and the category $Alg_{\mathcal{C}f}$ is Artinian, it follows from [20], Sect. 3.1 that F is strictly pro-representable. This means that there exists a projective system of finite-dimensional algebras $(A_i)_{i \in I}$ such that, for any morphism $i \to j$ the corresponding morphsim $A_j \to A_i$ is a categorical epimorphism and for any $A \in Ob(Alg_{\mathcal{C}f})$ one has

$$F(A) = \varinjlim_I Hom_{Alg_{\mathcal{C}f}}(A_i, A).$$

Equivalently,

$$F(A) = \varinjlim_I Hom_{Coalg_{\mathcal{C}f}}(A_i^*, A^*),$$

where $(A_i^*)_{i \in I}$ is an inductive system of finite-dimensional coalgebras and for any morphism $i \to j$ in I we have a categorical monomorphism $g_{ji} \colon A_i^* \to A_j^*$.

All what we need is to replace the projective system of algebras $(A_i)_{i \in I}$ by another projective system of algebras $(\overline{A}_i)_{i \in I}$ such that

(a) functors "$\varprojlim$"h_{A_i} and "$\varprojlim$"$h_{\overline{A}_i}$ are isomorphic (here h_X is the functor defined by the formula $h_X(Y) = Hom(X, Y)$);

(b) for any morphism $i \to j$ the corresponding homomorphism of algebras $\overline{f}_{ij} \colon \overline{A}_j \to \overline{A}_i$ is surjective.

Let us define $\overline{A}_i = \bigcap_{i \to j} Im(f_{ij})$, where $Im(f_{ij})$ is the image of the homomorphism $f_{ij} \colon A_j \to A_i$ corresponding to the morphism $i \to j$ in I. In order to prove a) it suffices to show that for any unital algebra B in $\mathcal{C}^f$ the natural map of sets

$$\varinjlim_I Hom_{\mathcal{C}^f}(A_i, B) \to \varinjlim_I Hom_{\mathcal{C}^f}(\overline{A}_i, B)$$

(the restriction map) is well-defined and bijective.

The set $\varinjlim_I Hom_{\mathcal{C}^f}(A_i, B)$ is isomorphic to $(\bigsqcup_I Hom_{\mathcal{C}^f}(A_i, B))/equiv$, where two maps $f_i : A_i \to B$ and $f_j : A_j \to B$ such that $i \to j$ are equivalent if $f_i f_{ij} = f_j$. Since $\mathcal{C}^f$ is an Artinian category, we conclude that there exists A_m such that $f_{im}(A_m) = \overline{A}_i$, $f_{jm}(A_m) = \overline{A}_j$. From this observation one easily deduces that $f_{ij}(\overline{A}_j) = \overline{A}_i$. It follows that the morphism of functors in (a) is well-defined and (b) holds. The proof that morphisms of functors biejectively correspond to homomorphisms of coalgebras is similar. This completes the proof of the theorem. ∎

12.3 Proof of Proposition 2.1.2

The result follows from the fact that any $x \in B$ belongs to a finite-dimensional subcoalgebra $B_x \subset B$ and if B was counital then B_x would be also counital. Let us describe how to construct B_x. Let Δ be the coproduct in B. Then one can write

$$\Delta(x) = \sum_i a_i \otimes b_i,$$

where a_i (resp. b_i) are linearly independent elements of B.

It follows from the coassociativity of Δ that

$$\sum_i \Delta(a_i) \otimes b_i = \sum_i a_i \otimes \Delta(b_i).$$

Therefore one can find constants $c_{ij} \in k$ such that

$$\Delta(a_i) = \sum_j a_j \otimes c_{ij},$$

and

$$\Delta(b_i) = \sum_j c_{ji} \otimes b_j.$$

Applying $\Delta \otimes id$ to the last equality and using the coassociativity condition again we get

$$\Delta(c_{ji}) = \sum_n c_{jn} \otimes c_{ni}.$$

Let B_x be the vector space spanned by x and all elements a_i, b_i, c_{ij}. Then B_x is the desired subcoalgebra. ∎

12.4 Formal Completion Along a Subscheme

Here we present a construction which generalizes the definition of a formal neighborhood of a k-point of a non-commutative smooth thin scheme.

Let $X = Spc(B_X)$ be such a scheme and $f \colon X \to Y = Spc(B_Y)$ be a closed embedding, i.e., the corresponding homomorphism of coalgebras $B_X \to B_Y$ is injective. We start with the category $\mathcal{N}_X$ of nilpotent extensions of X, i.e., homomorphisms $\phi \colon X \to U$, where $U = Spc(D)$ is a non-commutative thin scheme, such that the quotient $D/f(B_X)$ (which is always a non-counital coalgebra) is locally conilpotent. We recall that the local conilpotency means that for any $a \in D/f(B_X)$ there exists $n \geq 2$ such that $\Delta^{(n)}(a) = 0$, where $\Delta^{(n)}$ is the nth iterated coproduct Δ. If (X, ϕ_1, U_1) and (X, ϕ_2, U_2) are two nilpotent extensions of X then a morphism between them is a morphism of non-commutative thin schemes $t \colon U_1 \to U_2$, such that $t\phi_1 = \phi_2$ (in particular, $\mathcal{N}_X$ is a subcategory of the naturally-defined category of non-commutative relative thin schemes).

Let us consider the functor $G_f \colon \mathcal{N}_X^{op} \to Sets$ such that $G(X, \phi, U)$ is the set of all morphisms $\psi \colon U \to Y$ such that $\psi\phi = f$.

Proposition 12.7 Functor G_f is represented by a triple $(X, \pi, \widehat{Y}_X)$ where the non-commutative thin scheme denoted by $\widehat{Y}_X$ is called the formal neighborhood of $f(X)$ in Y (or the completion of Y along $f(X)$).

Proof. Let $B_f \subset B_X$ be the counital subcoalgebra which is the pre-image of the (non-counital) subcoalgebra in $B_Y/f(B_X)$ consisting of locally-conilpotent elements. Notice that $f(B_X) \subset B_f$. It is easy to see that taking $\widehat{Y}_X := Spc(B_f)$ we obtain the triple which represents the functor G_f. ∎

Notice that $\widehat{Y}_X \to Y$ is a closed embedding of non-commutative thin schemes.

Proposition 12.8 If Y is smooth then $\widehat{Y}_X$ is smooth and $\widehat{Y}_X \simeq \widehat{Y}_{\widehat{Y}_X}$.

Proof. Follows immediately from the explicit description of the coalgebra B_f given in the proof of the previous Proposition. ∎

Acknowledgments

We thank Vladimir Drinfeld for useful discussions and Victor Ginzburg and Kevin Costello for comments on the manuscript. Y.S. thanks Clay Mathematics Institute for supporting him as a Fellow during a part of the work. His work was also partially supported by an NSF grant. He is especially grateful to IHES for providing excellent research and living conditions for the duration of the project.

References

1. M. Artin, B. Mazur, Etale Homotopy, Lect. Notes Math., 100 (1969).
2. A. Beilinson, V. Drinfeld, Chiral algebras (in preparation)..
3. A. Beilinson, V. Drinfeld, Quantization of Hitchin's integrable system and Hecke eigensheaves (in preparation).
4. A. Bondal, M. Kapranov, Enhanced triangulated categories, Math. USSR Sbornik, 70:1, 93–107 (1991).
5. A. Bondal, M. Van den Bergh, Generators and representability of functors in commutative and noncommutative geometry, math.AG/0204218.
6. J. Boardman, R. Vogt, Homotopy invariant algebraic structures on topological spaces, Lect. Notes Math., 347 (1973).
7. A. Connes, Non-commutative geometry. Academic Press, 1994.
8. K. Costello, Topological conformal field theories, Calabi-Yau categories and Hochschild homology, preprint (2004).
9. J. Cuntz, D. Quillen, Cyclic homology and non-singularity, J. Amer. Math. Soc., 8:2, 373–442 (1995).
10. J. Cuntz, D. Quillen, Algebra extensions and non-singularity, J. Amer. Math. Soc., 8:2, 251–289 (1995).
11. J. Cuntz, G. Skandalis, B. Tsygan, Cyclic homology in noncommutative geometry, Encyclopaedia of Mathematical Sciences, v. 121, Springer Verlag, p. 74–113.
12. P. Deligne, L. Illusie, Relevements modulo p^2 et decomposition du complex de de Rham, Invent. Math. 89, 247–270 (1987).
13. P. Deligne, J.S. Milne, Tannakian categories, Lect. Notes Math., 900, 101–228 (1982).
14. V. Drinfeld, D.G. quotients of DG categories, math.KT/0210114.
15. K. Fukaya, Y.G. Oh, H. Ohta, K. Ono, Lagrangian intersection Floer theory-anomaly and obstruction. Preprint, 2000.
16. E. Getzler, Cartan homotopy formulas and Gauss-Manin connection in cyclic homology, Israel Math. Conf. Proc., 7, 65–78.
17. V. Ginzburg, Non-commutative symplectic geometry, quiver varieties and operads, math.QA/0005165.
18. V. Ginzburg, Lectures on Noncommutative Geometry, math.AG/0506603.
19. V. Ginzburg, Double derivations and cyclic homology, math.KT/0505236.
20. A. Grothendieck, Technique de descente.II. Sem. Bourbaki, 195 (1959/60).
21. M. Kashiwara, P. Schapira, Ind-sheaves, Asterisque 271 (2001).
22. L.Katzarkov, M. Kontsevich, T. Pantev, Calculating Gromov-Witten invariants from Fukaya category, in preparation.
23. H. Kajiura, Noncommutative homotopy algebras associated with open strings, arXiv:math/0306332.
24. D. Kaledin, Non-commutative Cartier operator and Hodge-to-de Rham degeneration, math.AG/0511665.
25. R. Kaufmann,Moduli space actions on the Hochschild co-chains of a Frobenius algebra I: Cell Operads, math.AT/0606064.
26. R. Kaufmann,Moduli space actions on the Hochschild co-chains of a Frobenius algebra II: Correlators, math.AT/0606065.

27. B. Keller, Introduction to A-infinity algebras and modules, Homology, Homotopy and Applications 3, 1–35 (2001).
28. B. Keller, On differential graded categories, math.KT/0601185.
29. M. Kontsevich, Deformation quantization of Poisson manifolds, math.QA/ 9709040.
30. M. Kontsevich, Formal non-commutative symplectic geometry. In: Gelfand Mathematical Seminars, 1990–1992, p. 173–187. *Birkhävser* Boston, MA, (1993).
31. M. Kontsevich, Feynman diagrams and low-dimensional topology. Proc. Europ. Congr. Math., 1 (1992).
32. M. Kontsevich, Notes on motives in finite characteristic, in preparation.
33. M. Kontsevich, Lecture on triangulated A_∞-categories at Max-Planck Institut für Mathematik, (2001).
34. M. Kontsevich, Yu.Manin, Gromov-Witten classes, quantum cohomology and enumerativ geometry, Comm. Math. Phys., 164:3, 525–562 (1994).
35. M. Kontsevich, Y. Soibelman, Deformations of algebras over operads and Deligne conjecture, math.QA/0001151, published in Lett. Math. Phys., 21:1, 255–307 (2000).
36. M. Kontsevich, Y. Soibelman, Deformation theory, (book in preparation).
37. L. Korogodski, Y. Soibelman, Algebras of functions on quantum groups.I. Math. Surveys Monogr. 56, AMS (1998).
38. L. Le Bruyn, Non-commutative geometry an n (book in preparation).
39. V. Lyubashenko, Category of A_∞-categories, math.CT/0210047.
40. V. Lyubashenko, S. Ovsienko, A construction of A_∞-category, math.CT/ 0211037.
41. S. Mac Lane, Categories for the working mathematician. Springer-Verlag (1971).
42. D. Orlov,Triangulated categories of singularities and equivalences between Landau-Ginzburg models, math.AG/0503630.
43. R. Rouquier, Dimensions of triangulated categories, math.CT/0310134.
44. S. Sanablidze, R. Umble, A diagonal of the associahedra, math. AT/0011065.
45. P. Seidel, Homological mirror symmetry for the quartic surface, math.SG/0310414.
46. Y. Soibelman, Non-commutative geometry and deformations of A_∞-algebras and A_∞-categories, Arbeitstagung 2003, Preprint Max-PLanck Institut für Mathematik, MPIM2003-60h, (2003).
47. Y. Soibelman, Mirror symmetry and non-commutative geometry of A_∞-categories, J. Math. Phys., 45:10, 3742–3757 (2004).
48. D. Tamarkin, B. Tsygan, Non-commutative differential calculus, homotopy BV-algebras and formality conjectures, arXiv:math/0010072.
49. D. Tamarkin, B. Tsygan, The ring of differential operators on forms in non-commutative calculus. Proceedings of Symposia in Pure Math., 73, 105–131 (2005).
50. B. Toen, M. Vaquie, Moduli of objects in dg-categories, math.AG/0503269.
51. J. L. Verdier, Categories derivees, etat 0. Lect. Notes in Math., 569, 262–312 (1977).

On Non-Commutative Analytic Spaces
Over Non-Archimedean Fields

Y. Soibelman

Department of Mathematics, Kansas State University, Manhattan, KS 66506, USA
soibel@math.ksu.edu

Abstract. We discuss examples of non-commutative spaces over non-archimedean fields. Those include non-commutative and quantum affinoid algebras, quantized K3 surfaces and quantized locally analytic p-adic groups. In the latter case we found a quantization of the Schneider–Teitelbaum algebra of locally analytic distributions by using the ideas of representation theory of quantized function algebras.

1 Introduction

1.1

Let A be a unital algebra over the field of real or complex numbers. Following [5] one can think of A as of the algebra of *smooth* functions $C^\infty(X_{NC})$ on some "non-commutative real smooth manifold X_{NC}". Differential geometry on X_{NC} has been developed by Connes and his followers. By adding extra structures to A one can define new classes of spaces. For example if A carries an antilinear involution one can try define a C^*-algebra $C(X_{NC})$ of "continuous functions on X_{NC} as a completion of A with respect to the norm $|f| = sup_\pi ||\pi(f)||$, where π runs through the set of all topologically irreducible $*$-representations in Hilbert spaces. By analogy with the commutative case, $C(X_{NC})$ corresponds to the non-commutative *topological* space X_{NC}. Similarly, von Neumann algebras correspond to non-commutative measurable spaces, etc. Main source of new examples for this approach are "bad" quotients and foliations.

Another class of non-commutative spaces consists of "non-commutative schemes" and their generalizations. Here we treat A as the algebra of *regular* functions on the "non-commutative affine scheme $Spec(A)$". The ground field can be arbitrary (in fact one can speak about rings, not algebras). Naively defined category of non-commutative affine schemes is the one opposite to the category of associative unital rings. But then one can ask what is the "*sheaf* of regular functions on $Spec(A)$". This leads to the question about

Soibelman, Y.: *On Non-Commutative Analytic Spaces Over Non-Archimedean Fields.* Lect. Notes Phys. **757**, 221–247 (2009)
DOI 10.1007/978-3-540-68030-7_7   © Springer-Verlag Berlin Heidelberg 2009

localization in the non-commutative framework which is a complicated task. An attempt to glue general non-commutative schemes from the affine ones, leads to a replacement of the naively-defined category of non-commutative affine schemes by a more complicated category (see e.g. [13]). Main source of examples for this approach is the representation theory (e.g. theory of quantum groups).

There is an obvious contradiction between the two points of view discussed above. Namely, associative algebras over $\mathbf{R}$ or $\mathbf{C}$ are treated as algebras of functions on different types of non-commutative spaces (smooth manifolds in non-commutative differential geometry and affine schemes in non-commutative algebraic geometry). There is no coherent approach to non-commutative geometry which resolves this contradiction. In other words, one cannot start with, say, non-commutative smooth algebraic variety over $\mathbf{C}$, make it into a non-commutative complex manifold and then define a non-commutative version of a smooth structure, so that it becomes a non-commutative real smooth manifold. Maybe this is a sign of a general phenomenon: there are many more types of non-commutative spaces than the commutative ones. Perhaps the traditional terminology (schemes, manifolds, algebraic spaces, etc.) has to be modified in the non-commutative world. Although non-commutative spaces resist an attempt to classify them, it is still interesting to study non-commutative analogues of "conventional" classes of commutative spaces. Many examples arise if one considers algebras which are close to commutative (e.g. deformation quantization, quantum groups), or algebras which are very far from commutative ones (like free algebras). In a sense these are two "extreme" cases and for some reason the corresponding non-commutative geometry is richer than the "generic" one.

1.2

In this chapter we discuss non-commutative analytic spaces over non-archimedean fields. The list of "natural" examples is non-empty (see e.g. [16]). Analytic non-commutative tori (or elliptic curves) from [16] are different from C^∞ non-commutative tori of Connes and Rieffel. Although a non-commutative elliptic curve over $\mathbf{C}$ (or over a non-archimedean valuation field, see [16]) appears as a "bad" quotient of an analytic space, it carries more structures than the corresponding "smooth" bad quotient which is an object of Connes theory. Non-commutative deformations of a non-archimedean K3 surface were mentioned in [8] as examples of a "quantization", which is not formal with respect to the deformation parameter. It seems plausible that a natural class of "quantum" non-archimedean analytic spaces can be derived from cluster ensembles (see [6]).

Present chapter is devoted to examples of non-commutative spaces which can be called *non-commutative rigid analytic spaces*. General theory is far from being developed. We hope to discuss some of its aspects elsewhere (see [14]).

Almost all examples of non-commutative analytic spaces considered in the present chapter are treated from the point of view of the approach to rigid analytic geometry offered in [2, 3]. The notion of spectrum of a commutative Banach ring plays a key role in the approach of Berkovich. Recall that the spectrum $M(A)$ introduced in [2] has two equivalent descriptions:

(a) the set of multiplicative continuous seminorms on a unital Banach ring A;

(b) the set of equivalence classes of continuous representations of A in 1-dimensional Banach spaces over complete Banach fields (i.e. continuous characters).

The space $M(A)$ carries a natural topology so that it becomes a (non-empty) compact Hausdorff space. There is a canonical map $\pi\colon M(A) \to Spec(A)$ which assigns to a multiplicative seminorm its kernel, a prime ideal. The spectrum $M(A)$ is a natural generalization of the Gelfand spectrum of a unital commutative C^*-algebra.

Berkovich's definition is very general and does not require A to be an affinoid algebra (i.e. an admissible quotient of the algebra of analytic functions on a non-archimedean polydisc). In the affinoid case one can make $M(A)$ into a ringed space (affinoid space). General analytic spaces are glued from affinoid ones similarly to the gluing of general schemes from affine schemes. Analytic spaces in the sense of Berkovich have better local properties than classical rigid analytic spaces (e.g. they are locally arcwise connected, see [4], while in the classical rigid analytic geometry the topology is totally disconnected).

Affinoid spaces play the same role of "building blocks" in non-archimedean analytic geometry as affine schemes play in the algebraic geometry. For example, localization of a finite Banach A-module M to an affinoid subset $V \subset M(A)$ is achieved by the topological tensoring of M with an affinoid algebra A_V, which is the localization of A on V (this localization *is not* an essentially surjective functor, differently from the case of algebraic geometry). In order to follow this approach one needs a large supply of "good" multiplicative subsets of A. If A is commutative this is indeed the case. It is natural to ask what has to be changed in the non-commutative case.

1.3

If A is a non-commutative unital Banach ring (or non-commutative affinoid algebra, whatever this means) then there might be very few "good" multiplicatively closed subsets of A. Consequently, the supply of affinoid sets can be insufficient to produce a rich theory of non-commutative analytic spaces. This problem is already known in non-commutative algebraic geometry and one can try to look for a possible solution there. One way to resolve the difficulty was suggested in [14]. Namely, instead of localizing *rings*, one should localize *categories of modules over rings*, e.g. using the notion of *spectrum* of an abelian category introduced in [14]. Spectrum is a topological space equipped with Zariski-type topology. For an associative unital ring A the category of left modules $A - mod$ gives rise to a sheaf of local categories on the spectrum of $A - mod$. If A is commutative, then the spectrum of $A - mod$ coincides with the usual spectrum $Spec(A)$. In the commutative case the fibre of the sheaf of categories over $p \in Spec(A)$ is the category of modules over a local ring A_p which is the localization of A at p. In the non-commutative case the fibre is not a category of modules over a ring. Nevertheless one can glue general non-commutative spaces from "affine" ones and call them *non-commutative schemes*. Thus non-commutative schemes are topological spaces equipped with sheaves of local categories (see [14] for details).

There are more general classes of non-commutative spaces than non-commutative schemes (see e.g. [10]). In particular, there might be no underlying topological space (i.e. no "spectrum"). Then one axiomatizes the notions of covering and descent. Main idea is the following. If $X = \cup_{i \in I} U_i$ is a "good" covering of a scheme, then the algebra of functions $C := \mathcal{O}(\sqcup_{i,j \in I}(U_i \cap U_j)$ is a coalgebra in the monoidal category of $A - A$-bimodules, where $A := \mathcal{O}(\sqcup_{i \in I} U_i)$. In order to have an equivalence of the category of descent data with the category of quasi-coherent sheaves on X one deals with the flat topology, which means that C is a (right) flat A-module. In this approach the category of non-commutative spaces is defined as a localization of the category of coverings with respect to a class of morphisms called refinements of coverings. This approach can be generalized to non-commutative case ([10]). One problem mentioned in [10] is the absence of a good definition of morphism of non-commutative spaces defined by means of coverings. Then one can use another approach based on the same idea, which deals with derived categories of quasi-coherent sheaves rather than with the abelian categories of quasi-coherent sheaves. Perhaps this approach can be generalized in the framework of non-commutative analytic spaces discussed in this chapter.

1.4

Let us briefly discuss some difficulties one meets trying to construct a theory of non-commutative analytic spaces (some of them are technical but other are conceptual).

(1) It is typical in non-commutative geometry to look for a point-independent (e.g. categorical) description of an object or a structure in the commutative case and then take it as a definition in the non-commutative case. For example, an affine morphism of schemes can be characterized by the property that the direct image functor is faithful and exact. This is taken as a definition (see [13], VII.1.4) of an affine morphism of non-commutative schemes. Another example is the algebra of regular functions on a quantized simple group which is defined via Peter–Weyl theorem (i.e. it is defined as the algebra of matrix elements of finite-dimensional representations of the quantized enveloping algebra, see [11]). Surprisingly many natural "categorical" questions do not have satisfactory (from the non-commutative point of view) answers

in analytic case. For example: how to characterize categorically an embedding $V \to X = M(A)$, where V is an affinoid domain?

(2) In the theory of analytic spaces all rings are topological (e.g. Banach). Topology should be involved already in the definition of the non-commutative version of Berkovich spectrum $M(A)$ as well as in the localization procedure (question: having a category of coherent sheaves on the Berkovich spectrum $M(A)$ of an affinoid algebra A how to describe categorically its stalk at a point $x \in M(A)$?).

(3) It is not clear what is a non-commutative analogue of the notion of affinoid algebra. In the commutative case a typical example of an affinoid algebra is the Tate algebra, i.e. the completion T_n of the polynomial algebra $K[x_1, ..., x_n]$ with respect to the Gauss norm $|| \sum_{l \in \mathbf{Z}_+} a_l x^l || = sup_l |a_l|$, where K is a valuation field. It is important (at least at the level of proofs) that T_n is noetherian. If we relax the condition that variables $x_i, 1 \le i \le n$ commute, then the noetherian property can fail. This is true, in particular, if one starts with the polynomial algebra $K \langle x_1, ..., x_n \rangle$ of free variables, equipped with the same Gauss norm as above (it is easy to see that the norm is still multiplicative). In "classical" rigid analytic geometry many proofs are based on the properties of T_n (e.g. all maximal ideals have finite codimension, Weierstrass division theorem, Noether normalization theorem, etc.). There is no "universal" replacement of T_n in non-commutative world which enjoys the same properties. On the other hand, there are some candidates which are good for particular classes of examples. We discuss them in the main body of the paper.

1.5

About the content of the paper. We start with elementary considerations, e.g. a non-commutative analog of the Berkovich spectrum $M_{NC}(A)$ (here A is a unital Banach ring). Our definition is similar to the algebro-geometric definition of the spectrum $Spec(A) := Spec(A - mod)$ from [13]. Instead of the category of A-modules in [13] we consider here the category of continuous A-modules complete with respect to a seminorm. We prove that $M_{NC}(A)$ is non-empty. There is a natural map $\pi \colon M_{NC}(A) \to Spec(A)$. Then we equip $M_{NC}(A)$ with the natural Hausdorff topology. The set of bounded multiplicative seminorms $M(A)$ is the usual Berkovich spectrum. Even if A is commutative, the space $M_{NC}(A)$ is larger than Berkovich spectrum $M(A)$. This phenomenon can be illustrated in the case of non-commutative algebraic geometry. Instead of considering k-points of a commutative ring A, where k is a field, one can consider matrix points of A, i.e. homomorphisms $A \to Mat_n(k)$. Informally speaking, such homomorphisms correspond to morphisms of a non-commutative scheme $Spec(Mat_n(k))$ into $Spec(A)$. Only the case $n = 1$ is visible in the "conventional" algebraic geometry. Returning to unital Banach algebras we observe that $M(A) \subset M_{NC}(A)$ regardless of commutativity of A. In this chapter we are going to use $M(A)$ only

Most of the chapter is devoted to examples, discussed in Sects. 3–7 below. We start with the non-commutative polydiscs and quantum tori. A non-commutative analytic K3 surface is the most non-trivial example considered in the paper. It can be also called *quantum* K3 surface, because it is a flat analytic deformation of the non-archimedean analytic K3 surface constructed in [8]. In the construction we follow the approach of [8], where the "commutative" analytic K3 surface was built by gluing it from "flat" pieces, each of which is a non-archimedean analytic analogue of a Lagrangian torus fibration in symplectic geometry. We recall the construction in Sect. 7.1. Main idea is based on the relationship between non-archimedean analytic Calabi–Yau manifolds and real manifolds with integral affine structure discussed in [8, 9]. Roughly speaking, to such a Calabi–Yau manifold X we associate a PL manifold $Sk(X)$, its skeleton. There is a continuous map $\pi : X \to Sk(X)$ such that the generic fibre is a non-archimedean analytic torus. Moreover, there is an embedding of $Sk(X)$ into X, so that π becomes a retraction. For the elliptically fibered K3 surface the skeleton $Sk(X)$ is a two-dimensional sphere S^2. It has an integral affine structure which is non-singular outside of the set of 24 points. It is analogous to the integral affine structure on the base of a Lagrangian torus fibration in symplectic geometry (Liouville theorem). Fibres of π are Stein spaces. Hence in order to construct the sheaf $\mathcal{O}_X$ of analytic functions on X it suffices to construct $\pi_*(\mathcal{O}_X)$.

In fact almost all examples considered in this chapter should be called *quantum non-commutative analytic spaces*. They are based on the version of Tate algebra in which the commutativity of variables $z_i z_j = z_j z_i$ is replaced by the q-commutativity $z_i z_j = q z_j z_i, i < j, q \in K^\times, |q| = 1$. In particular our non-commutative analytic K3 surface is defined as a ringed space, with the underlying topological space being an ordinary K3 surface equipped with the natural Grothendieck topology introduced in [4] and the sheaf of non-commutative algebras which is locally isomorphic to a quotient of the above-mentioned "quantum" Tate algebra. The construction of a quantum K3 surface uses a non-commutative analog of the map π. In other words, the skeleton $Sk(X)$ survives under the deformation procedure. We will point out a more general phenomenon, which does not exist in "formal" deformation quantization. Namely, as the deformation parameter q gets closer to 1 we recover more and more of the Berkovich spectrum of the undeformed algebra.

1.6

The theory of non-commutative non-archimedean analytic spaces should have applications to mirror symmetry in the spirit of [8, 9]. More precisely, it looks plausible that certain deformation of the Fukaya category of a maximally degenerate (see [8, 9]) hyperkahler manifold can be realized as the derived category of coherent sheaves on the non-commutative deformation of the dual

Calabi–Yau manifold (which is basically the same hyperkahler manifold). This remark was one of the motivations of this chapter. Another motivation is the theory of p-adic quantum groups. We will discuss it elsewhere.

2 Non-Commutative Berkovich Spectrum

2.1 Preliminaries

We refer to [2], Chap. 1 for the terminology of seminormed groups, etc. Here we recall few terms for convenience of the reader.

Let A be an associative unital Banach ring. Then, by definition, A carries a norm $|\bullet|_A$ and moreover, A is complete with respect to this norm. The norm is assumed to be submultiplicative, i.e. $|ab|_A \leq |a|_A |b|_A, a, b \in A$ and unital, i.e. $|1|_A = 1$. We call the norm *non-archimedean* if, instead of the usual inequality $|a + b|_A \leq |a|_A + |b|_A, a, b \in A$, we have a stronger one $|a + b|_A \leq max\{|a|_A, |b|_A\}$. A *seminormed module* over A is (cf. with [2]) a left unital (i.e. 1 acts as id_M) A-module M which carries a seminorm $|\bullet|$ such that $|am| \leq C|a|_A|m|$ for some $C > 0$ and all $a \in A, m \in M$. Seminormed A-modules form a category $A - mod^c$, such that a morphism $f : M \to N$ is a homomorphism of A-modules satisfying the condition $|f(m)| \leq const |m|$ (i.e. f is bounded). Clearly the kernel $Ker(f)$ is closed with respect to the topology defined by the seminorm on M. A morphism $f : M \to N$ is called *admissible* if the quotient seminorm on $Im(f) \simeq M/Ker(f)$ is equivalent to the one induced from N. We remark that the category $A - mod^c$ is not abelian in general. Following [2] we call *valuation field* a commutative Banach field K whose norm is multiplicative, i.e. $|ab| = |a||b|$. If the norm is non-archimedean, we will cal K a *non-archimedean* valuation field. In this case one can introduce a valuation map $val \colon K^\times \to \mathbf{R} \cup +\infty$, such that $val(x) = -log|x|$.

2.2 Spectrum of a Banach Ring

Let us introduce a partial order on the objects of $A - mod^c$. We say that $N \geq_c M$ if there exists a closed admissible embedding $i \colon L \to \oplus_{I-finite} N$ and an admissible epimorphism $pr \colon L \to M$. We will denote by $N \geq M$ a similar partial order on the category $A - mod$ of left A-modules (no admissibility condition is imposed). We will denote by $=_c$ (resp. $=$ for $A - mod$) the equivalence relation generated by the above partial order.

Let us call an object of $A - mod^c$ (resp. $A - mod$) *minimal* if it satisfies the following conditions (cf. [13]):

(1) if $i \colon N \to M$ is a closed admissible embedding (resp. any embedding in the case of $A - mod$) then $N \geq_c M$ (resp. $N \geq M$);

(2) there is an element $m \in M$ such that $|m| \neq 0$ (resp. $m \neq 0$ for $A-mod$).

We recall (see [13]) that the spectrum $Spec(A) := Spec(A - mod)$ consists of equivalence classes (w.r.t. $=$) of minimal objects of $A - mod$. It is known that $Spec(A)$ is non-empty and contains classes of A-modules A/m, where m is a left maximal ideal of A. Moreover, $Spec(A)$ can be identified with the so-called *left spectrum* $Spec_l(A)$, which can be also described in terms of a certain subset of the set of left ideals of A (for commutative A it is the whole set of prime ideals).

Let $A - mod^b$ be a full subcategory of $A - mod^c$ consisting of A-modules which are complete with respect to their seminorms (we call them *Banach modules* for short).

Definition 2.1 The non-commutative analytic spectrum $M_{NC}(A)$ consists of classes of equivalence (w.r.t. to $=_c$) of minimal (i.e. satisfying (1) and (2)) objects M of $A - mod^b$ which satisfy also the following property:

(3) if $m_0 \in M$ is such that $|m_0| \neq 0$ then the left module Am_0 is minimal, i.e. defines a point of $Spec(A)$ (equivalently, this means that $p := Ann(m_0) \in Spec_l(A)$) and this point does not depend on a choice of m_0.

The following easy fact implies that $M_{NC}(A) \neq \emptyset$.

Proposition 2.2 Every (proper) left maximal ideal $m \subset A$ is closed.

Proof. We want to prove that the closure $\overline{m}$ coincides with m. The ideal $\overline{m}$ contains m. Since m is maximal, then either $\overline{m} = m$ or $\overline{m} = A$. Assume the latter. Then there exists a sequence $x_n \to 1, n \to +\infty, x_n \in m$. Choose n so large that $|1 - x_n|_A < 1/2$. Then $x_n = 1 + (x_n - 1) := 1 + y_n$ is invertible in A, since $(1 + y_n)^{-1} = \sum_{l \geq 0} (-1)^l y_n^l$ converges. Hence $m = A$. Contradiction. ∎

Corollary 2.3 $M_{NC}(A) \neq \emptyset$

Proof. Let m be a left maximal ideal in A (it does exists because of the standard arguments which use Zorn lemma). It is closed by previous Proposition. Then $M := A/m$ is a cyclic Banach A-module with respect to the quotient norm. We claim that it contains no proper closed submodules. Indeed, let $N \subset M$ be a proper closed submodule. We may assume it contains an element n_0 such that annihilator $Ann(n_0)$ contains m as a proper subset. Since m is maximal we conclude that $Ann(n_0) = A$. But this cannot be true since $1 \in A$ acts without kernel on A/m, hence $1 \notin Ann(n_0)$. This contradiction shows that M contains no proper closed submodules. In order to finish the proof we recall that simple A-module A/m defines a point of $Spec(A)$. Hence the conditions (1)–(3) above are satisfied. ∎

Abusing terminology we will often say that an object belongs to the spectrum (rather than saying that its equivalence class belongs to the spectrum).

Proposition 2.4 If $M \in A - mod^b$ belongs to $M_{NC}(A)$ then its seminorm is, in fact, a Banach norm.

Proof. Let $N = \{m \in M | |m| = 0\}$. Clearly N is a closed submodule. Let L be a closed submodule of the finite sum of copies of N, such that there exists an admissible epimorphism $pr: L \to M$. Then the submodule L must carry trivial induced seminorm and, moreover, admissibility of the epimorphism $pr: L \to M$ implies that the seminorm on M is trivial. This contradicts to (2). Hence $N = 0$. ∎

Remark 2.5 (a) By condition (3) we have a map of sets $\pi: M_{NC}(A) \to Spec(A)$.

(b) If A is commutative then any bounded multiplicative seminorm on A defines a prime ideal p (the kernel of seminorm). Moreover, A/p is a Banach A-module, which belongs to $M_{NC}(A)$. Hence Berkovich spectrum $M(A)$ (see [2]) is a subset of $M_{NC}(A)$. Thus $M_{NC}(A)$ can be also called non-commutative Berkovich spectrum.

2.3 Topology on $M_{NC}(A)$

If a Banach A-module M belongs to $M_{NC}(A)$ then it is equivalent (with respect to $=_c$) to the closure of any cyclic submodule $M_0 = Am_0$. Then for a fixed $a \in A$ we have a function $\phi_a: (M, |\bullet|) \mapsto |am_0|$, which can be thought of as a real-valued function on $M_{NC}(A)$.

Definition 2.6 The topology on $M_{NC}(A)$ is taken to be the weakest one for which all functions $\phi_a, a \in A$ are continuous.

Proposition 2.7 The above topology makes $M_{NC}(A)$ into a Hausdorff topological space.

Proof. Let us take two different points of the non-commutative analytic spectrum $M_{NC}(A)$ defined by cyclic seminormed modules $(M_0, |\bullet|)$ and $(M_0', |\bullet|')$. If M_0 is not equivalent to M_0' with respect to $=$ (i.e. in $A - mod$), then there are exist two different closed left ideals p, p' such that $M_0 \simeq A/p$ and $M_0' \simeq A/p'$ (again, this is an isomorphism of A-modules only. Banach norms are not induced from A). Then there is an element $a \in A$ which belongs, to,say, p and does not belong to p' (if $p \subset p'$ we interchange p and p'). Hence the function ϕ_a is equal to zero on the closure of $(M_0, |\bullet|)$ and $\phi_a((\overline{M_0', |\bullet|'}) = c_a > 0$. Then open sets $U_0 = \{x \in M_{NC}(A) | \phi_a(x) < c_a/2\}$ and $U_0' = \{x \in M_{NC}(A) | 3c_a/4 < \phi_a(x) < c_a\}$ do not intersect and separate two given points of the analytic spectrum.

Suppose $M = M_0 = M_0' = Am_0$. Since the corresponding points of $Spec(A)$ coincide, we have the same cyclic A-module which carries two different norms $|\bullet|$ and $|\bullet|'$. Let $am_0 = m \in M$ be an element such that $|m| \neq |m|'$. Then the function ϕ_a takes different values at the corresponding points of $M_{NC}(A)$ (which are completions of M with respect to the above norms) and we can define separating open subsets as before. This concludes the proof. ∎

2.4 Relation to Multiplicative Seminorms

If $x\colon A \to \mathbf{R}_+$ is a multiplicative bounded seminorm on A (bounded means $x(a) \le |a|_A$ for all $a \in A$) then $Ker\, x$ is a closed two-sided ideal in A. If A is non-commutative, it can contain very few two-sided ideals. At the same time, a bounded multiplicative seminorm x gives rise to a point of $M_{NC}(A)$ such that the corresponding Banach A-module is $A/Ker\, x$ equipped with the left action of A.

There exists a class of *submultiplicative* bounded seminorms on A which is contained in $M(A)$, if A is commutative. More precisely, let us consider the set $P(A)$ of all submultiplicative bounded seminorms on A. By definition, an element of $P(A)$ is a seminorm such that $|ab| \le |a||b|, |1| = 1, |a| \le C|a|_A$ for all $a, b \in A$ and some $C > 0$ (C depends on the seminorm). The set $P(A)$ carries natural partial order: $|\bullet|_1 \le |\bullet|_2$ if $|a|_1 \le |a|_2$ for all $a \in A$. Let us call *minimal seminorm* a minimal element of $P(A)$ with respect to this partial order and denote by $P_{min}(A)$ the subset of minimal seminorms. The latter set is non-empty by Zorn lemma. Let us recall the following classical result (see e.g. [2]).

Proposition 2.8 If A is commutative then $P_{min}(A) \subset M(A)$, i.e. any minimal seminorm is multiplicative.

Since there exist multiplicative bounded seminorms which are not minimal (take, e.g. the trivial seminorm on the ring of integers $\mathbf{Z}$ equipped with the usual absolute value Banach norm) it is not reasonable to define $M(A)$ in the non-commutative case as a set of minimal bounded seminorms. On the other hand one can prove the following result.

Proposition 2.9 If A is left noetherian as a ring then a minimal bounded seminorm defines a point of $M_{NC}(A)$.

Proof. Let p be the kernel of a minimal seminorm v. Then p is a 2-sided closed ideal. The quotient $B = A/p$ is a Banach algebra with respect to the induced norm. It is topologically simple, i.e. does not contain non-trivial closed two-sided ideals. Indeed, let r be a such an ideal. Then B/r is a Banach algebra with the norm induced from B. The pullback of this norm to A gives rise to a bounded seminorm on A, which is smaller than v, since it is equal to zero on a closed two-sided ideal which contains $p = Ker\, v$. The remaining proof that $A/p \in M_{NC}(A)$ is similar to the one from [13]. Recall that it was proven in [13] that if n is a two-sided ideal in a noetherian ring R such that R/n contains no 2-sided ideals then $A/n \in Spec(A - mod)$. ∎

We will denote by $M(A)$ the subset of $M_{NC}(A)$ consisting of bounded multiplicative seminorms. It carries the induced topology, which coincides for a commutative A with the topology introduced in [13]. We will call the corresponding topological space *Berkovich spectrum* of A.

Returning to the beginning of this section we observe that the set $P(A)$ of submultiplicative bounded seminorms contains $M_{NC}(A)$. Indeed if $v \in P(A)$

then we have a left Banach A-module M_v, which is the completion of $A/Ker\,v$ with respect to the norm induced by v. Thus M_v is a cyclic Banach A-module. A submultiplicative bounded seminorm which defines a point of the analytic spectrum can be characterized by the following property: $v \in P(A)$ belongs to $M_{NC}(A)$ iff $A/Ker\,v \in Spec(A-mod)$ (equivalently, if $Ker\,v \in Spec_l(A)$).

2.5 Remark on Representations in a Banach Vector Space

Berkovich spectrum of a commutative unital Banach ring A can be understood as a set of equivalence classes of one-dimensional representations of A in Banach vector spaces over valuation fields. One can try to do a similar thing in the non-commutative case based on the following simple considerations.

Let A be as before, $Z \subset A$ its center. Clearly it is a commutative unital Banach subring of A. It follows from [2], 1.2.5(ii) that there exists a bounded seminorm on A such that its restriction to Z is multiplicative (i.e. $|ab| = |a||b|$). Any such a seminorm $x \in M(Z)$ gives rise to a valuation field Z_x, which is the completion (with respect to the induced multiplicative norm) of the quotient field of the domain $Z/Ker\,x$. Then the completed tensor product $A_x := Z_x \widehat{\otimes}_Z A$ is a Banach Z_x-algebra. For any valuation field F and a Banach F-vector space V we will denote by $B_F(V)$ the Banach algebra of all bounded operators on V. Clearly the left action of A_x on itself is continuous. Thus we have a homomorphism of Banach algebras $A_x \to B_{Z_x}(A_x)$. Combining this homomorphism with the homomorphism $A \to A_x, a \mapsto 1 \otimes a$ we see that the following result holds.

Proposition 2.10 For any associative unital Banach ring A there exists a valuation field F, a Banach F-vector space V and a representation $A \to B_F(V)$ of A in the algebra of bounded linear operators in V.

3 Non-Commutative Affinoid Algebras

Let K be a non-archimedean valuation field, $r = (r_1, ..., r_n)$, where $r_i > 0, 1 \le i \le n$. In the "commutative" analytic geometry an affinoid algebra A is defined as an admissible quotient of the unital Banach algebra $K\{T\}_r$ of formal series $\sum_{l \in \mathbf{Z}_+^n} a_l T^l$, such that $max\,|a_l| r^l \to 0, l = (l_1, ..., l_n)$. The latter algebra is the completion of the algebra of polynomials $K[T] := K[T_1, ..., T_n]$ with respect to the norm $|\sum_l a_l T^l|_r = max|a_l| r^l$. In the non-commutative case we start with the algebra $K\langle T \rangle := K\langle T_1, ..., T_n \rangle$ of polynomials in n free variables and consider its completion $K\langle\langle T \rangle\rangle_r$ with respect to the norm $|\sum_{\lambda \in P(\mathbf{Z}_+^n)} a_\lambda T^\lambda| = max_\lambda |a_\lambda| r^\lambda$. Here $P(\mathbf{Z}_+^n)$ is the set of finite paths in $\mathbf{Z}_+^n$ starting at the origin and $T^\lambda = T_1^{\lambda_1} T_2^{\lambda_2} ....$ for the path which moves λ_1 steps in the direction $(1, 0, 0, ...)$ then λ_2 steps in the direction $(0, 1, 0, 0...)$ and so on (repetitions are allowed, so we can have a monomial like $T_1^{\lambda_1} T_2^{\lambda_2} T_1^{\lambda_3}$). We say that a Banach unital algebra A is *non-commutative affinoid algebra* if there is an admissible surjective homomorphism $K\langle\langle T\rangle\rangle_r \to A$. In particular, $K\{T\}_r$ is such an algebra and hence all commutative affinoid algebras are such algebras. We will restrict ourselves to the class of *noetherian* non-commutative affinoid algebras, i.e. those which are noetherian as left rings. All classical affinoid algebras belong to this class.

For noetherian affinoid algebras one can prove the following result (the proof is similar to the proof of Prop. 2.1.9 from [2], see also [15]).

Proposition 3.1 Let A be a noetherian non-commutative affinoid algebra. Then the category $A - mod^f$ of (left) finite A-modules is equivalent to the category $A - mod^{fb}$ of (left) finite Banach A-modules.

An important class of noetherian affinoid algebras consists of *quantum affinoid algebras*. By definition they are admissible quotients of algebras $K\{T_1, ..., T_n\}_{q,r}$. The latter consists of formal power series $f = \sum_{l \in \mathbf{Z}_+^n} a_l T^l$ of q-commuting variables (i.e. $T_i T_j = q T_j T_i, i < j$ and $q \in K^\times, |q| = 1$) such that $|a_l| r^l \to 0$. Here $T^l = T_1^{l_1}...T_n^{l_n}$ (the order is important now). It is easy to see that for any $r = (r_1, ..., r_n)$ such that all $r_i > 0$ the function $f \mapsto |f|_r := max_l |a_l| r^l$ defines a multiplicative norm on the polynomial algebra $K[T_1, ..., T_n]_{q,r}$ in q-commuting variables $T_i, 1 \le i \le n$. Banach algebra $K\{T_1, ..., T_n\}_{q,r}$ is the completion of the latter with respect to the norm $f \mapsto |f|_r$. Similarly, let $Q = ((q_{ij}))$ be an $n \times n$ matrix with entries from K such that $q_{ij}q_{ji} = 1, |q_{ij}| = 1$ for all i, j. Then we define the quantum affinoid algebra $K\{T_1, ..., T_n\}_{Q,r}$ in the same way as above, starting with polynomials in variables $T_i, 1 \le i \le n$ such that $T_i T_j = q_{ij} T_j T_i$. One can think of this Banach algebra as of the quotient of $K\langle\langle T_i, t_{ij}\rangle\rangle_{r,\mathbf{1}_{ij}}$, where $1 \le i, j \le n$ and $\mathbf{1}_{ij}$ is the unit $n \times n$ matrix, by the two-sided ideal generated by the relations

$$t_{ij}t_{ji} = 1, T_i T_j = t_{ij} T_j T_i, t_{ij} a = a t_{ij},$$

for all indices i, j and all $a \in K\langle\langle T_i, t_{ij}\rangle\rangle_{r,\mathbf{1}_{ij}}$. In other words, we treat q_{ij} as variables which belong to the center of our algebra and have the norms equal to 1.

4 Non-Commutative Analytic Affine Spaces

Let k be a commutative unital Banach ring. Similarly to the previous section we start with the algebra $k\langle\langle x_1, ..., x_n\rangle\rangle$ of formal series in free variables $T_1, ..., T_n$. Then for each $r = (r_1, ..., r_n)$ we define a subspace $k\langle\langle T_1, ..., T_n\rangle\rangle_r$ consisting of series $f = \sum_{i_1,...,i_m} a_{i_1,...,i_m} T_{i_1}...T_{i_m}$ such that $\sum_{i_1,...,i_m} |a_{i_1,...,i_m}| r_{i_1}...r_{i_m} < +\infty$. Here the summation is taken over all sequences $(i_1, ..., i_m), m \ge 0$ and $|\bullet|$ denotes the norm in k. In the non-archimedean case the convergency condition is replaced by the following one:

$max\,|a_{i_1,...,i_m}|r_{i_1}...r_{i_m} < +\infty$. Clearly each $k\langle\langle x_1, ..., x_n\rangle\rangle_r$ is a Banach algebra (called the algebra of analytic functions on a non-commutative k-polydisc $E_{NC}(0, r)$, cf. [2], Sect. 1.5). We would like to define a non-commutative n-dimensional analytic k-affine space $\mathbf{A}^n_{NC}$ as the coproduct $\cup_r E_{NC}(0, r)$. By definition the algebra of analytic functions on the quantum affine space is given by the above series such that $max\,|a_{i_1,...,i_m}|r_{i_1}...r_{i_m} < +\infty$ for all $r = (r_1, ..., r_n)$. In other words, analytic functions are given by the series which are convergent in all non-commutative polydics with centers in the origin.

The algebra of analytic functions on the non-archimedean *quantum closed polydisc* $E_q(0, r)$ is, by definition, $k\{T_1, ..., T_n\}_{q,r}$. The algebra of analytic functions on non-archimedean *quantum affine space* $\mathbf{A}^N_q$ consists of the series f in q-commuting variables $T_1, ..., T_n$, such that for all r the norm $|f|_r$ is finite. Equivalently, it is the coproduct of *quantum closed polydiscs* $E_q(0, r)$. There is an obvious generalization of this example to the case when q is replaced by a matrix Q as in the previous section. We will keep the terminology for the matrix case as well.

5 Quantum Analytic Tori

Let K be a non-archimedean valuation field, L a free abelian group of finite rank d, $\varphi: L \times L \to \mathbf{Z}$ is a skew-symmetric bilinear form, $q \in K^*$ satisfies the condition $|q| = 1$. Then $|q^{\varphi(\lambda,\mu)}| = 1$ for any $\lambda, \mu \in L$. We denote by $A_q(T(L, \varphi))$ the *algebra of regular functions on the quantum torus* $T_q(L, \varphi)$. By definition, it is a K-algebra with generators $e(\lambda), \lambda \in L$, subject to the relation

$$e(\lambda)e(\mu) = q^{\varphi(\lambda,\mu)}e(\lambda + \mu).$$

Definition 5.1 The space $\mathcal{O}_q(T(L, \varphi, (1, ..., 1)))$ of analytic functions on the quantum torus of multiradius $(1, 1, ..., 1) \in \mathbf{Z}^d_+$ consists of series $\sum_{\lambda \in L} a(\lambda) e(\lambda), a(\lambda) \in K$ such that $|a(\lambda)| \to 0$ as $|\lambda| \to \infty$ (here $|(\lambda_1, ..., \lambda_d)| = \sum_i |\lambda_i|$).

It is easy to check (see [16]) the following result.

Lemma 5.2 *Analytic functions* $\mathcal{O}_q(T(L, \varphi, (1, ..., 1)))$ *form a Banach K-algebra. Moreover, it is a noetherian quantum affinoid algebra.*

This example admits the following generalization. Let us fix a basis $e_1, ..., e_n$ of L and positive numbers $r_1, ..., r_n$. We define $r^\lambda = r_1^{\lambda_1}...r_n^{\lambda_n}$ for any $\lambda = \sum_{1 \le i \le n} \lambda_i e_i \in L$. Then the algebra $\mathcal{O}_q(T(L, \varphi, r))$ of analytic functions on the *quantum torus of multiradius* $r = (r_1, ..., r_n)$ is defined by same series as in the above definition, with the only change that $|a(\lambda)|r^\lambda \to 0$ as $|\lambda| \to \infty$. We are going to denote the corresponding non-commutative analytic space by $T^{an}_q(L, \varphi, r)$. It is a quantum affinoid space. The coproduct $T^{an}_q(L, \varphi) = \cup_r T^{an}_q(L, \varphi, r)$ is called the *quantum analytic torus*. The algebra of analytic functions $\mathcal{O}_q(T(L, \varphi))$ on it consists, by definition, of the above series such that for all $r = (r_1, ..., r_d)$ one has: $|a(\lambda)|r^\lambda \to 0$ as $|\lambda| \to \infty$. To be consistent with the notation of the previous subsection we will often denote the dual to e_i by T_i.

5.1 Berkovich Spectra of Quantum Polydisc and Analytic Torus

Assume for simplicity that the pair (L, φ) defines a simply laced lattice of rank d, i.e. for some basis $(e_i)_{1 \le i \le d}$ of L one has $\varphi(e_i, e_j) = 1, i < j$. In the coordinate notation we have $T_i T_j = q T_j T_i, i < j, |q| = 1$. Any $r = (r_1, ..., r_d) \in \mathbf{R}_{>0}^d$ gives rise to a point $\nu_r \in M(\mathcal{O}_q(T(L, \varphi)))$ such that $\nu_r(f) = max_{\lambda \in L}|a(\lambda)|r^\lambda$. In this way we obtain a (continuous) embedding of $\mathbf{R}_{>0}^d$ into Berkovich spectrum of quantum torus. This is an example of a more general phenomenon. In fact $\mathbf{R}_{>0}^d$ can be identified with the so-called *skeleton* $S((G_m^{an})^d)$ (see [2, 4]) of the d-dimensional (commutative) analytic torus $(G_m^{an})^d$. Skeleta can be defined for more general analytic spaces (see [4]). For example, the skeleton of the d-dimensional Drinfeld upper-half space Ω_K^d is the Bruhat–Tits building of $PGL(d, K)$. There is also a different notion of skeleton, which makes sense for so-called maximally degenerate Calabi–Yau manifolds (see [8], Sect. 6.6 for the details). In any case a skeleton is a PL-space (polytope) naturally equipped with the sheaf of affine functions. One can expect that the notion of skeleton (either in the sense of Berkovich or in the sense of [8]) admits a generalization to the case of quantum analytic spaces modeled by quantum affinoid algebras.

We have constructed above an embedding of $S((G_m^{an})^d)$ into $M(\mathcal{O}_q(T (L, \varphi))$. If $q = 1$ then there is a retraction $(G_m^{an})^d \to S((G_m^{an})^d)$. The pair $((G_m^{an})^d, S((G_m^{an})^d))$ is an example of *analytic torus fibration* which plays an important role in mirror symmetry (see [8]). One can expect that this picture admits a quantum analogue.

The skeleton of the analytic quantum torus survives in q-deformations as long as $|q| = 1$. Another example of this sort is a quantum K3 surface considered in Sect. 7. One can expect that the skeleton survives under q-deformations with $|q| = 1$ for all analytic spaces which have a skeleton. Then it is natural to ask whether the Berkovich spectrum of a quantum non-archimedean analytic space contains more than just the skeleton. Surprisingly, as q gets sufficiently close to 1, the answer is positive. Let $\rho = (\rho_1, ..., \rho_d), r = (r_1, ..., r_d) \in \mathbf{R}_{\ge 0}^d$ and $a = (a_1, ..., a_d) \in K^d$. We assume that $|1 - q| < 1$ and $|a| \le |\rho| < r$ (i.e. $a_i \le \rho_i, < r_i, 1 \le i \le d$). Let $f = \sum_{n \in \mathbf{Z}_+^d} c_n T^n$ be a polynomial in q-commuting variables. Set $t_i = T_i - a_i, 1 \le i \le d$. Then f can be written as $f = \sum_{n \in \mathbf{Z}_+^d} b_n t^n$ (although t_i are no longer q-commute).

Proposition 5.3 The seminorm $\nu_{a,\rho}(f) := max_n|b_n|\rho^n$ defines a point of the Berkovich spectrum of the quantum closed polydisc disc $E_q(0, r)$.

Proof. It suffices to show that $\nu_{a,r}(fg) = \nu_{a,r}(f)\nu_{a,r}(g)$ for any two polynomials $f, g \in K[T_1, ..., T_d]_q$.

Let us introduce new variables t_i by the formulas $T_i - a_i = t_i, 1 \leq i \leq d$. Clearly $\nu_{a,\rho}(t_i) = \rho_i, 1 \leq i \leq d$. By definition $t_i t_j = q t_j t_i + (q-1)(a_i t_j + a_j t_i) + (q-1)a_i a_j$. Therefore, for any multi-indices α, β one has $t^\alpha t^\beta = q^{\varphi(\alpha,\beta)} t^\beta t^\alpha + D$, where $\nu_{a,\rho}(D) < \rho^{|\alpha|+|\beta|}$. Here $t^{(\alpha_1,...,\alpha_d)} := t_1^{\alpha_1}...t_d^{\alpha_d}$. Note that $\nu_{a,\rho}(t_i t_j) = \nu_{a,\rho}(t_j t_i) = \nu_{a,\rho}(t_i)\nu_{a,\rho}(t_j)$. Any polynomial $f = \sum_{n \in \mathbf{Z}_+^d} c_n T^n \in K[T_1,...,T_d]_q$ can be written as a finite sum $f = \sum_{n \in \mathbf{Z}_+^d} c_n t^n + B$ where $\nu_{a,\rho}(B) < \nu_{a,\rho}(\sum_{n \in \mathbf{Z}_+^d} a_n t^n)$. We see that $\nu_{a,\rho}(f) = max|c_n|\rho^n$. It follows that $\nu_{a,\rho}$ is multiplicative. This concludes the proof. ∎

We will say that the seminorm $\nu_{a,\rho}$ corresponds to the closed quantum polydisc $E_q(a, \rho) \subset E_q(0, r)$. Since the quantum affine space is the union of quantum closed discs (and hence the Berkovich spectrum of the former is by definition the union of the Berkovich spectra of the latter) we see that the Berkovich spectrum of $\mathbf{A}_q^d$ contains all points $\nu_{a,r}(f)$ with $|a| < r$.

Proposition 5.4 Previous Proposition holds for the quantum torus $T_q(L, \varphi, r)$.

Proof. The proof is basically the same as in the case $q = 1$. Let $f = \sum_{n \in \mathbf{Z}^d} c_n T^n$ be an analytic function on $T_q^{an}(L, \varphi, r)$. In the notation of the above proof we can rewrite f as $f = \sum_{n \in \mathbf{Z}^d} c_n(t+a)^n$, where $(t_i + a_i)^{-1} := t_i^{-1} \sum_{m \geq 0}(-1)^m a_i^m t_i^{-m}$. Therefore $f = \sum_{m \in \mathbf{Z}_+^d} d_m t^m + \sum_{m \in \mathbf{Z}_-^d} d_m t^m$. Suppose we know that $|d_m|\rho^m \to 0$ as $|m| \to +\infty$. Then similarly to the proof of the previous proposition one sees that $\nu_{a,\rho}(f) := max_m|d_m|\rho^m$ is a multiplicative seminorm. Let us estimate $|d_m|\rho^m$. For $m \in \mathbf{Z}_+^d$ we have: $d_m = \sum_{l \in \mathbf{Z}_+^d} b_l^{(m)} c_{l+m} a^l$, where $|b_l^{(m)}| \leq 1$. Since $|a| \leq \rho$ we have $|d_m| \leq max_{l \in \mathbf{Z}_+^d}|c_{l+m}|\rho^l \leq max_{l \in \mathbf{Z}_+^d}|c_{l+m}|\rho^{l+m}/\rho^m \leq p_m \rho^{-m}$, where $p_m \to 0$ as $|m| \to \infty$. Similar estimate holds for $m \in \mathbf{Z}_-^d$. Therefore $\nu_{a,\rho}(f)$ is a multiplicative seminorm. Berkovich spectrum of the quantum torus is, by definition, the union of the Berkovich spectra of quantum analytic tori $T_q(L, \varphi, r)$ of all multiradii r. Thus we see that the pair (a, ρ) as above defines a point of Berkovich spectrum of the quantum analytic torus. ∎

Remark 5.5 It is natural to ask whether any point of the Berkovich spectrum of the "commutative" analytic space $(G_m^{an})^d$ appears as a point of the Berkovich spectrum of the quantum analytic torus, as long as we choose q sufficiently close to 1. More generally, let us imagine that we have two quantum affinoid algebras A and A' which are admissible quotients of the algebras $K\{T_1,...,T_n\}_{Q,r}$ and $K\{T_1,...,T_m\}_{Q',r}$ respectively, where Q and Q' are matrices as in Sect. 3. One can ask the following question: for any closed subset $V \subset M(A)$ is there $\varepsilon > 0$ such that if $||Q - Q'|| < \varepsilon$ then there is a closed subset $V' \subset M(A')$ homeomorphic to V? Here the norm of the matrix $S = ((s_{ij}))$ is defined as $max_{i,j}|s_{ij}|$. If the answer is positive then taking $Q = id$ we see that the Berkovich spectrum of an affinoid algebra is a limit of the Berkovich spectrum of its quantum analytic deformation.

The above Proposition shows in a toy-model the drastic difference with formal deformation quantization. In the latter case deformations A_q of a commutative algebra A_1 are "all the same" as long as $q \neq 1$. In particular, except of few special values of q, they have the same "spectra" in the sense of non-commutative algebraic geometry or representation theory. In analytic case Berkovich spectrum can contain points which are "far" from the commutative ones (e.g. we have seen that the discs $E(a, \rho) \subset E(0, r)$ can be quantized as long as the radius $|\rho|$ is not small). Thus the quantized space contains "holes" (non-archimedean version of "discretization" of the space after quantization).

6 Non-Commutative Stein Spaces

This example is borrowed from [15].

Let K be a non-archimedean valuation field and A be a unital Frechet K-algebra. We say that A is *Frechet-Stein* if there is a sequence $v_1 \leq v_2 \leq \ldots \leq v_n \leq \ldots$ of continuous seminorms on A which define the Frechet topology and:

(a) the completions A_{v_n} of the algebras $A/Ker\, v_n$ are left notherian for all $n \geq 1$;

(b) each A_{v_n} is a flat $A_{v_{n+1}}$-module, $n \geq 1$.

Here we do not require that v_n are submultiplicative, only the inequalities $v_n(xy) \leq const\, v_n(x)v_n(y)$. Clearly the sequence $(A_{v_n})_{n \geq 1}$ is a projective system of algebras and its projective limit is isomorphic to A.

A *coherent sheaf* for a Frechet-Stein algebra $(A, v_n)_{n \geq 1}$ is a collection $M = (M_n)_{n \geq 1}$ such that each M_n is a finite Banach A_{v_n}-module and for each $n \geq 1$ one has natural isomorphism $A_{v_n} \otimes_{A_{v_{n+1}}} M_{n+1} \simeq M_n$.

The inverse limit of the projective system M_n is an A-module called the module of *global sections* of the coherent sheaf M. Coherent sheaves form an abelian category. It is shown in [15] that the global section functor from coherent sheaves to finite Banach A-modules is exact (this is an analogue of A and B theorems of Cartan).

One can define the non-commutative analytic spectrum of Frechet–Stein algebras in the same way as we did for Banach algebras, starting with the category of coherent sheaves.

It is shown in [15] that if G is a compact locally analytic group then the strong dual $D(G, K)$ to the space of K-valued locally analytic functions is a Frechet–Stein algebra. In the case $G = \mathbf{Z}_p$ it is isomorphic to the (commutative) algebra of power series converging in the open unit disc in the completion of the algebraic closure of K. In general, the Frechet–Stein algebra structure on $D(G, K)$ is defined by a family of norms which are submultiplicative only. Coherent sheaves for the algebra $D(G, K)$ should give rise to coherent sheaves on the non-commutative analytic spectrum $M_{NC}(D(G, K))$, rather than on the Berkovich spectrum $M(D(G, K))$.

7 Non-Commutative Analytic K3 Surfaces

7.1 General Scheme

The following way of constructing a non-commutative analytic K3 surface X over the field $K = \mathbf{C}((t))$ was suggested in [8].

(1) We start with a two-dimensional sphere $B := S^2$ equipped with an integral affine structure outside of a finite subset $B^{sing} = \{x_1, ..., x_{24}\}$ of 24 distinct points (see [8] for the definitions and explanation why $|B^{sing}| = 24$). We assume that the monodromy of the affine structure around each point x_i is conjugate to the 2×2 unipotent Jordan block (it is proved in [8] that this restriction enforces the cardinality of B^{sing} to be equal to 24).

(2) In addition to the above data we have an infinite set of "trees" embedded in S^2, called *lines* in [8]. Precise definition and the existence of such a set satisfying certain axioms can be found in [8], Sects. 7.9, 11.5.

(3) The non-commutative analytic space X_q^{an} will be defined defined by a pair $(X_0, \mathcal{O}_{X_0,q})$, where $q \in K^*$ is an arbitrary element satisfying the condition $|q| = 1$, X_0 is a topological K3 surface and $\mathcal{O}_{X_0,q}$ is a sheaf on (a certain topology on) X_0 of non-commutative noetherian algebras over the field K.

(4) There is a natural continuous map $\pi \colon X_0 \to S^2$ with the generic fibre being a two-dimensional torus. The sheaf $\mathcal{O}_{X_0,q}$ is uniquely determined by its direct image $\mathcal{O}_{S^2,q} := \pi_*(\mathcal{O}_{X_0,q})$. Hence the construction of X_q is reduced to the construction of the sheaf $\mathcal{O}_{S^2,q}$ on the sphere S^2.

(5) The sheaf $\mathcal{O}_{S^2,q}$ is glued from two sheaves: a sheaf $\mathcal{O}_q^{sing}$ defined in a neighbourhood W of the "singular" subset B^{sing} and a sheaf $\mathcal{O}_q^{nonsing}$ on $S^2 \setminus W$.

(6) The sheaf $\mathcal{O}_q^{sing}$ is defined by an "ansatz" described below, while the sheaf $\mathcal{O}_q^{nonsing}$ is constructed in two steps. First, with any integral affine structure and an element $q \in K^*, |q| = 1$ one can associate canonically a sheaf $\mathcal{O}_q^{can}$ of non-commutative algebras over K. Then, with each line l one associates an automorphism φ_l of the restriction of $\mathcal{O}_q^{can}$ to l. The sheaf $\mathcal{O}_q^{nonsing}$ is obtained from the restriction of $\mathcal{O}_q^{can}$ to the complement to the union of all lines by the gluing procedure by means of φ_l.

The above scheme was realized in [8] in the case $q = 1$. In that case one obtains a sheaf $\mathcal{O}_{X_0,1}$ of *commutative* algebras which is the sheaf of analytic functions on the non-archimedean analytic K3 surface. In this section we will explain what has to be changed in [8] in order to handle the case $q \neq 1, |q| = 1$. As we mentioned above, in this case $\mathcal{O}_{X_0,q}$ will be a sheaf of non-commutative algebras, which is a flat deformation of $\mathcal{O}_{X_0,0}$. It was observed in [8] that $\mathcal{O}_{X_0,1}$ is a sheaf of Poisson algebras. The sheaf $\mathcal{O}_{X_0,q}$ is a *deformation quantization* of $\mathcal{O}_{X_0,1}$. It is an analytic (not a formal) deformation quantization with respect to the parameter $q - 1$. The topology on X_0 will be clear from the construction.

7.2 Z-Affine Structures and the Canonical Sheaf

Let K be a non-archimedean valuation field. Fix an element $q \in K^*, |q| = 1$. Let us introduce invertible variables ξ, η such that

$$\eta\xi = q\xi\eta.$$

Then we define a sheaf $\mathcal{O}_q^{can}$ on $\mathbf{R}^2$ such that for any open connected subset U one has

$$\mathcal{O}_q^{can}(U) = \left\{ \sum_{n,m \in \mathbf{Z}} c_{n,m}\xi^n\eta^m \,|\, \forall (x,y) \in U \sup_{n,m} \left(\log(|c_{n,m}|) + nx + my\right) < \infty \right\}.$$

The above definition is motivated by the following considerations.

Recall that an integral affine structure (**Z**-affine structure for short) on an n-dimensional topological manifold Y is given by a maximal atlas of charts such that the change of coordinates between any two charts is described by the formula

$$x_i' = \sum_{1 \le j \le n} a_{ij}x_j + b_i,$$

where $(a_{ij}) \in GL(n, \mathbf{Z}), (b_i) \in \mathbf{R}^n$. In this case one can speak about the sheaf of **Z**-affine functions, i.e. those which can be locally expressed in affine coordinates by the formula $f = \sum_{1 \le i \le n} a_i x_i + b, a_i \in \mathbf{Z}, b \in \mathbf{R}$. Another equivalent description: **Z**-affine structure is given by a covariant lattice $T^{\mathbf{Z}} \subset TY$ in the tangent bundle (recall that an affine structure on Y is the same as a torsion free flat connection on the tangent bundle TY).

Let Y be a manifold with **Z**-affine structure. The sheaf of **Z**-affine functions $Aff_{\mathbf{Z}} := Aff_{\mathbf{Z},Y}$ gives rise to an exact sequence of sheaves of abelian groups

$$0 \to \mathbf{R} \to Aff_{\mathbf{Z}} \to (T^*)^{\mathbf{Z}} \to 0,$$

where $(T^*)^{\mathbf{Z}}$ is the sheaf associated with the dual to the covariant lattice $T^{\mathbf{Z}} \subset TY$.

Let us recall the following notion introduced in [8], Sect. 7.1.

Definition 7.1 A K-affine structure on Y compatible with the given **Z**-affine structure is a sheaf Aff_K of abelian groups on Y, an exact sequence of sheaves

$$1 \to K^\times \to Aff_K \to (T^*)^{\mathbf{Z}} \to 1,$$

together with a homomorphism Φ of this exact sequence to the exact sequence of sheaves of abelian groups

$$0 \to \mathbf{R} \to Aff_{\mathbf{Z}} \to (T^*)^{\mathbf{Z}} \to 0,$$

such that $\Phi = id$ on $(T^*)^{\mathbf{Z}}$ and $\Phi = val$ on $K^\times$.

Since Y carries a $\mathbf{Z}$-affine structure, we have the corresponding $GL(n, \mathbf{Z}) \ltimes \mathbf{R}^n$-torsor on Y, whose fibre over a point x consists of all $\mathbf{Z}$-affine coordinate systems at x.

Then one has the following equivalent description of the notion of K-affine structure.

Definition 7.2 A K-affine structure on Y compatible with the given $\mathbf{Z}$-affine structure is a $GL(n, \mathbf{Z}) \ltimes (K^\times)^n$-torsor on Y such that the application of $val^{\times n}$ to $(K^\times)^n$ gives the initial $GL(n, \mathbf{Z}) \ltimes \mathbf{R}^n$-torsor.

Assume that Y is oriented and carries a K-affine structure compatible with a given $\mathbf{Z}$-affine structure. Orientation allows us to reduce to $SL(n, \mathbf{Z}) \ltimes (K^\times)^n$ the structure group of the torsor defining the K-affine structure. One can define a higher-dimensional version of the sheaf $\mathcal{O}_q^{can}$ in the following way. Let $z_1, ..., z_n$ be invertible variables such that $z_i z_j = q z_j z_i$, for all $1 \le i < j \le n$. We define the sheaf $\mathcal{O}_q^{can}$ on $\mathbf{R}^n, n \ge 2$ by the same formulas as in the case $n = 2$:

$$\mathcal{O}_q^{can}(U) = \left\{ \sum_{I = (I_1, ..., I_n) \in \mathbf{Z}^n} c_I z^I, \mid \forall (x_1, ..., x_n) \right.$$

$$\left. \in U \ \sup_I \left(\log(|c_I|) + \sum_{1 \le m \le n} I_m x_m \right) < \infty \right\},$$

where $z^I = z_1^{I_1} \ldots z_n^{I_n}$. Since $|q| = 1$ the convergency condition does not depend on the order of variables.

The sheaf $\mathcal{O}_q^{can}$ can be lifted to Y (we keep the same notation for the lifting). In order to do that it suffices to define the action of the group $SL(n, \mathbf{Z}) \ltimes (K^\times)^n$ on the canonical sheaf on $\mathbf{R}^n$. Namely, the inverse to an element $(A, \lambda_1, ..., \lambda_n) \in SL(n, \mathbf{Z}) \ltimes (K^\times)^n$ acts on monomials as

$$z^I = z_1^{I_1} \ldots z_n^{I_n} \mapsto \left(\prod_{i=1}^n \lambda_i^{I_i} \right) z^{A(I)} .$$

The action of the same element on $\mathbf{R}^n$ is given by a similar formula:

$$x = (x_1, \ldots, x_n) \mapsto A(x) - (val(\lambda_1), \ldots, val(\lambda_n)) .$$

Note that the stalk of the sheaf $\mathcal{O}_q^{can}$ over a point $y \in Y$ is isomorphic to a direct limit of algebras of functions on quantum analytic tori of various multiradii.

Let $Y = \cup_\alpha U_\alpha$ be an open covering by coordinate charts $U_\alpha \simeq V_\alpha \subset \mathbf{R}^n$ such that for any α, β we are given elements $g_{\alpha, \beta} \in SL(n, \mathbf{Z}) \ltimes (K^\times)^n$ satisfying the 1-cocycle condition for any triple α, β, γ. Then the lifting of $\mathcal{O}_h^{can}$ to Y is obtained via gluing by means of the transformations $g_{\alpha, \beta}$.

Let G_m^{an} be the analytic space corresponding to the multiplicative group G_m and $(T^n)^{an} := (G_m^{an})^n$ the n-dimensional analytic torus. Then one has a canonically defined continuous map $\pi_{can} : (T^n)^{an} \to \mathbf{R}^n$ such that $\pi_{can}(p) = (-val_p(z_1), ..., -val_p(z_n)) = (log|z_1|_p, ..., log|z_n|_p)$, where $|a|_p$ (resp. $val_p(a)$) denotes the seminorm (resp. valuation) of an element a corresponding to the point p.

For an open subset $U \in \mathbf{R}^n$ we have a topological K-algebra $\mathcal{O}_q^{can}(U)$ defined by the formulas above. Note that a point $x = (x_1, ..., x_n) \in \mathbf{R}^n$ defines a multiplicative seminorm $|\sum_I c_I z^I|_{exp(x)}$ on the algebra of formal series of q-commuting variables $z_1, ..., z_n$ (here $exp(x) = (exp(x_1), ..., exp(x_n))$).

Let $M(\mathcal{O}_q^{can}(U))$ be the set of multiplicative seminorms ν on $\mathcal{O}_q^{can}(U)$ extending the norm on K. We have defined an embedding $U \to M(\mathcal{O}_q^{can}(U))$, such that $(x_1, ..., x_n)$ corresponds to a seminorm with $|z_i| = exp(x_i), 1 \le i \le n$. The map $\pi_{can} : |\bullet| \mapsto (log|z_1|, ..., log|z_n|)$ is a retraction of $M(\mathcal{O}_q^{can}(U))$ to the image of U.

Let $S_1^n \subset (K^\times)^n$ be the set of such $(s_1, ..., s_n)$ that $|s_i| = 1, 1 \le i \le n$. The group S_1^n acts on $\mathcal{O}_q^{can}(U)$ in such a way that $z_i \mapsto s_i z_i$. Clearly the map π_{can} is S_1^n-invariant. For this reason we will call π_{can} a *quantum analytic torus fibration* over U. More precisely, we reserve this name for a pair $(U, \mathcal{O}_q^{can}(U))$, where the algebra is equipped with the S_1^n-action. We suggest to think about such a pair as of the algebra $\mathcal{O}_q(\pi_{can}^{-1}(U))$ of analytic functions on the open subset $\pi_{can}^{-1}(U)$ of the non-commutative analytic torus $(T_q^n(\mathbf{Z}^n)^{an}, \varphi_0)$, where $\varphi_0((a_1, ..., a_n), (b_1, ..., b_n)) = a_1 b_1 + ... + a_n b_n$.

We can make a category of the above pairs, defining a morphism $(U, \mathcal{O}_q^{can}(U)) \to (V, \mathcal{O}_q^{can}(V))$ as a pair (f, ϕ) where $f : U \to V$ is a continuous map and $\phi : \mathcal{O}_q^{can}(f^{-1}(V)) \to \mathcal{O}_q^{can}(V)$ is a S_1^n-equivariant homomorphism of algebras. In particular we have the notion of isomorphism of quantum analytic torus fibrations.

Let $U \subset \mathbf{R}^n$ be an open set and A be a non-commutative affinoid K-algebra equipped with a S_1^n-action. We say that a pair (U, A) defines a quantum analytic torus fibration over U if it is isomorphic to the pair $(U, \mathcal{O}_q^{can}(U))$. Notice that morphisms of quantum analytic torus fibrations are compatible with the restrictions on the open subsets. Therefore we can introduce a topology on $(T^n)^{an}$ taking $\pi_{can}^{-1}(U), U \subset \mathbf{R}^2$ as open subsets and make a ringed space assigning the algebra $\mathcal{O}_q(U)$ to the open set $\pi_{can}^{-1}(U)$. We will denote the sheaf by $(\pi_{can})_*(\mathcal{O}_{(T^n)^{an},q})$. Its global sections (for $U = \mathbf{R}^n$) coincides with the projective limit of algebras of analytic functions on quantum tori of all possible multiradii. Slightly abusing the terminology we will call the above ringed space a quantum analytic torus.

Remark 7.3 The above definition is a toy-model of a q-deformation of the natural retraction $X^{an} \to Sk(X)$ of the analytic space X^{an} associated with the maximally degenerate Calabi–Yau manifold X onto its skeleton $Sk(X)$ (see [8]). Analytic torus fibrations introduced in [8] are "rigid analytic" analogues of Lagrangian torus fibrations in symplectic geometry. Moreover, the mirror symmetry functor (or rather its incarnation as a Fourier–Mukai transform) interchanges these two types of torus fibrations for mirror dual Calabi–Yau manifolds.

It turns out that the sheaf $\mathcal{O}_q^{can}$ is not good for construction of a non-commutative analytic K3 surface. We will explain later how it should be modified. Main reason for the complicated modification procedure comes from Homological Mirror Symmetry, as explained in [8]. In a few words, the derived category of coherent sheaves on a non-commutative analytic K3 surface should be equivalent to a certain deformation of the Fukaya category of the mirror dual K3 surface. If the K3 surface is realized as an elliptic fibration over $\mathbf{CP}^1$ then there are fibers (they are 2-dimensional Lagrangian tori) which contain boundaries of holomorphic discs. Those discs give rise to an infinite set of lines on the base of the fibration. In order to have the above-mentioned categorical equivalence one should modify the canonical sheaf for each line.

7.3 Model Near a Singular Point

Let us fix $q \in K^*, |q| = 1$.

We start with the open covering of $\mathbf{R}^2$ by the following sets $U_i, 1 \leq i \leq 3$. Let us fix a number $0 < \varepsilon < 1$ and define

$$U_1 = \{(x,y) \in \mathbf{R}^2 | x < \varepsilon|y| \}$$
$$U_2 = \{(x,y) \in \mathbf{R}^2 | x > 0, y < \varepsilon x \}$$
$$U_3 = \{(x,y) \in \mathbf{R}^2 | x > 0, y > 0\}$$

Clearly $\mathbf{R}^2 \setminus \{(0,0)\} = U_1 \cup U_2 \cup U_3$. We will also need a slightly modified domain $U_2' \subset U_2$ defined as $\{(x,y) \in \mathbf{R}^2 | x > 0, y < \frac{\varepsilon}{1+\varepsilon}x \}$.

Recall that one has a canonical map $\pi_{can} : (T_q^2)^{an} \to \mathbf{R}^2$.

We define $T_i := \pi_{can}^{-1}(U_i), i = 1,3$ and $T_2 := \pi_{can}^{-1}(U_2')$ (see the explanation below). Then the projections $\pi_i : T_i \to U_i$ are given by the formulas

$$\pi_i(\xi_i, \eta_i) = \pi_{can}(\xi_i, \eta_i) = (\log|\xi_i|, \log|\eta_i|), \quad i = 1,3$$

$$\pi_2(\xi_2, \eta_2) = \begin{cases} (\log|\xi_2|, \log|\eta_2|) & \text{if } |\eta_2| < 1 \\ (\log|\xi_2| - \log|\eta_2|, \log|\eta_2|) & \text{if } |\eta_2| \geq 1 \end{cases}$$

In these formulas (ξ_i, η_i) are coordinates on $T_i, 1 \leq i \leq 3$. More pedantically, one should say that for each T_i we are given an algebra $\mathcal{O}_q(T_i)$ of series $\sum_{m,n} c_{mn}\xi_i^m \eta_i^n$ such that $\xi_i\eta_i = q\eta_i\xi_i$ and for a seminorm $| \bullet |$ corresponding to a point of T_i (which means that $(log|\xi_i|, log|\eta_i|) \in U_i$) one has: $sup_{m,n}(m \, log|\xi_i| + n \, log|\eta_i|) < +\infty$. In this way we obtain a sheaf of non-commutative algebras on the set U_i, which is the subset of the Berkovich spectrum of the algebra $\mathcal{O}_q(T_i)$. We will denote this sheaf by $\mathcal{O}_{T_i,q}$.

Let us introduce the sheaf $\mathcal{O}_q^{can}$ on $\mathbf{R}^2 \setminus \{(0,0)\}$. It is defined as $(\pi_i)_* (\mathcal{O}_{T_i,q})$ on each domain U_i, with identifications

$$(\xi_1, \eta_1) = (\xi_2, \eta_2) \quad \text{on } U_1 \cap U_2$$
$$(\xi_1, \eta_1) = (\xi_3, \eta_3) \quad \text{on } U_1 \cap U_3$$
$$(\xi_2, \eta_2) = (\xi_3 \eta_3, \eta_3) \quad \text{on } U_2 \cap U_3$$

Let us introduce the sheaf $\mathcal{O}_q^{sing}$ on $\mathbf{R}^2 \setminus \{(0,0)\}$. On the sets U_1 and $U_2 \cup U_3$ this sheaf is isomorphic to $\mathcal{O}_q^{can}$ (by identifying of coordinates (ξ_1, η_1) and of glued coordinates (ξ_2, η_2) and (ξ_3, η_3) respectively). On the intersection $U_1 \cap (U_2 \cup U_3)$ we identify two copies of the canonical sheaf by an automorphism φ of $\mathcal{O}_q^{can}$. More precisely, the automorphism is given (we skip the index of the coordinates) by

$$\varphi(\xi, \eta) = \begin{cases} (\xi(1 + \eta), \eta) & \text{on } U_1 \cap U_2 \\ (\xi(1 + 1/\eta), \eta) & \text{on } U_1 \cap U_3 \end{cases}$$

7.4 Lines and Automorphisms

We refer the reader to [8] for the precise definition of the set of lines and axioms this set is required to obey. Roughly speaking, for a manifold Y which carries a $\mathbf{Z}$-affine structure a line l is defined by a continuous map $f_l : (0, +\infty) \to Y$ and a covariantly constant nowhere vanishing integer-valued 1-form $\alpha_l \in \Gamma((0, +\infty), f_l^*((T^*)^{\mathbf{Z}})$. A set $\mathcal{L}$ of lines is required to be decomposed into a disjoint union $\mathcal{L} = \mathcal{L}_{in} \cup \mathcal{L}_{com}$ of *initial* and *composite* lines. Each composite line is obtained as a result of a finite number of "collisions" of initial lines. A collision is described by a Y-shape figure, where the leg of Y is a composite line, while two other segments are "parents" of the leg. A construction of the set $\mathcal{L}$ satisfying the axioms from [8] was proposed in [8], Section 9.3.

With each line l we can associate a continuous family of automorphisms of stalks of sheaves of algebras $\varphi_l(t) : (\mathcal{O}_q^{can})_{Y, f_l(t)} \to (\mathcal{O}_q^{can})_{Y, f_l(t)}$.

Automorphisms φ_l can be defined in the following way (see [8], Sect. 10.4).

First we choose affine coordinates in a neighborhood of a point $b \in B \setminus B^{sing}$, identifying b with the point $(0, 0) \in \mathbf{R}^2$. Let $l = l_+ \in \mathcal{L}_{in}$ be (in the standard affine coordinates) a line in the half-plane $y > 0$ emerging from $(0, 0)$ (there is another such line l_- in the half-plane $y < 0$, see [8] for the details). Assume that t is sufficiently small. Then we define $\varphi_l(t)$ on topological generators ξ, η by the formula

$$\varphi_l(t)(\xi, \eta) = (\xi(1 + 1/\eta), \eta).$$

In order to extend $\varphi_l(t)$ to the interval $(0, t_0)$, where t_0 is not small, we cover the corresponding segment of l by open charts. Then a change of affine coordinates transforms η into a monomial multiplied by a constant from $K^\times$. Moreover, one can choose the change of coordinates in such a way that $\eta \mapsto C\eta$ where $C \in K^\times, |C| < 1$ (such change of coordinates preserve the 1-form dy. Constant C is equal to $exp(-L)$, where L is the length of the segment of l between two points in different coordinate charts). Therefore η extends analytically in a unique way to an element of $\Gamma((0, +\infty), f_l^*((\mathcal{O}_q^{can})^\times))$. Moreover the norm $|\eta|$ strictly decreases as t increases and remains strictly smaller than 1. Similarly to [8], Sect. 10.4 one deduces that $\varphi_l(t)$ can be extended for all $t > 0$. This defines $\varphi_l(t)$ for $l \in \mathcal{L}_{in}$.

Next step is to extend $\varphi_l(t)$ to the case when $l \in \mathcal{L}_{com}$, i.e. to the case when the line is obtained as a result of a collision of two lines belonging to $\mathcal{L}_{in}$. Following [8], Sect. 7.10, we introduce a group G which contains all the automorphisms $\varphi_l(t)$ and then prove the factorization theorem (see [8], Theorem 6) which allows us to define $\varphi_l(0)$ in the case when l is obtained as a result of a collision of two lines l_1 and l_2. Then we extend $\varphi_l(t)$ analytically for all $t > 0$ similarly to the case $l \in \mathcal{L}_{in}$.

More precisely, the construction of G goes such as follows. Let $(x_0, y_0) \in \mathbf{R}^2$ be a point, $\alpha_1, \alpha_2 \in (\mathbf{Z}^2)^*$ be 1-covectors such that $\alpha_1 \wedge \alpha_2 > 0$. Denote by $V = V_{(x_0,y_0),\alpha_1,\alpha_2}$ the closed angle

$$\{(x,y) \in \mathbf{R}^2 | \langle \alpha_i, (x,y) - (x_0, y_0) \rangle \geq 0, i = 1, 2\}.$$

Let $\mathcal{O}_q(V)$ be a K-algebra consisting of series $f = \sum_{n,m \in \mathbf{Z}} c_{n,m} \xi^n \eta^m$, such that $\xi \eta = q \eta \xi$ and $c_{n,m} \in K$ satisfy the condition that for all $(x, y) \in V$ we have:

1. if $c_{n,m} \neq 0$ then $\langle (n, m), (x, y) - (x_0, y_0) \rangle \leq 0$, where we identified $(n, m) \in \mathbf{Z}^2$ with a covector in $(T_p^*Y)^{\mathbf{Z}}$;
2. $\log |c_{n,m}| + nx + my \to -\infty$ as long as $|n| + |m| \to +\infty$.

For an integer covector $\mu = a \, dx + b \, dy \in (\mathbf{Z}^2)^*$ we denote by R_μ the monomial $\xi^a \eta^b$. Then we consider a pro-unipotent group $G := G(q, \alpha_1, \alpha_2, V)$ of automorphisms of $\mathcal{O}_q(V)$ having the form

$$g = \sum_{n_1, n_2 \geq 0, n_1 + n_2 > 0} c_{n_1, n_2} R_{\alpha_1}^{-n_1} R_{\alpha_2}^{-n_2}$$

where

$$\log |c_{n,m}| - n_1 \langle \alpha_1, (x, y) \rangle - n_2 \langle \alpha_2, (x, y) \rangle \leq 0 \quad \forall (x, y) \in V$$

The latter condition is equivalent to $\log |c_{n,m}| - \langle n_1 \alpha_1 + n_2 \alpha_2, (x_0, y_0) \rangle \leq 0$.

Fixing the ratio $\lambda = n_2/n_1 \in [0, +\infty]_{\mathbf{Q}} := \mathbf{Q}_{\geq 0} \cup \infty$ we obtain a subgroup $G_\lambda := G_\lambda(q, \alpha_1, \alpha_2, V) \subset G$. There is a natural map $\prod_\lambda G_\lambda \to G$, defined as in [8], Sect. 10.2. The above-mentioned factorization theorem states that this map is a bijection of sets.

Let us now assume that lines l_1 and l_2 collide at $p = f_{l_1}(t_1) = f_{l_2}(t_2)$, generating the line $l \in \mathcal{L}_{com}$. Then $\varphi_l(0)$ is defined with the help of factorization theorem. More precisely, we set $\alpha_i := \alpha_{l_i}(t_i)$, $i = 1, 2$ and the angle V is the intersection of certain half-planes $P_{l_1,t_1} \cap P_{l_2,t_2}$ defined in [8], Sect. 10.3. The half-plane $P_{l,t}$ is contained in the region of convergence of $\varphi_l(t)$. By construction, the elements $g_0 := \varphi_{l_1}(t_1)$ and $g_{+\infty} := \varphi_{l_2}(t_2)$ belong respectively to G_0 and $G_{+\infty}$. The we have:

$$g_{+\infty}g_0 = \prod_{\rightarrow} \left((g_\lambda)_{\lambda \in [0,+\infty]_{\mathbf{Q}}} \right) = g_0 \cdots g_{1/2} \cdots g_1 \cdots g_{+\infty}.$$

Each term g_λ with $0 < \lambda = n_1/n_2 < +\infty$ corresponds to the newborn line l with the direction covector $n_1\alpha_{l_1}(t_1) + n_2\alpha_{l_2}(t_2)$. Then we set $\varphi_l(0) := g_\lambda$. This transformation is defined by a series which is convergent in a neighbourhood of p and using the analytic continuation we obtain $\varphi_l(t)$ for $t > 0$, as we said above. Recall that every line carries an integer 1-form $\alpha_l = adx + bdy$. By construction, $\varphi_l(t) \in G_\lambda$, where λ is the slope of α_l.

Having automorphisms φ_l assigned to lines $l \in \mathcal{L}$ we proceed as in [8], Section 7.11, modifying the sheaf $\mathcal{O}_q^{can}$ along each line. We denote the resulting sheaf $\mathcal{O}_q^{nonsing}$. By construction it is isomorphic to the sheaf $\mathcal{O}_q^{sing}$ in a neighborhood of the point $(0,0)$.

Let us now consider the manifold $Y = S^2 \setminus B^{sing}$, i.e. the complement of 24 points on the sphere S^2 equipped with the $\mathbf{Z}$-affine structure, which has standard singularity at each point $x_i \in B^{sing}, 1 \le i \le 24$ (see Sect. 7.1). Using the above construction (with any choice of set of lines on S^2) we define the sheaf $\mathcal{O}_{S^2,q}^{nonsing}$ on Y. Notice that in a small neighbourhood of each singular point x_i the sheaf $\mathcal{O}_{S^2,q}^{nonsing}$ is isomorphic to the sheaf $\mathcal{O}_q^{sing}$ (in fact they become isomorphic after identification of the punctured neighbourhood of x_i with the punctured neighbourhood of $(0,0) \in \mathbf{R}^2$ equipped with the standard singular $\mathbf{Z}$-affine structure (see Sect. 7.1 and [8], Sect. 6.4 for the description of the latter). In the next subsection we will give an alternative description of the sheaf $\mathcal{O}_q^{sing}$. It follows from that description that $\mathcal{O}_q^{sing}$ can be extended to the point $(0,0)$. It gives a sheaf $\mathcal{O}_{S^2,q}^{sing}$ in the neighbourhood of B^{sing}. As a result we will obtain the sheaf $\mathcal{O}_{S^2,q}$ of non-commutative K-algebras on the whole sphere S^2 such that it is isomorphic to $\mathcal{O}_{S^2,q}^{nonsing}$ on the complement of B^{sing} and isomorphic to $\mathcal{O}_{S^2,q}^{sing}$ in a neighborhood of B^{sing}.

7.5 About the Sheaf $\mathcal{O}_q^{sing}$

We need to check that the sheaf $\mathcal{O}_{S^2,q}$ is a flat deformation of the sheaf $\mathcal{O}_{S^2}$ constructed in [8]. For the sheaf $\mathcal{O}_{S^2,q}^{nonsing}$ this follows from the construction. Indeed, the algebra of analytic functions on the quantum analytic torus (of any multiradius) is a flat deformation of the algebra of analytic functions on the corresponding "commutative" torus, equipped with the Poisson bracket $\{x,y\} = xy$. The group $G = G(q)$ described in the previous subsection is a flat deformation of its "commutative" limit $G(1) := G(q = 1)$ defined in [8], Sect. 7.10. The group $G(1)$ preserves the above Poisson bracket. In order to complete the construction of the non-commutative space analytic K3 surface X_q we need to prove that the sheaf $\mathcal{O}_q^{sing}$ is a flat deformation with respect to $q - 1$ of the sheaf $\mathcal{O}^{model}$ introduced in [8], Sect. 7.8. First we recall the definition of the latter.

Let $S \subset \mathbf{A}^3$ be an algebraic surface given by equation $(\alpha\beta - 1)\gamma = 1$ in coordinates (α, β, γ) and S^{an} be the corresponding analytic space. We define a continuous map $f : S^{an} \to \mathbf{R}^3$ by the formula $f(\alpha, \beta, \gamma) = (a, b, c)$ where $a = \max(0, \log |\alpha|_p), b = \max(0, \log |\beta|_p), c = \log |\gamma|_p = -\log |\alpha\beta - 1|_p$. Here $| \cdot |_p = \exp(-val_p(\cdot))$ denotes the multiplicative seminorm corresponding to the point $p \in S^{an}$.

Let us consider the embedding $j : \mathbf{R}^2 \to \mathbf{R}^3$ given by formula

$$j(x, y) = \begin{cases} (-x, \max(x + y, 0), -y) & \text{if } x \leq 0 \\ (0, x + \max(y, 0), -y) & \text{if } x \geq 0 \end{cases}$$

One can easily check that the image of j coincides with the image of f. Let us denote by $\pi : S^{an} \to \mathbf{R}^2$ the map $j^{(-1)} \circ f$. Finally, we denote $\pi_*(\mathcal{O}_{S^{an}})$ by $\mathcal{O}^{model} := \mathcal{O}_{\mathbf{R}^2}^{model}$. It was shown in [8], Sect. 7.8, that $\mathcal{O}^{model}$ is canonically isomorphic to the sheaf $\mathcal{O}_{q=1}^{sing}$ (the latter is defined as a modification of the sheaf $\mathcal{O}_{q=1}^{can}$ by means of the automorphism φ, given by the formula at the end of Sect. 7.3 for commuting variables ξ and η).

Let us consider a non-commutative K-algebra $A_q(S)$ generated by generators α, β, γ subject to the following relations:

$$\alpha\gamma = q\gamma\alpha, \beta\gamma = q\gamma\beta,$$

$$\beta\alpha - q\alpha\beta = 1 - q,$$

$$(\alpha\beta - 1)\gamma = 1.$$

For $q = 1$ this algebra coincides with the algebra of regular functions on the surface $X \subset \mathbf{A}_K^3$ given by the equation $(\alpha\beta - 1)\gamma = 1$ and moreover, it is a flat deformation of the latter with respect to the parameter $q - 1$.

Recall that in Sect. 7.3 we defined three open subsets $T_i, 1 \leq i \leq 3$ of the two-dimensional quantum analytic torus $(T^2)^{an}$. The subset T_i is defined as a ringed space $(\pi_i^{-1}(U_i), \mathcal{O}_{T_i, q})$, where U_i are open subsets of $\mathbf{R}^2$ and $\mathcal{O}_{T_i, q}$ is a sheaf of non-commutative algebras, uniquely determined by the K-algebra $\mathcal{O}_q(T_i)$ of its global sections.

We define morphisms $g_i : T_i \hookrightarrow S, 1 \leq i \leq 3$ by the following formulas

$$g_1(\xi_1, \eta_1) = (\xi_1^{-1}, \xi_1(1 + \eta_1), \eta_1^{-1})$$
$$g_2(\xi_2, \eta_2) = ((1 + \eta_2)\xi_2^{-1}, \xi_2, \eta_2^{-1})$$
$$g_3(\xi_3, \eta_3) = ((1 + \eta_3)(\xi_3\eta_3)^{-1}, \xi_3\eta_3, (\eta_3)^{-1})$$

More precisely this means that for each $1 \leq i \leq 3$ we have a homomorphism of K-algebras $A_q(S) \to \mathcal{O}_q(T_i)$ such that α is mapped to the first coordinate of g_i, β is mapped to the second coordinate and γ is mapped to the third coordinate. One checks directly that three coordinates obey the relations between α, β, γ. Modulo $(q - 1)$ these morphisms are inclusions. In the non-commutative case they induce embeddings of $M(\mathcal{O}_q(T_i)), 1 \leq i \leq 3$ into the set $M(A_q(S))$ of multiplicative seminorms on $A_q(S)$.

Notice that in the commutative case we have: $j \circ \pi_i = f \circ g_i$ and $f^{-1}(j(U_i)) = g_i(T_i)$ for all $1 \leq i \leq 3$. Using this observation we can decompose a neighbourhood V of $\pi^{-1}(0, 0)$ in S^{an} into three open analytic subspaces and describe explicitly algebras of analytic functions as series in coordinates (α, β) or (β, γ) or (α, γ) (choice of the coordinates depend on the domain) with certain grows conditions on the coefficients of the series. This gives explicit description of the algebra $\pi_*(\mathcal{O}_{S^{an}}(\pi(V)))$. Then we *declare* the same description in the non-commutative case to be the answer for the direct image. Non-commutativity does not affect the convergency condition because $|q| = 1$. This description, perhaps, can be obtained from the "general theory" which will developed elsewhere. The direct check, as in the commutative case, shows the compatibility of this description of the direct image sheaf with the description of $\mathcal{O}_q^{nonsing}$ in the neighborhood of $(0,0)$. Therefore we can glue both sheaves together, obtaining $\mathcal{O}_{S^2,q}$. This completes the construction.

Acknowledgements

I am grateful to IHES and Max-Planck Institut für Mathematik for excellent research conditions. I thank Vladimir Berkovich, Vladimir Drinfeld, Maxim Kontsevich and Alexander Rosenberg for useful discussions. I thank Hanfeng Li for comments on the paper. I am especially grateful to Lapchik (a.k.a Lapkin) for encouragement, constant interest and multiple remarks on the preliminary drafts (see [12]). The main object of the present chapter should be called "lapkin spaces". Only my inability to discover the property "sweetness" (predicted by Lapchik, cf. with the "charm" of quarks) led me to the (temporary) change of terminology to "non-commutative non-archimedean analytic spaces".

References

1. S. Bosch, U. Güntzer, R. Remmert, Non-archimedean analysis. Springer-Verlag, (1984).
2. V. Berkovich, Spectral theory and analytic geometry over non-archimedean fields. AMS Math. Surveys and Monographs, n. 33, 1990.
3. V. Berkovich, Etale cohomology for non-Archimedean analytic spaces, Publ. Math. IHES 78 (1993), 5–161.
4. V. Berkovich, Smooth p-adic analytic spaces are locally contractible, Inv. Math., 137 (1999), 1–84.
5. A. Connes, Non-commutative geometry, Academic Press, 1994.
6. V. Fock, A. Goncharov, Cluster ensembles, quantization and the dilogarithm, math.AG/0311245.
7. J. Fresnel, M. van der Put, Rigid analytic geometry and applications. Birkhauser, (2003).
8. M. Kontsevich, Y. Soibelman, Affine structures and non-archimedean analytic spaces, math.AG/0406564.
9. M. Kontsevich, Y. Soibelman, Homological mirror symmetry and torus fibrations, math.SG/0011041.

10. M. Kontsevich, A. Rosenberg, Non-commutative smooth spaces, math.AG/ 9812158.
11. L. Korogodsky, Y. Soibelman, Algebras of functions on quantum groups. I, Amer. Math. Soc., (1997).
12. Lapchik, personal communications, 2004–2005, see also www.allalapa.net
13. A. Rosenberg, Non-commutative algebraic geometry and representations of quantized algebras. Kluwer Academic Publishers, (1995).
14. A. Rosenberg, Y. Soibelman, Non-commutative analytic spaces, in preparation.
15. P. Schneider, J. Teitelbaum, Algebras of p-adic distributions and admissible representations, Invent. math. 153, 145–196 (2003).
16. Y. Soibelman, V. Vologodsky, Non-commutative compactifications and elliptic curves, math.AG/0205117, published in Int. Math.Res. Notes, 28 (2003).

Derived Categories and Stacks in Physics

E. Sharpe

Departments of Physics, Mathematics, University of Utah Salt Lake City, UT
84112, USA
ersharpe@rt.edu

Abstract. We review how both derived categories and stacks enter physics. The
physical realization of each has many formal similarities. For example, in both the
cases, equivalences are realized via renormalization group flow: in the case of derived
categories, (boundary) renormalization group flow realizes the mathematical procedure of localization on quasi-isomorphisms and, in the case of stacks, worldsheet
renormalization group flow realizes presentation-independence. For both, we outline
current technical issues and applications.

1 Introduction

For many years, much mathematics relevant to physics (Gromov–Witten theory, Donaldson theory, quantum cohomology, etc.) has appeared physically in
correlation function computations in supersymmetric field theories. Typically
one can see all aspects of the mathematics encoded somewhere in the physics,
if one takes the time to work through the details. In this fashion we have been
able to understand the relation of these parts of mathematics to physics very
concretely.

However, more recently we have begun to see a more complicated dictionary, in which mathematical ideas of homotopy and categorical equivalences
map to the physical notion of the renormalization group. The renormalization
group is a very powerful idea in physics, but unlike the correlation function
calculations alluded to in the last paragraph, it is not currently technically
feasible to follow the renormalization group explicitly and concretely a finite
distance along its flow. Unlike what has happened in the past, we can no
longer see all details of the mathematics explicitly and directly in the physics,
and instead have to appeal to indirect arguments to make the connection.

In this note we will outline two such recent examples of pieces of mathematics in which important components map to the renormalization group in
physics. Specifically, we will briefly discuss how derived categories and stacks
enter physics. For information on the mathematics of derived categories, see

Sharpe, E.: *Derived Categories and Stacks in Physics.* Lect. Notes Phys. **757**, 249–272 (2009)
DOI 10.1007/978-3-540-68030-7_8 © Springer-Verlag Berlin Heidelberg 2009

for example [1, 2, 3] and for a more extensive description of how derived categories enter physics, see the review article [4]. For information on the mathematics of stacks see [5, 6].

2 The Renormalization Group

For readers not[1] familiar with the notion, the renormalization group is a semigroup operation on an abstract space of physical theories. Given one quantum field theory, the semigroup operation constructs new quantum field theories which are descriptions valid at longer and longer distance scales.

In particular, under the renormalization group two distinct theories can sometimes become the same (the semigroup operation is not invertible). We have schematically illustrated such a process in the two pictures below. Although the two patterns look very different, at long distances the checkerboard on the right becomes a better and better approximation to the square on the left, until the two are indistinguishable.

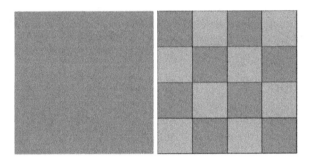

Two theories that flow to the same theory under the renormalization group are said to be in the same "universality class" of renormalization group flow.

The renormalization group is a powerful tool but unfortunately we cannot follow it completely explicitly in general. The best we can typically do is construct an asymptotic series expansion to the tangent vector of the flow at any given point. Thus, ordinarily we cannot really prove in any sense that two theories will flow under the renormalization group to the same point.

On the one hand, a mathematical theory that makes predictions for how different theories will flow, as happens in the application of both derived categories and stacks to physics, is making strong statements about physics. On the other hand, we can only check such statements indirectly, by performing many consistency tests in numerous examples.

---

[1] The talk was given to an audience of both mathematicians and physicists, and we have attempted to make these notes accessible to both audiences as well.

3 Derived Categories in Physics

3.1 History

Derived categories entered physics gradually through a succession of developments. Before describing the modern understanding, let us take a moment to review the history.

One of the original motivations was work of Kontsevich, his "homological mirror symmetry" approach to mirror symmetry [7]. Ordinary mirror symmetry is a relation between two Calabi–Yau's, but Kontsevich's Homological mirror symmetry relates derived categories of coherent sheaves on one Calabi–Yau to Fukaya categories on the other. At the time it was proposed, no physicist had any idea how or even if derived categories entered physics – his proposal predated even D-branes – a testament to Kontsevich's insight.

Shortly after Kontsevich's work, in 1995 Polchinski reminded everyone of his work on D-branes and explained their relevance to string duality [8]. Although neither derived categories nor sheaves appeared in [8], they would later play a role.

About a year later, Harvey and Moore speculated in [9] that coherent sheaves might be a good mathematical description for some D-branes. Given that impetus, other authors soon discovered experimentally that, indeed, mathematical properties of coherent sheaves at least often computed physical quantities of corresponding D-branes. For example, it was discovered empirically that massless states between D-branes were counted by Ext groups between the corresponding sheaves, though the complete physical understanding of why that was the case was not worked out until [10, 11, 12].

Having understood how sheaves could at least often be relevant to physics was an important step, but derived categories of sheaves are more complicated than just sheaves – derived categories of sheaves involve complexes of sheaves, not just sheaves. The next intellectual step was Sen's introduction of antibranes [13] and Witten's realization that Sen's work amounted to a physical realization of K theory [14] in 1998.

With that insight, the first proposals for how derived categories enter physics became possible. Shortly after Witten's introduction of K theory, it was proposed in [15] that the same notion of antibranes, when applied to sheaf models, could be used to give a physical realization of derived categories. Specifically, a complex of branes representing any given object in a derived category should correspond physically to a set of branes and antibranes, with the maps in the complex defining a set of tachyons. Two objects in the derived category related by quasi-isomorphism should correspond physically to two sets of branes and antibranes that are in the same universality class of renormalization group flow. In this fashion one could finally begin to have a physical understanding of Kontsevich's homological mirror symmetry.

Although several papers were written and talks given, this physical understanding of derived categories languished in obscurity until Douglas popularized the same notion 2 years later in [16]. Douglas also introduced the notion of pi-stability, which has seen some interest in the mathematics community (see e.g. [17]).

3.2 D-Branes and Sheaves

To lowest order, a D-brane is a pair, consisting of a submanifold of spacetime together with a vector bundle on that submanifold. In fact, we will specialize to D-branes in the open string topological B model, in which case the submanifolds are complex submanifolds and the vector bundles are holomorphic. Such data we can describe by the sheaf $i_* \mathcal{E}$, where i is the inclusion map of the submanifold into the spacetime and $\mathcal{E}$ is a vector bundle on the submanifold.

In what sense are sheaves a good model for D-branes? After all, physically D-branes are specified by a set of boundary conditions on open strings together with Chan–Paton data, which is not the same thing as a sheaf. However, we can compute physical quantities (such as massless spectra) using mathematical operations on sheaves (such as Ext groups) and it is for this reason that we consider sheaves to be a good model for D-branes.

In particular, mathematical deformations of a sheaf match physical deformations of the corresponding D-brane. This is ordinarily one of the first tests that one performs given some new mathematical model of part of physics.

Let us next briefly outline how these sorts of computations are performed.

Massless states in the topological B model are BRST-closed combinations of the fields ϕ^i, $\phi^{\bar{i}}$, $\eta^{\bar{i}}$, θ_i, modulo BRST-exact combinations. The fields ϕ^i, $\phi^{\bar{i}}$ are local coordinates on the target space and $\eta^{\bar{i}}$, θ_i are Grassman-valued. In simple cases, the BRST operator acts as follows:

$$Q_{BRST} \cdot \phi^i = 0, \quad Q_{BRST} \cdot \phi^{\bar{i}} \neq 0$$
$$Q_{BRST} \cdot \eta^{\bar{i}} = 0, \quad Q_{BRST} \cdot \theta_i = 0$$

States are then of the form

$$b(\phi)^{\alpha\beta \; j_1 \cdots j_m}_{\bar{i}_1 \cdots \bar{i}_n} \eta^{\bar{i}_1} \cdots \eta^{\bar{i}_n} \theta_{j_1} \cdots \theta_{j_m}$$

where α, β are "Chan–Paton" indices, coupling to two vector bundles $\mathcal{E}$, $\mathcal{F}$. We can understand these states mathematically by applying the dictionary (for an open string both of whose ends lie on a submanifold $S \subseteq X$)

$$Q_{BRST} \sim \bar{\partial}, \quad \eta^{\bar{i}} \sim d\bar{z}^{\bar{i}} \sim TS, \quad \theta_i \sim \mathcal{N}_{S/X}$$

Then the states above can be identified with elements of the sheaf cohomology group

$$H^n(S, \mathcal{E}^\vee \otimes \mathcal{F} \otimes \Lambda^m \mathcal{N}_{S/X}).$$

The analysis above is a bit quick and only applies in special cases. In general, one must take into account complications such as:

1. The Freed–Witten anomaly [18], which says that to the sheaf $i_* \mathcal{E}$ one associates a D-brane on S with 'bundle' $\mathcal{E} \otimes \sqrt{K_S}$ instead of $\mathcal{E}$ [10].

2. The open string analogue of the Calabi–Yau condition, which for two D-branes with trivial bundles wrapped on submanifolds S, T, becomes the constraint [10]

$$\Lambda^{top}\mathcal{N}_{S\cap T/S} \otimes \Lambda^{top}\mathcal{N}_{S\cap T/T} \cong \mathcal{O}$$

3. When the Chan–Paton bundles have nonzero curvature, the boundary conditions on the fields are modified [19]; for example, for line bundles, the constraint can be written

$$\theta_i = F_{i\bar{j}}\eta^{\bar{j}}$$

Taking into account these complications will in general physically realize a spectral sequence [10]; for example, for D-branes wrapped on the same submanifold S,

$$H^n\left(S, \mathcal{E}^\vee \otimes \mathcal{F} \otimes \Lambda^m\mathcal{N}_{S/X}\right) \implies \text{Ext}_X^{n+m}\left(i_*\mathcal{E}, i_*\mathcal{F}\right)$$

Not all sheaves are of the form $i_*\mathcal{E}$ for some vector bundle $\mathcal{E}$ on S. How can one handle more general cases?

A partial answer was proposed in [20] (before the publication of [16]) and later worked out in more detail in [21], using mathematics appearing in [22]. The proposal made there was that some more general sheaves can be understood as mathematical models of D-branes with non-zero "Higgs fields". For each direction perpendicular to the D-brane worldvolume, there is a Higgs field, which one can interpret as a holomorphic section of $\mathcal{E}^\vee \otimes \mathcal{E} \otimes \mathcal{N}_{S/X}$. The idea is that we can interpret such a section as defining a deformation of the ring action on the module for $\mathcal{E}$, yielding a more general module, meaning a more general sheaf.

A trivial example of this is as follows. Start with a skyscraper sheaf at the origin of the complex line, corresponding to a single D-brane at the origin. The module corresponding to that sheaf is $\mathbf{C}[x]/(x)$. Now, consider the Higgs vev a ($\mathcal{E}^\vee \otimes \mathcal{E} \otimes \mathcal{N}_{S/X} \cong \mathcal{O}$ here). Describe the original module as a generator α subject to the relation $x\cdot\alpha = 0$, then define the new module by $x\cdot\alpha = a\alpha$. That relation is the same as $(x - a)\cdot\alpha = 0$ and so the new module is $\mathbf{C}[x]/(x - a)$, which describes a D-brane shifted from the origin to point a. A Higgs field in such a simple case should translate the D-brane, so this result is exactly what one would expect.

Similarly, the sheaf $\mathbf{C}[x]/(x^2)$ (the structure sheaf of a non-reduced scheme) corresponds to a pair of D-branes over the origin of the complex line with Higgs field

$$\begin{bmatrix} 0 & 1 \\ 0 & 0 \end{bmatrix}$$

We believe this mathematics has physical meaning because massless states in theories with non-zero Higgs fields can be shown to match Ext groups between the sheaves obtained by the process above. Physically, a non-zero Higgs field deforms the BRST operator to the form

$$Q_{BRST} = \bar{\partial} + \Phi_1^i \theta_i - \Phi_2^i \theta_i$$

where Φ_1^i, Φ_2^i are Higgs fields on either side of the open string. A necessary condition for the topological field theory to be sensible is that the BRST operator square to zero, which imposes the constraints that the Higgs fields be holomorphic and that the different Higgs fields commute with one another (which ordinarily would be an F term condition in the target-space theory). When one computes massless states using the deformed BRST operator above, one gets Ext groups between the sheaves dictated by the dictionary above. See [21] for more information.

3.3 Derived Categories

There is more to derived categories than just, sheaves. For example, where does the structure of complexes come from? Not to mention, where does the renormalization group enter?

First, in addition to D-branes, there are also anti-D-branes. An anti-D-brane is specified by the same data as a D-brane, but dynamically a D-brane and an anti-D-brane will try to annihilate one another.

Furthermore, in addition to antibranes, there are also "tachyons" between branes and antibranes, represented by maps between the sheaves representing the branes and antibranes.

The dictionary between derived categories and physics can now be stated. Given a complex

$$\cdots \longrightarrow \mathcal{E}_0 \longrightarrow \mathcal{E}_1 \longrightarrow \mathcal{E}_2 \longrightarrow \cdots$$

we map it to a brane/antibrane system in which the $\mathcal{E}_i$ for i odd, say, define branes, the other sheaves define antibranes, and the maps are tachyons [15, 16].

The first problem with this dictionary is that we do not know how to associate branes to every possible sheaf: we can map branes to sheaves but not necessarily the reverse.

The solution to this problem is as follows. So long as we are on a smooth complex manifold, every equivalence class of objects has a representative in terms of a complex of locally free sheaves, i.e. a complex of bundles, and we do know how to associate branes to those.

So, for any given equivalence class of objects, we pick a physically realizable representative complex (at least one exists) and map it to branes/antibranes/tachyons.

The next problem is that such representatives are not unique and different representatives lead to different physics. For example, the sheaf 0, describing no branes or antibranes, is equivalent in a derived category to the complex

$$0 \longrightarrow \mathcal{E} \xrightarrow{=} \mathcal{E} \longrightarrow 0$$

which is described by an unstable set of equivalent branes and antibranes. However, although these two systems are physically distinct, we believe that after a long time they will evolve to the same configuration – the branes and antibranes will completely annihilate. Such time evolution corresponds to worldsheet boundary renormalization group flow.

Thus, the proposal is that any two brane/antibrane systems representing quasi-isomorphic complexes flow to the same physical theory under the renormalization group. In other words, the mathematics of derived categories is providing a classification of universality classes of open strings.

A proposal of this form cannot be checked explicitly – it is not technically possible to explicitly follow renormalization group flow. Thus, we must perform numerous indirect tests, to accumulate evidence to determine whether the proposal is correct.

One test we can perform is to calculate massless spectra in the nonconformal theory describing brane/antibrane/tachyon systems and check that, again, one gets Ext groups. Let us work through those details.

On the worldsheet, to describe tachyons, we add a term to the boundary, which has the effect of modifying the BRST operator, which becomes

$$Q_{BRST} = \overline{\partial} + \sum_i \phi_i^{\alpha\beta}$$

schematically. A necessary condition for the topological field theory to remain well-defined is that $Q_{BRST}^2 = 0$, which implies that [23]

1. $\overline{\partial}\phi^{\alpha\beta} = 0$, i.e. the maps are holomorphic
2. $\phi_i^{\alpha\beta}\phi_{i+1}^{\beta\gamma} = 0$, i.e. the composition of successive maps vanishes, the condition for a complex.

Furthermore, if $f. : C. \to D.$ is a chain homotopy between complexes, i.e. if

$$f = \phi_D s - s\phi_C$$

for $s_n : C_n \to D_{n-1}$, then $f = Q_{BRST}s$ and so is BRST exact. So, modding out BRST exact states will have the effect of modding out chain homotopies.

As an example, let us compute $\mathrm{Ext}_{\mathbf{C}}^n(\mathcal{O}_D, \mathcal{O})$ in this language, for D a divisor on the complex line $\mathbf{C}$.

$$0 \longrightarrow \mathcal{O}(-D) \xrightarrow{\phi} \mathcal{O} \longrightarrow \mathcal{O}_D \longrightarrow 0$$

Relevant boundary states are of the form

$$b_{0\overline{i}_1\cdots\overline{i}_n}^{\alpha\beta}\eta^{\overline{i}_1}\cdots\eta^{\overline{i}_n} \sim H^n\left(\mathcal{O}(-D)^\vee \otimes \mathcal{O}\right)$$
$$b_{1\overline{i}_1\cdots\overline{i}_n}^{\alpha\beta}\eta^{\overline{i}_1}\cdots\eta^{\overline{i}_n} \sim H^n\left(\mathcal{O}^\vee \otimes \mathcal{O}\right)$$

In this language, degree one states are of the form $b_0 + b_{1\overline{i}}\eta^{\overline{i}}$. The BRST closure conditions are

$$\overline{\partial}b_0 = -\phi(b_{1\overline{i}}d\overline{z}^{\overline{i}}$$
$$\overline{\partial}\left(b_{1\overline{i}}d\overline{z}^{\overline{i}}\right) = 0$$

and the state is BRST exact if

$$b_0 = \phi a$$
$$b_{1\bar{i}}d\bar{z}^{\bar{i}} = \bar{\partial}a$$

for some a. These conditions imply that

$$b_0 \bmod \operatorname{Im} \phi \in H^0\left(D, \mathcal{O}(-D)^{\vee}|_D \otimes \mathcal{O}|_D\right) = \operatorname{Ext}^1(\mathcal{O}_D, \mathcal{O}).$$

Conversely, given an element of

$$\operatorname{Ext}^1(\mathcal{O}_D, \mathcal{O}) = H^0\left(D, \mathcal{O}(-D)^{\vee}|_D \otimes \mathcal{O}_D\right)$$

we can define b_0 and b_1 using the long exact sequence

$$\cdots \longrightarrow H^0(\mathcal{O}) \longrightarrow H^0(\mathcal{O}(D)) \longrightarrow H^0(D, \mathcal{O}(D)|_D) \overset{\delta}{\longrightarrow} H^1(\mathcal{O}) \longrightarrow \cdots$$

from which we see b_1 is the image under δ and b_0 is the lift to an element of $C^{\infty}(\mathcal{O}(D))$.

More generally, it can be shown that Ext groups can be obtained in this fashion.

Thus, massless spectra can be counted in the non-conformal theory and they match massless spectra of the corresponding conformal theory: both are counted by Ext groups. This gives us a nice test of presentation-independence of renormalization group flow, of the claim that localization on quasi-isomorphisms is realized by the renormalization group.

3.4 Grading

Let us next take a few minutes to describe how the grading appears physically in terms of $U(1)_R$ charges.

For branes and antibranes wrapped on the entire space, the analysis is straightforward. The tachyon T is a degree zero operator. The term we add to the boundary to describe a tachyon is the descendant

$$\int_{\partial \Sigma} [G, T]$$

where G is the topologically twisted boundary supercharge. The operator G has $U(1)_R$ charge -1, so $[G, T]$ has charge -1. Now, a necessary condition to preserve supersymmetry is that boundary terms must be neutral under $U(1)_R$ (otherwise the $U(1)_R$ is broken, which breaks the $\mathcal{N} = 2$ boundary supersymmetry). Thus, the Noether charge associated to the $U(1)_R$ symmetry must have boundary conditions on either side of the boundary-condition-changing operator above such that the grading shifts by one.

For lower-dimensional sheaves, on the other hand, the relationship between the $U(1)_R$ charge and the grading is more subtle. In particular, the state corresponding to an element of $\operatorname{Ext}^n(\mathcal{S}, \mathcal{T})$ for two sheaves $\mathcal{S}, \mathcal{T}$ need not have

$U(1)_R$ charge equal to n – the degree of the Ext group will not match the charge of the state. If we build the states as combinations of fields acting on a vacuum, then the $U(1)_R$ charge of the field combinations will be the same as the degree of the Ext group, but the vacuum will make an additional contribution to the $U(1)_R$ charge which will spoil the relationship. In particular, if the two sheaves do not correspond to mutually supersymmetric branes, then the charge of the vacuum need not even be integral. This mismatch is very unlike closed strings, where typically the vacuum charge contribution precisely insures that the total $U(1)_R$ charge *does* match the degree of corresponding cohomology. This mismatch was known at the time of [15] and has been verified more thoroughly since (see e.g. [4]), though it is often misstated in the literature.

3.5 Generalized Complexes

Another question the reader might ask is, why should maps between branes and antibranes unravel into a linear complex as opposed to a more general set of maps? For example, why can one not have a configuration that unravels to something of the form

$$\mathcal{E}_0 \longrightarrow \mathcal{E}_1 \rightleftharpoons \mathcal{E}_2 \longrightarrow \mathcal{E}_3 \longrightarrow \mathcal{E}_4 \longrightarrow \mathcal{E}_5$$

In fact, such configurations are allowed physically and also play an important mathematical role.

Physically, if we add a boundary operator $\mathcal{O}$ of $U(1)_R$ charge n, then $[G, \mathcal{O}]$ has charge $n - 1$, so the boundaries it lies between must have relative $U(1)_R$ charge $1 - n$ and so give rise to the 'wrong-way' maps displayed above.

Adding such operators deforms the BRST operator

$$Q_{BRST} = \overline{\partial} + \sum_i \phi_i^{\alpha\beta}$$

and demanding that $Q_{BRST}^2 = 0$ now merely implies

$$\sum_i \overline{\partial}\phi_i + \sum_{i,j} \phi_i \circ \phi_j = 0.$$

Complexes with wrong-way arrows of the form above, such that the maps obey the condition stated above, are examples of "generalized complexes" used in [24] to define a technical improvement of ordinary derived categories. The relevance of [24] to physics was first described in [25, 26, 27].

3.6 Cardy Condition and Hirzebruch–Riemann–Roch

Other aspects of the physics of the open string B model have also been shown to have a mathematical understanding. For example, the Cardy condition, which says that interpreting the annulus diagram in terms of either closed or open string propagation, has been shown by A. Caldararu to be the same mathematically as the Hirzebruch–Riemann–Roch index theorem:

$$\int_M \text{ch}(\mathcal{E})^* \wedge \text{ch}(\mathcal{F}) \wedge \text{td}(TM) = \sum_i (-)^i \dim \text{Ext}_M^i(\mathcal{E}, \mathcal{F})$$

3.7 Open Problems

Lest the reader get the impression that the connection between derived categories and physics is well-understood, there are still some very basic open problems that have never been solved. For example:

1. One of the most basic problems is that we have glossed over technical issues in dealing with bundles of rank greater than one. For such bundles, it is not completely understood how their curvature modifies the boundary conditions on the open strings. Such boundary conditions will modify the arguments we have given for tachyon vevs and although we are optimistic that the modification will not make essential changes to the argument, no one knows for certain.

2. Anomaly cancellation in the open string B model implies that one can only have open strings between some D-branes and not others. At the moment, we can only describe the condition in special circumstances, as the (unknown) boundary conditions above play a crucial role. On the one hand, this condition seems to violate the spirit of the arguments we have presented so far and we would hope that some additional physical effect (anomaly inflow, perhaps) modifies the conclusion. On the other hand, this anomaly cancellation condition plays a role in understanding how Ext groups arise.

3. We understand how to associate D-branes to some sheaves, but not to other sheaves. A more comprehensive dictionary would be useful.

Although we now have most of the puzzle pieces, so to speak, a complete comprehensive physical understanding still does not exist.

4 Stacks in Physics

4.1 Introduction

So far we have discussed the physical understanding of derived categories as one application of the renormalization group: the mathematical process of localization on quasi-isomorphisms is realized physically by worldsheet boundary renormalization group flow.

Another application is to the physical understanding of stacks,[2] where the renormalization group will play an analogous role in washing out potential presentation-dependence.

Although in the mathematics literature the words "stack" and "orbifold" are sometimes used interchangeably, in the physics community the term "orbifold" has a much more restrictive meaning: global quotients by finite effectively acting groups. Most stacks cannot be understood as global quotients by finite effectively acting groups.

Understanding physical properties of string orbifolds has led physicists through many ideas regarding orbifolds (in the sense used by physicists, global quotients by finite effectively acting groups). For example, for a time, many thought that the properties of orbifolds were somehow intrinsic to string theory or CFT. Later, the fact that string orbifolds are well-behaved CFT's unlike sigma models on quotient spaces was attributed to a notion of "B fields at quotient singularities" [28], a notion that only made sense in special cases. Later still, because of the properties of D-branes in orbifolds [29], many claimed that string orbifolds were the same as strings propagating on resolutions of quotient spaces, a notion that can not make sense for terminal singularities such as $\mathbf{C}^4/\mathbf{Z}_2$ or in some nonsupersymmetric orbifolds such as $[\mathbf{C}/\mathbf{Z}_k]$.

It was proposed in [30] that a better way of understanding the physical properties of string orbifolds lie just in thinking of terms of the geometry of stacks. For readers who are not well acquainted with the notion, the quotient stack $[X/G]$ encodes properties of orbifolds. For example, a function on the stack $[X/G]$ is a G-invariant function on X, a metric on the stack $[X/G]$ is a G-invariant metric on X, a bundle on the stack $[X/G]$ is a G-equivariant bundle on X and so forth. Roughly, the stack $[X/G]$ is the same as the space X/G except over the singularities of X/G, where the stack has additional structure that results in it being smooth while X/G is singular. For G finite, a map $Y \to [X/G]$ is a pair consisting of a principal G bundle P over Y together with a G-equivariant map $P \to X$. The reader should recognize this as a twisted sector (defined by P) together with a twisted map in that sector. In other words, summing over maps into the stack $[X/G]$, for G finite, is the same as summing over the data in the path integral description of a string orbifolds.

Part of the proposal of [30] was that the well-behavedness of string orbifold CFT's, as opposed to sigma models on quotient spaces, could be understood geometrically as stemming from the smoothness of the corresponding stack. In particular, the stack $[X/G]$ is smooth even when the space X/G is singular because of fixed-points of G. Another part of the proposal of [30] was a conjecture for how the "B fields at quotient singularities" could be understood mathematically in terms of stacks, though that particular conjecture has since been contradicted.

---

[2] In this contribution by "stack" we mean Deligne–Mumford stacks and their smooth analogues.

However, at the time of [30], there were many unanswered questions. First and foremost was a mismatch of moduli. One of the first tests of any proposed mathematical model of part of physics is whether the physical moduli match mathematical moduli – this was one of the reasons described for why sheaves are believed to be a good model of D-branes, for example. Unfortunately, physical moduli of string orbifolds are not the same as mathematical moduli of stacks. For example, the stack $[\mathbf{C}^2/\mathbf{Z}_2]$ is rigid, it has no moduli, whereas physically the $\mathbf{Z}_2$ string orbifold of $\mathbf{C}^2$ does have moduli, which are understood as deformations and resolutions of the quotient space $\mathbf{C}^2/\mathbf{Z}_2$. If string orbifolds can be understood in terms of stacks, then it will be the first known case in which mathematical moduli do not match physical moduli.

Understanding this moduli mismatch was the source of much work over the next few years.

Another problem at the time of [30] was the issue of understanding how to describe strings on more general stacks. Most stacks cannot be understood as, global quotients by finite effectively acting groups and so cannot be understood in terms of orbifolds in the sense used by physicists. One of the motivations for thinking about stacks was to introduce a potentially new class of string compactifications, so understanding strings on more general stacks was of interest. At a more basic level, if the notion of strings on more general stacks did not make sense, then that would call into question whether string orbifolds can be understood meaningfully in terms of (special) stacks.

4.2 Strings on More General Stacks

How to understand strings on more general stacks? Although most stacks can not be understood as global orbifolds by finite groups, locally in patches they look like quotients by finite not-necessarily effectively acting groups. Unfortunately, that does not help us physically: only[3] global quotients are known to define CFTs and only effectively acting quotients are well-understood.

So, we are back to the question of describing strings on more general stacks. It is a mathematical result that most[4] stacks can be presented as a global quotient $[X/G]$ for some group, not necessarily finite and not necessarily effectively acting. To such a presentation, we can associate a G-gauged sigma model.

---

[3] Very recently some progress has been made understanding CFT's perturbatively in terms of local data on the target space [31], but the work described there is only perturbative, not non-perturbative and does not suffice to define a CFT.

[4] Experts are referred to an earlier footnote where we define our usage of 'stack'. For the exceptions to the rule described above, i.e. for those few stacks not presentable in the form $[X/G]$ for some not necessarily finite, not necessarily effectively acting G, it is not currently known whether they define a CFT. Even if it is possible to associate a CFT to them, it is certainly not known how one would associate a CFT to them.

So, given a presentation of the correct form, we can get a physical theory. Unfortunately, such presentations are not unique and different presentations lead to very different physics. For example, following the dictionary above, $[\mathbf{C}^2/\mathbf{Z}_2]$ defines a conformally invariant two-dimensional theory. However, that stack can also be presented as $[X/\mathbf{C}^\times]$ where $X = (\mathbf{C}^2 \times \mathbf{C}^\times)/\mathbf{Z}_2$ and to that presentation one associates a non-conformally invariant two-dimensional theory, a $U(1)$-gauged sigma model. As stacks, they are the same,

$$[\mathbf{C}^2/\mathbf{Z}_2] = [X/\mathbf{C}^\times]$$

but the corresponding physics is very different.

Thus, we have a potential presentation-dependence problem. These problems are, again, analogous to those in understanding the appearance of derived categories in physics. There, to a given object in a derived category, one picks a representative with a physical description (as branes/antibranes/tachyons), just as here, given a stack, we must first pick a physically realizable presentation. Every equivalence class of objects has at least one physically realizable representation; unfortunately, such representatives are not unique. It is conjectured that different representatives give rise to the same low-energy physics, via boundary renormalization group flow, but only indirect tests are possible.

Here also, we conjecture that worldsheet renormalization group flow takes different presentations of the same stack to the same CFT. Unfortunately, just as in the case of derived categories, worldsheet renormalization group flow cannot be followed explicitly and so we cannot explicitly check such a claim. Instead, we must rely on indirect tests.

At the least, one would like to check that the spectrum of massless states is presentation-independent. After all, in the case of derived categories, this was one of the important checks we outlined that the renormalization group respected localization on quasi-isomorphisms: spectra computed in non-conformal presentations matched spectra computed in conformal presentations believed to be the endpoint of renormalization group flow.

Unfortunately, no such test is possible here. Massless spectra are only explicitly computable for global quotients by finite groups and have only been well-understood in the past for global quotients by finite effectively acting groups. For global quotients by non-finite groups, there has been a longstanding unsolved technical question of how to compute massless spectra. Since the theory is not conformal, the spectrum cannot be computed in the usual fashion by enumerating vertex operators and although there exist conjectures for how to find massive representations of some such states in special cases (gauged *linear* sigma models), not even in such special cases does anyone have conjectures for massive representations of all states, much less a systematic computation method. The last (unsuccessful) attempt appearing in print was in [32] and there has been little progress since then.

Thus, there is no way to tell if massless spectra are the same across presentations. On the other hand, a presentation-independent ansatz for massless spectra makes predictions for massless spectra in situations where they are not explicitly computable.

As alluded to earlier, another of the first indirect tests of the presentation-independence claim is whether deformations of stacks match deformations of corresponding CFT's. In every other known example of geometry applied to physics, mathematical deformations match physical deformations. Unfortunately, stacks fail this test, which one might worry might be a signal of presentation-dependence. Maybe renormalization group flow does not respect stacky equivalences of gauged sigma models; maybe some different mathematics is relevant instead of stacks.

To justify that stacks are relevant physically, as opposed to some other mathematics, one has to understand this deformation theory issue, as well as conduct tests for presentation-dependence. This was the subject of several papers [33, 34, 35].

In the rest of these notes, we shall focus on special kinds of stacks known as gerbes (which are described physically by quotients by non-effectively acting groups).

4.3 Strings on Gerbes

Strings on gerbes, i.e. global quotients by non-effectively acting groups, have additional physical difficulties beyond those mentioned in the last subsection. For example, the naive massless spectrum calculation contains multiple dimension zero operators, which manifestly violates cluster decomposition, one of the foundational axioms of quantum field theory.

There is a single known loophole: if the target space is disconnected, in which case cluster decomposition is also violated, but in the mildest possible fashion. We believe that is more or less what is going on.

Consider $[X/H]$ where

$$1 \longrightarrow G \longrightarrow H \longrightarrow K \longrightarrow 1$$

G acts trivially, K acts effectively and neither H nor K need to be finite.

We claim [36]

$$\mathrm{CFT}([X/H]) = \mathrm{CFT}([(X \times \hat{G})/K])$$

(together with some B field), where $\hat{G}$ is the set of irreducible representations of G. We refer to this as our "decomposition conjecture". The stack $[(X \times \hat{G})/K]$ is not connected and so the CFT violates cluster decomposition but in the mildest possible fashion.

For banded gerbes, K acts trivially upon $\hat{G}$, so the decomposition conjecture reduces to

$$\mathrm{CFT}\,(G - \mathrm{gerbe\ on}\ X) = \mathrm{CFT}\left(\coprod_{\hat{G}}(X, B)\right)$$

where the B field on each component is determined by the image of the characteristic class of the gerbe under

$$H^2(X, Z(G)) \overset{Z(G) \to U(1)}{\longrightarrow} H^2(X, U(1)).$$

For our first example, consider $[X/D_4]$, where the $\mathbf{Z}_2$ centre of the dihedral group D_4 acts trivially:

$$1 \longrightarrow \mathbf{Z}_2 \longrightarrow D_4 \longrightarrow \mathbf{Z}_2 \times \mathbf{Z}_2 \longrightarrow 1$$

This example is banded and the decomposition conjecture above predicts

$$\mathrm{CFT}([X/D_4]) = \mathrm{CFT}\left([X/\mathbf{Z}_2 \times \mathbf{Z}_2] \coprod [X/\mathbf{Z}_2 \times \mathbf{Z}_2]\right)$$

One of the effective $\mathbf{Z}_2 \times \mathbf{Z}_2$ orbifolds has vanishing discrete torsion, the other has nonvanishing discrete torsion, using the relationship between discrete torsion and B fields first described in [37, 38].

One easy check of that statement lies in computing genus one partition functions. Denote the elements of the group D_4 by

$$D_4 = \{1, z, a, b, az, bz, ab, ba = abz\}$$

and the elements of $\mathbf{Z}_2 \times \mathbf{Z}_2$ by

$$\mathbf{Z}_2 \times \mathbf{Z}_2 = \{1, \overline{a}, \overline{b}, \overline{ab}\}$$

and the map $D_4 \to \mathbf{Z}_2 \times \mathbf{Z}_2$ proceeds as, for example, $a, az \mapsto \overline{a}$. The genus one partition function of the noneffective D_4 orbifold can be described as

$$Z(D_4) = \frac{1}{|D_4|} \sum_{g,h \in D_4, gh=hg} Z_{g,h}$$

Each of the $Z_{g,h}$ twisted sectors that appears is the same as a $\mathbf{Z}_2 \times \mathbf{Z}_2$ sector, appearing with multiplicity $|\mathbf{Z}_2|^2 = 4$ except for the

$$\overline{a}\,\boxed{}_{\overline{b}}, \quad \overline{a}\,\boxed{}_{\overline{ab}}, \quad \overline{b}\,\boxed{}_{\overline{ab}}$$

sectors, which have no lifts to the D_4 orbifold. The partition function can be expressed as

$$Z(D_4) = \frac{|\mathbf{Z}_2 \times \mathbf{Z}_2|}{|D_4|} |\mathbf{Z}_2|^2 \left(Z(\mathbf{Z}_2 \times \mathbf{Z}_2) - (\text{some twisted sectors}) \right)$$

$$= 2 \left(Z(\mathbf{Z}_2 \times \mathbf{Z}_2) - (\text{some twisted sectors}) \right)$$

The factor of 2 is important – in ordinary QFT, one ignores multiplicative factors in partition functions, but string theory is a two-dimensional QFT coupled to gravity and so such numerical factors are important.

Discrete torsion acts as a sign on the

$$\bar{a}\,\underset{\bar{b}}{\square}\,,\quad \bar{a}\,\underset{\overline{ab}}{\square}\,,\quad \bar{b}\,\underset{\overline{ab}}{\square}$$

$\mathbf{Z}_2 \times \mathbf{Z}_2$ twisted sectors, so adding partition functions with and without discrete torsion will have the effect of removing the sectors above and multiplying the rest by a factor of two. Thus, we see that

$$Z([X/D_4]) \;=\; Z\left([X/\mathbf{Z}_2 \times \mathbf{Z}_2]\coprod[X/\mathbf{Z}_2 \times \mathbf{Z}_2]\right)$$

with discrete torsion in one component. (The same computation is performed at arbitrary genus in [36].)

Another quick tests of this example comes from comparing massless spectra. Using the Hodge decomposition, the massless spectrum for $[T^6/D_4]$ can be expressed as

$$
\begin{array}{ccccccc}
 & & & 2 & & & \\
 & & 0 & & 0 & & \\
 & 0 & & 54 & & 0 & \\
2 & & 54 & & 54 & & 2 \\
 & 0 & & 54 & & 0 & \\
 & & 0 & & 0 & & \\
 & & & 2 & & & \\
\end{array}
$$

and the massless spectrum for each $[T^6/\mathbf{Z}_2 \times \mathbf{Z}_2]$, with and without discrete torsion, can be written

$$
\begin{array}{ccccccc}
 & & & 1 & & & \\
 & & 0 & & 0 & & \\
 & 0 & & 3 & & 0 & \\
1 & & 51 & & 51 & & 1 \\
 & 0 & & 3 & & 0 & \\
 & & 0 & & 0 & & \\
 & & & 1 & & & \\
\end{array}
$$

and

$$
\begin{array}{ccccccc}
 & & & 1 & & & \\
 & & 0 & & 0 & & \\
 & 0 & & 51 & & 0 & \\
1 & & 3 & & 3 & & 1 \\
 & 0 & & 51 & & 0 & \\
 & & 0 & & 0 & & \\
 & & & 1 & & & \\
\end{array}
$$

The sum of the states from the two $[T^6/\mathbf{Z}_2 \times \mathbf{Z}_2]$ factors matches that of $[T^6/D_4]$, precisely as expected.

Another example of the decomposition conjecture is given by $[X/\mathbf{H}]$, where $\mathbf{H}$ is the eight-element group of quaternions and a $\mathbf{Z}_4$ acts trivially:

$$1 \longrightarrow <i> (\cong \mathbf{Z}_4) \longrightarrow \mathbf{H} \longrightarrow \mathbf{Z}_2 \longrightarrow 1$$

The decomposition conjecture predicts

$$\text{CFT}([X/\mathbf{H}]) = \text{CFT}\left([X/\mathbf{Z}_2] \coprod [X/\mathbf{Z}_2] \coprod X\right).$$

It is straightforward to show that this statement is true at the level of partition functions, as before.

Another class of examples involves global quotients by non-effectively-acting non-finite groups. For example, the banded $\mathbf{Z}_k$ gerbe over $\mathbf{P}^{N-1}$ with characteristic class $-1 \mod k$ can be described mathematically as the quotient

$$\left[\frac{\mathbf{C}^N - \{0\}}{\mathbf{C}^\times}\right]$$

where the $\mathbf{C}^\times$ acts as rotations by k times rather than once. Physically this quotient can be described by a $U(1)$ supersymmetric gauge theory with N chiral fields all of charge k, rather than charge 1. The only difference between this and the ordinary supersymmetric $\mathbf{P}^{N-1}$ model is that the charges are non-minimal.

Now, how can this be physically distinct from the ordinary supersymmetric $\mathbf{P}^{N-1}$ model? After all, perturbatively having non-minimal charges makes no difference. The difference lies in non-perturbative effects. For example, consider the anomalous global $U(1)$ symmetries of these models. In the ordinary supersymmetric $\mathbf{P}^{N-1}$ model, the axial $U(1)$ is broken to $\mathbf{Z}_{2N}$ by anomalies, whereas here it is broken to $\mathbf{Z}_{2kN}$. The non-vanishing A model correlation functions of the ordinary supersymmetric $\mathbf{P}^{N-1}$ model are given by

$$< X^{N(d+1)-1} > = q^d$$

whereas here the non-zero A model correlation functions are given by

$$< X^{N(kd+1)-1} > = q^d$$

As a result, the quantum cohomology ring of the ordinary $\mathbf{P}^{N-1}$ model is given by

$$\mathbf{C}[x]/(x^N - q)$$

whereas the quantum cohomology ring of the current model is given by

$$\mathbf{C}[x]/(x^{kN} - q)$$

In short, having non-minimal charges does lead to different physics.

Why should having non-minimal charges make a difference non-perturbatively? On a compact worldsheet, this can be understood as follows. To specify a Higgs field completely on a compact space, we need to specify what bundle they couple to. Thus, if the gauge field couples to $\mathcal{L}$ then saying a Higgs field Φ has charge Q implies $\Phi \in \Gamma(\mathcal{L}^{\otimes Q})$. Different bundles implies fields have different zero modes, which implies different anomalies, which implies different physics.

On a non-compact worldsheet, the argument is different [39]. If electrons have charge k, then instantons have charge $1/k$ and the theory reduces to the minimal-charge case. Suppose we add massive fields of charge ± 1, of mass greater than the energy scale at which we are working. One can determine instanton numbers by periodicity of the theta angle, which acts like an electric field in two dimensions. If all fields have charge k, then the theta angle has periodicity $2\pi k$ and we reduce to the ordinary case. However, the existence of massive fields of unit charge means the theta angle has periodicity 2π, which is the new case. Thus, even on a non-compact worldsheet, having non-minimal charges can be distinguished from minimal charges.

There are four-dimensional analogues of this distinction. For example, $SU(n)$ and $SU(n)/\mathbf{Z}_n$ gauge theories are perturbatively equivalent (since their Lie algebras are identical), but have distinct nonperturbative corrections, a fact that is crucial to the analysis of [40]. Similarly, Spin(n) and $SO(n)$ gauge theories are perturbatively identical but non-perturbatively distinct. M. Strassler has studied Seiberg duality in this context [41] and has examples of Spin(n) gauge theories with $\mathbf{Z}_2$ monopoles (distinguishing Spin(n) from $SO(n)$ non-perturbatively) Seiberg dual to Spin(n) gauge theory with massive spinors (distinguishing Spin(n) from $SO(n)$ perturbatively).

The equivalence of CFT's implied by our decomposition conjecture implies a statement about K theory, thanks to D-branes. Suppose H acts on X with a trivially acting subgroup G:

$$1 \longrightarrow G \longrightarrow H \longrightarrow K \longrightarrow 1$$

Our decomposition conjecture predicts that the ordinary H-equivariant K theory of X is the same as the twisted K-equivariant K theory of $X \times \hat{G}$. This result can be derived just within K theory (see [36]) and provides a check of the decomposition conjecture.

Another check of the decomposition conjecture comes from derived categories. Our decomposition conjecture predicts that D-branes on a gerbe should be the same as D-branes on a disjoint union of spaces, together with flat B fields, and this corresponds to a known mathematics result. Specifically, a sheaf on a gerbe is the same as a twisted sheaf on $[X \times \hat{G}/K]$. A sheaf on a banded G-gerbe is the same thing as a twisted sheaf on the underlying space, twisted by the image of the characteristic class of the gerbe in $H^2(X, Z(G))$. Thus, sheaves on gerbes behave in exactly the fashion one would expect from D-branes according to our decomposition conjecture.

Similarly, massless states between D-branes also have an analogous decomposition. For D-branes on a disjoint union of spaces, there will only be massless states between D-branes which are both on the same connected component. Mathematically, in the banded case for example, sheaves on a banded G-gerbe decompose according to irreducible representations of G and sheaves associated with distinct irreducible representations have vanishing Ext groups between them. This is precisely consistent with the idea that sheaves associated with distinct irreducible representations should describe D-branes on different components of a disconnected space.

4.4 Mirror Symmetry for Stacks

There exist mirror constructions for any model realizable as a two-dimensional abelian gauge theory [42, 43]. There is a notion of toric stacks [44], generalizing toric varieties, which can be described physically via gauged linear sigma models [35]. Standard mirror constructions [42, 43] now produce [35] character-valued fields, a new effect, which ties into the stacky fan description of [44].

For example, the "Toda dual" of the supersymmetric $\mathbf{P}^N$ model is described by Landau–Ginzburg model with superpotential

$$W = \exp(-Y_1) + \cdots + \exp(-Y_N) + \exp(Y_1 + \cdots + Y_N)$$

The analogous duals to $\mathbf{Z}_k$ gerbes over $\mathbf{P}^N$, of characteristic class $-n$ mod k, are given by [35]

$$W = \exp(-Y_1) + \cdots + \exp(-Y_N) + \Upsilon^n \exp(Y_1 + \cdots + Y_N)$$

where Υ is a character-valued field, in this case valued in the characters of $\mathbf{Z}_k$.

In the same language, the Landau-Ginzburg point mirror to the quintic hypersurface in a $\mathbf{Z}_k$ gerbe over $\mathbf{P}^4$ is described by (an orbifold of) the superpotential

$$W = x_0^5 + \cdots + x_4^5 + \psi \Upsilon x_0 x_1 x_2 x_3 x_4$$

where ψ is the ordinary complex structure parameter (mirror to the Kähler parameter) and Υ is a discrete (character-)valued field as above.

In terms of the path integral measure,

$$\int [Dx_i, \Upsilon] = \int [Dx_i] \sum_\Upsilon = \sum_\Upsilon \int [Dx_i]$$

so having a discrete-valued field is equivalent to summing over contributions from different theories, or, equivalently, summing over different components of the target space.

In the case of the gerby quintic, the presence of the discrete-valued field Υ on the mirror means that the CFT is describing a target space with multiple components. Moreover, the mirror map for the ordinary quintic says

$$B + iJ = -\frac{5}{2\pi i} \log(5\psi) + \cdots$$

so shifting ψ by phases has precisely[5] the effect of shifting the B field, exactly as the decomposition conjecture predicts for this case.

---

[5] Higher order corrections invalidate geometric conclusions, so we are omitting them.

4.5 Applications

One of the original proposed applications of these ideas, described in [30], was to understand physical properties of string orbifolds. For example, the fact that string orbifolds define well-behaved CFTs, unlike sigma models on quotient spaces, might be attributable to the smoothness of the corresponding quotient stack, instead of traditional notions such as "B fields at quotient singularities" or "string orbifolds are strings on resolutions", which do not even make sense in general.

Another basic application is to give a concrete understanding of local orbifolds, i.e. stacks which are locally quotients by finite groups but which cannot be expressed globally as quotients by finite groups. We now have a concrete way to manufacture a corresponding CFT – by rewriting the local orbifold as a global quotient by a non-finite group – and we also understand what problems may arise – the construction might not be well-defined, as different rewritings might conceivably flow under the renormalization group to distinct CFTs.

Implicit here is that stacks give a classification of universality classes of worldsheet renormalization group flow in gauged sigma models, just as derived categories give a classification of universality classes of worldsheet boundary renormalization group flow in the open string B model.

Another application is the computation of massless spectra in cases where direct calculations are not currently possible, such as global quotients by nonfinite groups. A presentation-independent ansatz was described in [30, 33, 34, 35] which predicts massless spectra in cases where explicitly enumerating vertex operators is not possible.

Another application of these ideas is to the properties of quotients by noneffective group actions, i.e. group actions in which elements other than the identity act trivially. Such quotients correspond to strings propagating on special kinds of stacks known as gerbes.

Non-effective group actions play a crucial role in [40], the recent work on the physical interpretation of the geometric Langlands program. One application of the decomposition conjecture of the last section is to give a concrete understanding of some aspects of [40]. For example, related work [45] describes two-dimensional theories in the language of gerbes, whereas [40] deals exclusively with spaces. As a result of the decomposition conjecture, we see that the language of [45] is physically equivalent to that of [40], as sigma models on the gerbes of [45] define the same CFTs as sigma models on the disjoint unions of spaces of [40].

Our decomposition conjecture makes a prediction for Gromov–Witten invariants of gerbes, as defined in the mathematics literature in, for example, [46]. Specifically, the Gromov–Witten theory of $[X/H]$ should match that of $[(X \times \hat{G})/K]$. This prediction works in basic cases [47].

Another result of our work is quantum cohomology for toric stacks. Toric stacks are a stacky generalization of toric varieties [44]. Just as toric varieties can be described with gauged linear sigma models, so too can toric stacks [35], and so the technology of gauged linear sigma models can be applied to their understanding. In particular, Batyrev's conjecture for quantum cohomology rings can be extracted from the two-dimensional effective action of the gauge theory, without any explicit mention of rational curves [48].

In the present case, old results of [48] generalize from toric varieties to toric stacks. Let the toric stack be described in the form

$$\left[\frac{\mathbf{C}^N - E}{(\mathbf{C}^\times)^n}\right]$$

where E is some exceptional set, Q_i^a the weight of the ith vector under the ath $\mathbf{C}^\times$, then the analogue of Batyrev's conjecture for the quantum cohomology ring is of the form $\mathbf{C}[\sigma_1, \cdots, \sigma_n]$ modulo the relations [35]

$$\prod_{i=1}^{N} \left(\sum_{b=1}^{n} Q_i^b \sigma_b\right)^{Q_i^a} = q_a.$$

For example, the quantum cohomology ring of $\mathbf{P}^N$ is

$$\mathbf{C}[x]/(x^{N+1} - q)$$

and according to the formula above the quantum cohomology ring of a $\mathbf{Z}_k$ gerbe over $\mathbf{P}^N$ with characteristic class $-n$ mod k is

$$\mathbf{C}[x, y]/(y^k - q_2, \ x^{N+1} - y^n q_1).$$

As an aside, note the calculations above give us a check of the massless spectrum. In physics, we can derive quantum cohomology rings without knowing the massless spectrum and we are unable to calculate the massless spectrum directly for the gerbes above, hence we can use the quantum cohomology rings to read off the additive part of the massless spectrum.

Also note that we can see the decomposition conjecture for gerbes in the quantum cohomology rings of toric stacks. Consider for example, the quantum cohomology ring of a $\mathbf{Z}_k$ gerbe on $\mathbf{P}^N$, as above. In that ring, the y's index copies of the quantum cohomology ring of $\mathbf{P}^N$ with variable q's. The gerbe is banded, so this is exactly what we expect – copies of $\mathbf{P}^N$, variable B field.

More generally, a gerbe structure is indicated from the quotient description whenever the $\mathbf{C}^\times$ charges are non-minimal. In such a case, from our generalization of Batyrev's conjecture, at least one relation will have the form $p^k = q$, where p is a relation in the quantun cohomology ring of the toric variety and k is the greatest divisor in the non-minimal charges. We can rewrite that relation in the same form as for a gerbe on $\mathbf{P}^N$ and in this fashion can see our decomposition conjecture in our generalization of Batyrev's quantum cohomology.

Other applications of stacks to understanding D-branes and their derived categories model are discussed in [49, 50, 51].

5 Conclusions

In this contribution we have outlined how both derived categories and stacks enter physics, and the crucial role of the renormalization group. In both the cases, to physically realize either a derived category or stack, one picks physically-realizable presentations (which are guaranteed to exist, though not all presentations are physically realizable), which yield non-conformal theories. Remaining presentation-dependence is removed via renormalization group flow.

Acknowledgements

I have learned a great deal of both mathematics and physics from my collaborators on the papers we have written concerning derived categories and stacks. Listed alphabetically, they are M. Ando, A. Caldararu, R. Donagi, S. Hellerman, A. Henriques, S. Katz and T. Pantev.

References

1. C. Weibel, *An Introduction to Homological Algebra*, Cambridge studies in advanced mathematics 38, Cambridge University Press, 1994.
2. R. Hartshorne, *Residues and Duality*, Lecture Notes in Mathematics 20, Springer-Verlag, Berlin, 1966.
3. R. Thomas, "Derived categories for the working mathematician," `math.AG/0001045`.
4. E. Sharpe, "Lectures on D-branes and sheaves," lectures given at the twelfth Oporto meeting on "Geometry, topology, and physics," `hep-th/0307245`.
5. A. Vistoli, "Intersection theory on algebraic stacks and on their moduli spaces," Inv. Math. **97** (1989) 613–670.
6. T. Gomez, "Algebraic stacks," Proc. Indian Acad. Sci. Math. Sci. **111** (2001) 1–31, `math.AG/9911199`.
7. M. Kontsevich, "Homological algebra of mirror symmetry," in *Proceedings of the International Congress of Mathematicians*, pp. 120–139, Birkhäuser, (1995), `alg-geom/9411018`.
8. J. Polchinski, "Dirichlet branes and Ramond-Ramond charges," Phys. Rev. Lett. **75** (1995) 4724–4727, `hep-th/9510017`.
9. J. Harvey and G. Moore, "On the algebras of BPS states," Comm. Math. Phys. **197** (1998) 489–519, `hep-th/9609017`.
10. S. Katz and E. Sharpe, "D-branes, open string vertex operators, and Ext groups," Adv. Theor. Math. Phys. 6 (2003) 979–1030, `hep-th/0208104`.
11. S. Katz, T. Pantev, and E. Sharpe, "D-branes, orbifolds, and Ext groups," Nucl. Phys. **B673** (2003) 263–300, `hep-th/0212218`.
12. A. Caldararu, S. Katz, and E. Sharpe, "D-branes, B fields, and Ext groups," Adv. Theor. Math. Phys. **7** (2004) 381–404, `hep-th/0302099`.

13. A. Sen, "Tachyon condensation on the brane-antibrane system," J. High Energy Phys. **08** (1998) 012, `hep-th/9805170`.
14. E. Witten, "D-branes and K theory," J. High Energy Phys. **9812** (1998) 019, `hep-th/9810188`.
15. E. Sharpe, "D-branes, derived categories, and Grothendieck groups," Nucl. Phys. **B561** (1999) 433–450, `hep-th/9902116`.
16. M. Douglas, "D-branes, categories, and $\mathcal{N} = 1$ supersymmetry," J. Math. Phys. **42** (2001) 2818–2843, `hep-th/0011017`.
17. T. Bridgeland, "Stability conditions on triangulated categories," `math`. Annals of Math. 166 (2007) 317–345.
18. D. Freed and E. Witten, "Anomalies in string theory with D-branes," `hep-th/9907189`.
19. A. Abouelsaood, C. Callan, C. Nappi, and S. Yost, "Open strings in background gauge fields," Nucl. Phys. **B280** (1987) 599–624.
20. T. Gomez and E. Sharpe, "D-branes and scheme theory," `hep-th/0008150`.
21. R. Donagi, S. Katz, and E. Sharpe, "Spectra of D-branes with Higgs vevs," Adv. Theor. Math. Phys. **8** (2005) 813–259, `hep-th/0309270`.
22. R. Donagi, L. Ein, and R. Lazarsfeld, "A non-linear deformation of the Hitchin dynamical system," `alg-geom/9504017`, a.k.a. "Nilpotent cones and sheaves on K3 surfaces," pp. 51–61 in *Birational Algebraic Geometry*, Contemp. Math. 207, Amer. Math. Soc, Providence, Rhode Island, (1997).
23. P. Aspinwall and A. Lawrence, "Derived categories and zero-brane stability," J. High Energy Phys. **0108** (2001) 004, `hep-th/0104147`.
24. A. Bondal and M. Kapranov, "Enhanced triangulated categories," Math. USSR Sbornik **70** (1991) 93–107.
25. C. Lazaroiu, "Generalized complexes and string field theory," JHEP 0106 (2001) 052, `hep-th/0102122`.
26. C. Lazaroiu, "Graded lagrangians, exotic topological D-branes and enhanced triangulated categories," JHEP 0106 (2001) 064, `hep-th/0105063`.
27. E. Diaconescu, "Enhanced D-brane categories from string field theory," JHEP 0106 (2001) 016, `hep-th/0104200`.
28. E. Witten, "Some comments on string dynamics," contribution to proceedings of Strings '95, `hep-th/9507121`.
29. M. Douglas, B. Greene, and D. Morrison, "Orbifold resolution by D-branes," Nucl. Phys. **B506** (1997) 84–106, `hep-th/9704151`.
30. E. Sharpe, "String orbifolds and quotient stacks," Nucl. Phys. **B627** (2002) 445–505, `hep-th/0102211`.
31. E. Witten, "Two-dimensional models with (0,2) supersymmetry: perturbative aspects," `hep-th/0504078`.
32. E. Witten, "The N matrix model and gauged WZW models," Nucl. Phys. **B371** (1992) 191–245.
33. T. Pantev and E. Sharpe, "Notes on gauging noneffective group actions," `hep-th/0502027`.
34. T. Pantev and E. Sharpe, "String compactifications on Calabi-Yau stacks," Nucl. Phys. **B733** (2006) 233–296, `hep-th/0502044`.
35. T. Pantev and E. Sharpe, "GLSM's for gerbes (and other toric stacks)," Adv. Theor. Math. Phys. **10** (2006) 77–121, `hep-th/0502053`.
36. S. Hellerman, A. Henriques, T. Pantev, E. Sharpe, and M. Ando, "Cluster decomposition, T-duality, and gerby CFT's," Adv. Theor. Math. Phys. 11 (2007) 751–818, `hep-th/0606034`.

37. E. Sharpe, "Discrete torsion," Phys. Rev. **D68** (2003) 126003, hep-th/0008154.
38. E. Sharpe, "Recent developments in discrete torsion," Phys. Lett. **B498** (2001) 104–110, hep-th/0008191.
39. J. Distler and R. Plesser, private communication.
40. A. Kapustin, E. Witten, "Electric-magnetic duality and the geometric Langlands program," hep-th/0604151.
41. M. Strassler, "Duality, phases, spinors, and monopoles in $SO(n)$ and $\mathrm{Spin}(n)$ gauge theories," hep-th/9709081.
42. D. Morrison and R. Plesser, "Towards mirror symmetry as duality for two-dimensional abelian gauge theories," Nucl. Phys. Proc. Suppl. **46** (1996) 177–186, hep-th/9508107.
43. K. Hori and C. Vafa, "Mirror symmetry," hep-th/0002222.
44. L. Borisov, L. Chen, and G. Smith, "The orbifold Chow ring of toric Deligne-Mumford stacks," math.AG/0309229.
45. R. Donagi and T. Pantev, "Langlands duality for Hitchin systems," math.AG/0604617.
46. D. Abramovich, T. Graber, and A. Vistoli, "Algebraic orbifold quantum products," math.AG/0112004.
47. T. Graber and J. Bryan, private communication.
48. D. Morrison and R. Plesser, "Summing the instantons: quantum cohomology and mirror symmetry in toric varieties," Nucl. Phys. **B440** (1995) 279–354, hep-th/9412236.
49. R. Karp, "$\mathbf{C}^2/\mathbf{Z}_n$ fractional branes and monodromy," Comm. Math. Phys. 270 (2007) 163–196, hep-th/0510047.
50. R. Karp, "On the $\mathbf{C}^n/\mathbf{Z}_m$ fractional branes," hep-th/0602165.
51. C. Herzog and R. Karp, "On the geometry of quiver gauge theories (stacking exceptional collections)," hep-th/0605177.